NOTES, PROBLEMS AND SOLUTIONS IN DIFFERENTIAL EQUATIONS

A. K. Nandakumaran

P. S. Datti

Shaftesbury Road, Cambridge CB2 8EA, United Kingdom

One Liberty Plaza, 20th Floor, New York, NY 10006, USA

477 Williamstown Road, Port Melbourne, VIC 3207, Australia

314–321, 3rd Floor, Plot No. 3, Splendor Forum, Jasola District Centre, New Delhi – 110025, India

103 Penang Road, #05–06/07, Visioncrest Commercial, Singapore 238467

Cambridge University Press is part of Cambridge University Press & Assessment, a department of the University of Cambridge.

We share the University's mission to contribute to society through the pursuit of education, learning and research at the highest international levels of excellence.

www.cambridge.org
Information on this title: www.cambridge.org/9781009610025

© A. K. Nandakumaran and P. S. Datti 2025

This publication is in copyright. Subject to statutory exception and to the provisions of relevant collective licensing agreements, no reproduction of any part may take place without the written permission of Cambridge University Press & Assessment.

First published 2025

Printed in India by Repro India Ltd.

A catalogue record for this publication is available from the British Library

ISBN 978-1-1009-61002-5 Paperback

Cambridge University Press & Assessment has no responsibility for the persistence or accuracy of URLs for external or third-party internet websites referred to in this publication and does not guarantee that any content on such websites is, or will remain, accurate or appropriate.

For EU product safety concerns, contact us at Calle de José Abascal, 56, 1°, 28003 Madrid, Spain, or email eugpsr@cambridge.org.

NOTES, PROBLEMS AND SOLUTIONS IN DIFFERENTIAL EQUATIONS

This book is designed for senior undergraduate and graduate students pursuing courses in mathematics, physics, engineering and biology. The text begins with a study of ordinary differential equations, and concepts of first- and second-order equations. It includes discussion on linear systems, series solutions, regular Sturm–Liouville theory, boundary value problems and qualitative theory. With a detailed exploration of partial differential equations, it covers first order partial differential equations, their classification and Laplace and Poisson equations. It concludes with heat equation, one dimensional wave equation and wave equation in higher dimensions. The book highlights the importance of analysis, linear algebra and geometry in the study of the subject. It provides sufficient theoretical material at the beginning of each chapter, which will enable students to understand the concepts better and will initiate them into problem solving.

A. K. Nandakumaran is Professor and Chairman of Department of Mathematics at Indian Institute of Science, Bangalore. He is the co-author of the books *Ordinary Differential Equations: Principles and Applications* (2017) and *Partial Differential Equations: Classical Theory with a Modern Touch* (2020) published by Cambridge University Press.

P. S. Datti is a former faculty member at the Tata Institute of Fundamental Research Centre for Applicable Mathematics, Bangalore. He is the co-author of the books *Ordinary Differential Equations: Principles and Applications* (2017) and *Partial Differential Equations: Classical Theory with a Modern Touch* (2020) published by Cambridge University Press.

To our family members

Sujatha
Nileena Vivek
Vishnu

Madhumathi
Sarang Makiko Anika
Sharmishte Vaibhav

Contents

Preface xi

Acknowledgements xv

Notation xvii

List of Illustrations xxi

PART ONE | ORDINARY DIFFERENTIAL EQUATIONS 1

1 First- and Second-Order ODE 3
   1.1   Introduction 3
   1.2   Some Applications 8
        1.2.1   The Envelope of a Family of Curves 9
        1.2.2   Orthogonal and Isogonal Trajectories 11
   1.3   Exercises 12
   1.4   Solutions 20

2 Linear Systems 47
   2.1   Introduction 47
   2.2   Some Applications 52
        2.2.1   Kinematics 52
        2.2.2   Controllability and Observability of Linear Systems 53
   2.3   Exercises 58
   2.4   Solutions 62

3 Series Solutions: Frobenius Theory 85
   3.1   Introduction 85
   3.2   Real Analytic Functions 86
   3.3   Equations with Analytic Coefficients 88
   3.4   Regular Singular Points 91
        3.4.1   Equations with Regular Singular Points 92
        3.4.2   Singularity at Infinity 98
   3.5   Exercises 100
   3.6   Solutions 102

4 Regular Sturm–Liouville Theory and Boundary Value Problems 111
   4.1   Introduction 111
        4.1.1   Boundary Value Problems 113
        4.1.2   Existence and Uniqueness of Green's Function 113
        4.1.3   Green's Function for the Adjoint Equation (Symmetry) 116
   4.2   Prüfer Substitution 118

	4.3	Exercises	120
	4.4	Solutions	122
5	**Qualitative Theory**		**141**
	5.1	Introduction	141
		5.1.1 Liapunov Stability, Liapunov Function	143
		5.1.2 Linearization	144
		5.1.3 Liapunov Function and Stability	145
		5.1.4 Hamiltonian System	147
		5.1.5 Two-Dimensional Potential Flows	149
	5.2	Exercises	150
	5.3	Solutions	154

PART TWO | PARTIAL DIFFERENTIAL EQUATIONS — 183

6	**First-Order Partial Differential Equations**		**185**
	6.1	Method of Characteristics	185
	6.2	Hamilton–Jacobi Equation	188
	6.3	Conservation Laws	190
	6.4	Exercises	193
	6.5	Solutions	201
7	**Classification of Partial Differential Equations**		**235**
	7.1	Introduction	235
	7.2	Classification	237
	7.3	Second-Order Linear Equations in Two Variables	240
		7.3.1 Reduction to Canonical Form	240
	7.4	Higher-Order Equations	242
	7.5	Exercises	242
	7.6	Solutions	243
8	**Laplace and Poisson Equations**		**247**
	8.1	Introduction	247
	8.2	Exercises	255
	8.3	Solutions	262
9	**Heat Equation**		**285**
	9.1	Homogeneous Equation: Fourier–Poisson Formula	285
		9.1.1 Inhomogeneous Equation: Duhamel's Principle	286
		9.1.2 Heat Equation in a Finite Interval: Fourier Method	286
	9.2	Exercises	287
	9.3	Solutions	292
10	**One-Dimensional Wave Equation**		**307**
	10.1	Homogeneous Equation: D'Alembert's Formula	307
	10.2	Domain of Dependence and Other Concepts	308
	10.3	Discontinuous Initial Data: Propagation of Singularities	309
	10.4	Inhomogeneous Equation: Duhamel's Principle	310
	10.5	Equation with Two Distinct Speeds	311

10.6		Characteristic Parallelogram Property	312
10.7		Telegraph Equation	314
10.8		Exercises	314
10.9		Solutions	317

11 Wave Equation in Higher Dimensions — 329
11.1		Introduction	329
	11.1.1	Two-Dimensional Wave Equation: Method of Descent	331
	11.1.2	Telegraph Equation	332
11.2		Exercises	336
11.3		Solutions	339

References — 353

Index — 357

Preface

The present book, consisting of two parts, is essentially a collection of Exercises and their Solutions in Ordinary Differential Equations (ODEs) and Partial Differential Equations (PDEs). Its special nature, which makes it different from the usual books on the subject, consists of providing sufficient theoretical materials at the beginning of each chapter, pertaining to that chapter. This certainly helps students straight way dwell into solving the exercises, referring to the relevant textbooks only for understanding the proofs of the important theorems and additional topics. The topics are at (senior) undergraduate and beginning graduate levels.

The importance of the subjects of ODE and PDE in the modern science and engineering curricula is well-known. Many interesting and important real-life problems are modelled using ODE and PDE. These include, but not limited to, physics, chemistry, biology, engineering, economics, sociology and psychology. These topics require extensive mathematical techniques, which are dealt with in many existing books on the subjects, with varying degrees of difficulty and contents. Many books, especially those used as textbooks, also contain a comprehensive list of exercises. Certain books also have Solutions Manuals, usually made available to teachers only. Many other books contain solutions to a selected list of exercises. In our experience, as teachers of these subjects for many years, we find that many students do not attempt to solve the exercises, except routine straightforward ones. This could be partially due to the fact that some exercises turn out to be tricky or difficult, often requiring some advanced knowledge in analysis and linear algebra for their solutions. Furthermore, those students who study the subjects on their own find solving exercises more difficult, thus missing out an important aspect of learning the subject.

This book is an outcome of our experience of teaching and research, not only in our respective institutions but also in many other institutions and universities across India. We have extensively given lectures on differential equations at these institutions, for more than three decades, as part of workshops, schools and series of lectures. After the successful publication of our two co-authored books, one on ODEs (2017) (Ref.[46])[1] and the other on PDEs (2020) (Ref.[45]), there was a

---

[1] The numbers in the square brackets [ ] indicate the references listed at the end of the book.

constant inquiry about the Solutions Manuals of the problems in these books from the student and teacher community, as well as our well-wishers. A similar inquiry was also there to the problems that were discussed in our courses on First Course in PDEs on the online platform of the National Programme for Technology Enhanced Learning (NPTEL) (2021, 2022). One of the concerns was that many problems presented in these books are not straightforward, routine type but requiring for their solutions a deeper knowledge in analysis and linear algebra, which somehow are not well-connected in the study of differential equations in our university curricula. This was the main motivation for us to write a detailed "Problems and Solutions" book, including a detailed description of the theoretical results pertaining to each chapter, rather than just writing a Solutions Manual. This is reflected in the title of the book. Thus, the present book contains a large number of problems gathered from our lectures and courses we have taught for many years, including some from the above-mentioned two books. And some exercises are from our student days!

Many of these problems were also a part of the online courses we have given under the auspices of the NPTEL, Department of Science and Technology (DST), Government of India. Our way of presenting the solutions is also unique in the sense that many lengthy computations, though routine in nature, are left to the readers. Certain solutions require some non-trivial results from analysis and linear algebra; some brief proofs are provided wherever possible and others are indicated in footnotes. A diligent reader can work out them separately. Wherever it is felt essential, we have drawn figures for a better illustration of a solution. The figures are drawn using the *tikz* and *pgfplots* software. We hope this book will be useful to the students and teachers in various institutions.

The contents in the book are arranged in two parts; the first one is on ODEs, containing five chapters, and the second on PDEs, containing six chapters. The chapters have been numbered consecutively. Each chapter begins with Notes followed by Exercises and Answers, which appear at the end after listing all the exercises. This gives an opportunity to the readers to work out a problem independently before looking at the solution. Of course, some of the problems are important theorems! Some of the highlights of the book, in addition to the standard topics of a first course, are as follows:

- Determining the numerical range of a real or complex matrix of order 2, as an application of the envelope of a family of curves.
- Small oscillations with many degrees of freedom near a position of a stable equilibrium.
- Linear finite dimensional control systems.

- Discussion on the construction of a Lyapunov function for a linear system of the first-order ODE.
- Discussion of a particular two-dimensional autonomous system called *potential flow*.
- A brief discussion of Hamiltonian systems and the canonical transformation.
- Widder's theorem on uniqueness of non-negative solution to the heat equation is included.
- Examples of the fourth-order hyperbolic equations with simple or multiple characteristics.
- Examples of systems of the second-order wave-like equations. Typical examples of such a system are the system of Maxwell's equations of electrodynamics and the system of equations of linear elasticity.

We hope that the present book will serve as a good addition to differential equations community across the globe.

Bangalore
February 2025

A. K. Nandakumaran
P. S. Datti

Acknowledgements

We wish to acknowledge the encouragement we have received from our respective institutions and the moral support from our colleagues, during the preparation of the manuscript. We sincerely thank the Cambridge University Press (CUP) for their continued support. Our earlier two coauthored books on *Ordinary Differential Equations: Principles and Applications* (2017) and *Partial Differential Equations: Classical Theory with a Modern Touch* (2020) were also published by them in the Cambridge-IISc Series. It was indeed a pleasant experience to work with CUP, in particular Mr Ankush Kumar for his meticulous coordination during the preparation of the manuscript and Mr Karan Gupta for his timely coordination during the proofreading. We also wish to thank the anonymous reviewers for their constructive criticism and suggestions, improving the overall presentation of the book. We thank our academic fraternity, especially our students, for making valuable suggestions while reading parts of the manuscript.

Notation

The notations used in the book are the standard ones extensively used in the literature. Below is a list of these notations.

- Abbreviations ODE(s) and PDE(s) mean ordinary differential equation(s) and partial differential equation(s) respectively, and PDO(s) means partial differential operator(s).
- Points in the Euclidean space $\mathbb{R}^n$ are denoted by $x = (x_1, \ldots, x_n)$; in some places, they are also treated as column vectors and written as $\mathbf{x} = \begin{bmatrix} x_1 & \cdots & x_n \end{bmatrix}^\mathrm{T}$, using bold face.
- For $x, y \in \mathbb{R}^n$, their *dot product* or *scalar product* or *inner product* is defined by $x \cdot y = \sum_{i=1}^n x_i y_i$. We also write $(x, y) = x \cdot y$. The *standard norm* in $\mathbb{R}^n$ is denoted by $|x| = \sqrt{(x,x)}$.
- For $z, w \in \mathbb{C}^n$, their *dot product* or *scalar product* or *inner product* is defined by $(z, w) = \sum_{i=1}^n z_i \bar{w}_i$. We also write $(x, y) = x \cdot y$. The *standard norm* in $\mathbb{C}^n$ is denoted by $|z|^2 = (z, z) = |z_1|^2 + \cdots + |z_n|^2$.
- If A and B are subsets of $\mathbb{R}^n$, we write $A \Subset B$ if $\bar{A} \subset B$, where $\bar{A}$ denotes the *closure* of A.
- For $x \in \mathbb{R}^n$ and $r > 0$, the *open ball* with centre at x and radius r is denoted by $B_r(x)$; the *sphere* with centre at x and radius r is denoted by $S_r(x)$ or $\partial B_r(x)$, which is the boundary of $B_r(x)$. Thus,
$$B_r(x) = \{y \in \mathbb{R}^n : |x - y| < r\}$$
and
$$S_r(x) = \{y \in \mathbb{R}^n : |x - y| = r\}.$$
- The *volume* of the unit ball $B_1(0)$ in $\mathbb{R}^n$ is denoted by ω_n, and the *surface area* of the unit sphere $S_1(0)$ in $\mathbb{R}^n$ is denoted by σ_n. Thus, $\sigma_n = \dfrac{2\pi^{n/2}}{\Gamma(n/2)}$ and

$\omega_n = \dfrac{\sigma_n}{n}$, where Γ is the Euler *gamma function*. The *volume* of $B_r(x)$ and the *surface area* of $S_r(x)$ are denoted by $|B_r(x)|$ and $|S_r(x)|$, respectively. Thus, $|B_r(x)| = \omega_n r^n$ and $|S_r(x)| = \sigma_n r^{n-1}$.

- The (first) derivative of a function of a single variable is denoted by $'$ and the second derivative by $''$. These are frequently used in the ODE part.

- The partial derivatives are denoted by

$$D_j = \partial_j = \frac{\partial}{\partial x_j}, \; j = 1, 2, \ldots, n.$$

- A multi-index is an n-tuple $\alpha = (\alpha_1, \ldots, \alpha_n)$ with α_j being all non-negative integers. The *order* of α is the non-negative integer $|\alpha| = \alpha_1 + \cdots + \alpha_n$.

- If α is a multi-index, we write

$$D^\alpha = D_1^{\alpha_1} \cdots D_n^{\alpha_n} = \frac{\partial^{|\alpha|}}{\partial x_1^{\alpha_1} \cdots \partial x_n^{\alpha_n}}.$$

- If α is a multi-index and $x \in \mathbb{R}^n$, we write $x^\alpha = x_1^{\alpha_1} \cdots x_n^{\alpha_n}$.

- Unless otherwise stated, Ω denotes a bounded open set in $\mathbb{R}^n$ with smooth boundary $\partial\Omega$. The surface measure on $\partial\Omega$ or, more generally, on a smooth surface S is denoted by $d\sigma$ or $d\sigma(y)$ to stress the variable y in $\partial\Omega$ or S.

- The space $C^k(\Omega)$ denotes the set of all real-valued functions defined on Ω which have continuous partial derivatives of order up to k, for a non-negative integer k. We also write $C = C^0$ when $k = 0$.

- The set $C^k(\overline{\Omega}) \subset C^k(\Omega)$ denotes the set of all real-valued functions defined on Ω whose partial derivatives up to order k are continuous in $\overline{\Omega}$.

- For $0 < \alpha \leq 1$, the set $C^{k,\alpha}(\Omega)$ denotes the subset of $C^k(\Omega)$ consisting of all those u such that $D^\beta u$, $|\beta| = k$ are Hölder continuous of order α in Ω (Lipschitz continuous if $\alpha = 1$).

- For a function $f : \Omega \to \mathbb{R}$, its *support* is defined as the closure of the set $\{x \in \Omega : f(x) \neq 0\}$ and is denoted by $\mathrm{supp} f$.

- The set of functions in $C^k(\Omega)$ having compact support in Ω is denoted by $C_c^k(\Omega)$.

- The set of all infinitely differentiable functions defined on Ω and having compact support in Ω is denoted by $C_c^\infty(\Omega)$. The space of test functions $\mathcal{D}(\Omega)$ is the set $C_c^\infty(\Omega)$, with a specified topology.

- If $u : \Omega \to \mathbb{R}$, we write $u_{ij} = D_i D_j u$ for $i, j = 1, 2, \ldots, n$. The *Hessian* of u is the real symmetric matrix $[u_{ij}]$ and is denoted by $D^2 u$.

- For functions of $(x, t) \in \mathbb{R} \times \mathbb{R}$, we use the notation $D_x = \partial_x = \dfrac{\partial}{\partial x}$ and $D_t = \partial_t = \dfrac{\partial}{\partial t}$ and similarly for mixed derivatives.

- The operator ∇ denotes the *grad* operator: $\nabla u = (\partial_1 u, \ldots, \partial_n u)$. When needed, we also stress the variable, as in ∇_x, ∇_y.

- For a vector-valued function $u = (u_1, \ldots, u_n)$, its *divergence* is defined by $\operatorname{div} u = \nabla \cdot u = \partial_1 u_1 + \cdots + \partial_n u_n$.

- The *Laplace operator* or the *Laplacian* in $\mathbb{R}^n$ is the PDO defined by

$$\Delta = \nabla \cdot \nabla = D_1^2 + \cdots + D_n^2 = \frac{\partial^2}{\partial x_1^2} + \cdots \frac{\partial^2}{\partial x_n^2}.$$

- The *wave operator* or the *D'Alembertian* in $\mathbb{R}^{n+1}$ is the PDO defined by

$$\Box_c = \partial_t^2 - c^2 \Delta_x = \partial_{tt} - c^2 \Delta_x,$$

where $c > 0$ is a constant, $t \in \mathbb{R}$ and $x \in \mathbb{R}^n$; Δ_x denotes the Laplacian with respect to the x variables.

- The *heat operator* is $\partial_t - a^2 \Delta_x$.

- For an $m \times n$ real matrix A, its *transpose* is denoted by A^{T}, which is an $n \times m$ matrix. For an $m \times n$ complex matrix A, its *conjugate transpose* is denoted by A^*, which is an $n \times m$ matrix. For a (column) vector $\mathbf{x} \in \mathbb{R}^n$, its transpose $\mathbf{x}^{\mathrm{T}}$ denotes a (row) vector and vice versa. At some places, matrices are written in bold face.

- For $z \in \mathbb{C}$, $\Re z$ and $\Im z$ denote, respectively, the real and imaginary parts of z; $\bar{z}$ denotes its conjugate.

Illustrations

1.1	Solution graphs of Exercise 1.5: (i) $0 < y_0 < a/b$, (ii) $y_0 > a/b$, (iii) $y_0 < 0$.	24
1.2	Solution graphs of Exercise 1.9.	28
2.1	Phase portraits of Exercise 2.2.	64
2.2	Phase portrait in $y_1 - y_2$ and $x_1 - x_2$ planes in Exercise 2.4.	67
2.3	Orbits in the Exercise 2.7.	69
2.4	Orbits in Exercise 2.10.	72
2.5	Schematic description in Exercise 2.15.	77
4.1	Graphical solution of equation (4.3.11).	134
5.1	Functions z and w in Exercise 5.2.	155
5.2	Typical orbits in Exercise 5.2.	156
5.3	Orbits of Exercise 5.5: $c = 0$.	158
5.4	Orbits of Exercise 5.5: $0 < c < 1$.	160
5.5	Orbits of Exercise 5.5: $c = 1$.	161
5.6	Homoclinic orbits and solution curves of Exercise 5.7.	164
5.7	Potential function (a) and an orbit for $E < -2/3$ (b) in Exercise 5.9.	166
5.8	Typical orbits in Exercise 5.9: (a) $-2/3 < E < 2/3$; (b) $E = 2/3$.	167
5.9	Limit cycle in Exercise 5.14.	171
5.10	Phase portrait of potential flow for $\Re(z^4)$ (*top*) and $\Re(z^{-4})$ (*bottom*).	177
5.11	(a) Joukowsky airfoil and (b) phase portrait for Joukowsky function.	178
5.12	The auxiliary energy function h and the graphs of $f_\pm$ with the parabola x^2.	180
5.13	Phase portrait of the Hamiltonian system in Exercise 5.23.	182
6.1	Initial curve and characteristics.	202
6.2	Smooth, rarefaction and shock formation.	210
6.3	Exit times.	219
6.4	(a) Brachistochrone and (b) catenary.	221
6.5	Schematic description of $x - t$ plane of Exercise 6.34.	228
8.1	Reflection points through (a) upper semi-ball and (b) quarter-ball.	269
8.2	The sets K_1 and K_2 in Exercise 8.35.	282
9.1	Temperature profiles in Exercise 9.8.	298
10.1	Domain of dependence and range of influence.	308
10.2	(a) Range of influence and (b) domain of determinacy of $[a, b]$.	308
10.3	Discontinuous data: for u_0 and u_1.	309
10.4	Characteristics of the wave equation in the first quadrant.	313
10.5	Domain of integration in Exercise 10.1.	317
10.6	Characteristics of the wave equation in a finite interval.	322
10.7	The function u_0 in Exercise 10.6.	324
10.8	The characteristics for Exercise 10.6.	324

Illustrations

PART ONE

ORDINARY DIFFERENTIAL EQUATIONS

1
First- and Second-Order ODE

1.1 Introduction

The first-order equations–linear and non-linear–and the second-order linear equations, with constant or variable coefficients, are considered in this chapter. For solving the first-order equations, familiar methods such as the method of separation of variables or a method that can be reduced to this are used. Also discussed are the exact differential equations or those equations that can be reduced to this form using a suitable *integrating factor* (*IF*). We also emphasize the peculiarities that may arise in an initial value problem (IVP) when sufficient conditions imposed in the Cauchy–Peano existence theorem or in the method of Picard's iterations are not satisfied. Many exercises deal with the maximal interval of existence of a solution to an IVP.

Only second-order linear equations are considered here. The non-linear equations or, more generally, the two-dimensional systems of first-order equations are treated in Chapter 5 on qualitative analysis. The treatment of equations with constant coefficients is straightforward. The equations with variable coefficients are more difficult to deal with, and, in general, it is not possible to obtain the solution in explicit form. However, the structure of solutions to the homogeneous and inhomogeneous equations is well-understood.

A general first-order ordinary differential equation (ODE) takes the form $f(t, x(t), x'(t)) = 0$, where f is a given function and $x = x(t)$ is the unknown function to be determined. A general theory for the above equation is rather difficult. A more realistic equation for which a general theory can be developed is given by the regular form $x'(t) = f(t, x(t))$, and the corresponding IVP is

$$x'(t) = f(t, x(t)), \, t \in I, \, x(t_0) = x_0. \qquad (1.1.1)$$

It should be noted that the initial time t_0 and the initial value x_0 come from the domain of the definition of f. If $\mathbf{f} = (f_1, \cdots, f_n)$ is a vector-valued (n-dimensional) and $\mathbf{x} = \mathbf{x}(t) \in \mathbb{R}^n$ is a vector, then equation (1.1.1) is written in boldface as $\mathbf{x}'(t) = \mathbf{f}(t, \mathbf{x}(t)), t \in I, \mathbf{x}(t_0) = \mathbf{x}_0$ and is a system of n equations in n unknowns $\mathbf{x}(t) = (x_1(t), \cdots x_n(t))$. It is readily seen that any nth-order equation $x^{(n)}(t) = f(t, x(t), x'(t), \cdots x^{(n-1)}(t))$ can be reduced to the first-order system by substituting $x_1(t) = x(t), x_2(t) = x_1'(t), \cdots, x_{n-1}(t) = x_{n-2}'(t)$ and $x_n(t) = x_{n-1}'(t) = f(t, x_1(t), \cdots, x_n(t))$, that is, $\mathbf{x}'(t) = \mathbf{F}(t, \mathbf{x}(t))$ with

$$\mathbf{F}(t, \mathbf{x}(t)) = (x_2(t), \cdots, x_n(t), f(t, x_1(t), \cdots, x_n(t)).$$

First-Order Linear Equations: A general regular form of the first-order equation is given by

$$x'(t) + p(t)x(t) = q(t), \qquad (1.1.2)$$

where we assume that p and q are given continuous functions defined on an interval I. This equation can be solved using an IF $\mu(t) = \exp\left(\int^t p(\tau)\,d\tau\right)$. Multiplying equation (1.1.2) by μ, the equation reduces to an integrable (exact) form as

$$(\mu x)' = \mu q,$$

and the solution is given by $x(t) = (\mu(t))^{-1}\left(\int^t \mu q\,d\tau + C\right)$.

The ODE $x'(t) = f(t, x(t))$ is said to be an *exact equation* if it can be reduced to an integrable (exact) form that can be directly integrated to get the solution. That is, the equation can be written as $\dfrac{d}{dt}\phi(t, x(t)) = 0$ for some two-variable function ϕ. If we write the first-order equation in a general form as $M(t, x) + N(t, x)x'(t) = 0$, then it is exact if and only if $\dfrac{\partial M}{\partial x} = \dfrac{\partial N}{\partial t}$, assuming the functions M and N are C^1 in a domain D in the (t, x) plane. If it is not exact, we look for a function μ so that the equation $\mu M + \mu N x' = 0$ is exact. Such a μ, when exists, is called an IF.

Thus, we have a reasonable complete theory for solving the first order linear equations. Such a complete procedure is not available for the second- (and higher-) order linear equations of the form

$$x''(t) + p(t)x'(t) + q(t)x(t) = r(t),\ t \in I. \qquad (1.1.3)$$

Nevertheless, we have the following result regarding the existence theory and structure of the solutions.

Theorem 1.1. Assume p, q and r are continuous functions in a compact interval I. Then, the following statements hold:

(i) For arbitrary $x_0, x_1 \in \mathbb{R}$, equation (1.1.3) has a unique solution x in some sub-interval $I(t_0)$ of I containing t_0 satisfying the initial conditions $x(t_0) = x_0$, $x'(t_0) = x_1$.

(ii) Let S be the set of all solutions of the homogeneous equation (1.1.3) (i.e., with $r \equiv 0$). Then, S is a vector space and $\dim(S) = 2$. In other words, the homogeneous equation has two linearly independent solutions.

(iii) Let $\tilde{S}$ be the set of all solutions of equation (1.1.3). Then, $\tilde{S} = S + x_p$, where x_p is any particular solution of equation (1.1.3).

Thus, $\tilde{S}$ is a hyperplane. In general, there is no specific procedure to determine two linearly independent solutions of the homogeneous equation. However, if one (non-trivial) solution $x_1(t)$ is known, then the second linearly independent solution can be obtained by the method of order reduction (also called the method of variation of parameters or constants). More precisely, we look for a second solution of the form $x_2(t) = c(t)x_1(t)$, where the function $c(t)$ needs to be determined. It is readily seen that c satisfies the equation $x_1(t)c''(t) + (2x_1'(t) + p(t)x_1(t))c'(t) = 0$, as x_1 is a solution. Thus, c can be obtained by performing two integrations.

On the other hand, if the homogeneous equation has constant coefficients, then the solutions can be explicitly written down. More precisely, consider the equation $ax''(t) + bx'(t) + cx(t) = 0$, where $a, b, c \in \mathbb{R}$ and $a \neq 0$. Then, the general solution is given by

$$x(t) = \begin{cases} Ae^{r_1 t} + Be^{r_2 t} & \text{if } b^2 - 4ac > 0, \\ (A + Bt)e^{\alpha t} & \text{if } b^2 - 4ac = 0, \\ e^{\alpha t}(A \cos \beta t + B \sin \beta t) & \text{if } b^2 - 4ac < 0. \end{cases}$$

Here, A and B are arbitrary constants, $\alpha = -b/(2a)$, $\beta = \sqrt{4ac - b^2}/(2a)$ and

$$r_1 = (-b + \sqrt{b^2 - 4ac})/(2a), \quad r_2 = (-b - \sqrt{b^2 - 4ac})/(2a).$$

Returning to the IVP (1.1.1), we have general results regarding the existence, uniqueness and continuous dependence of solutions. For existence, the continuity of f is a sufficient condition (Peano's theorem), whereas the uniqueness requires a stronger assumption of Lipschitz continuity of f. We state these results in the sequel. First, we recall Gronwall's inequality, a powerful and interesting result in establishing uniqueness result.

Theorem 1.2 (Gronwall's Inequality). Suppose that p and q are continuous real-valued functions defined on $[a, b]$ with $q(t) \geqslant 0$ on $[a, b]$. Assume p and q satisfy the inequality

$$p(t) \leqslant C + k \int_{t_0}^{t} q(s)p(s)ds,$$

for all $t \in [a,b]$ and $t \geqslant t_0$, where $t_0 \in [a,b]$ is fixed and C and k are constants with $k \geqslant 0$. Then,

$$p(t) \leqslant C \exp\left(k \int_{t_0}^{t} q(s)\, ds\right),$$

for all $t \in [a,b]$, $t \geqslant t_0$; for $t \leqslant t_0$, interchange the limits in the above integrals.

It is easy to verify that a differentiable function $x = x(t)$ defined on $I = (a,b)$ is a solution of IVP (1.1.1) if and only if x is a solution of the corresponding integral equation

$$x(t) = x_0 + \int_{t_0}^{t} f(\tau, x(\tau))d\tau,\ t \in (a,b). \tag{1.1.4}$$

This is termed as the integral formulation. However, we remark that to define a solution of equation (1.1.4), we do not require the differentiability of x. Thus, it is possible to define the solution in a weaker sense that a continuous function is a solution if it satisfies the integral formulation. This weaker notion of a solution even allows to relax the condition of continuity on the function f. This has many applications including control theory, where the function f need not be a continuous function when discontinuous controls are employed. Nevertheless, it is an interesting fact that if we start with a continuous function f and continuous solution x, then we can establish that x is differentiable and it is a classical solution, that is, a solution of equation (1.1.1).

By the above observation, it suffices to prove the existence and uniqueness of a continuous solution to the integral equation, when f is a continuous function.

Definition 1.3 (Lipschitz Continuity). A function $f : D \subseteq \mathbb{R} \to \mathbb{R}$ is said to be locally Lipschitz continuous in D if, for any $x_0 \in D$, there exists neighbourhood N_{x_0} of x_0 and an $\alpha = \alpha(x_0) > 0$ such that

$$|f(x) - f(y)| \leqslant \alpha |x - y|,\ \text{for all } x, y \in N_{x_0}.$$

The function $f : D \subset \mathbb{R} \to \mathbb{R}$ is said to be Lipschitz continuous (or globally Lipschitz continuous) in D if there is a constant $\alpha > 0$ such that $|f(x) - f(y)| \leqslant \alpha\, |x - y|$, for all $x, y \in D$.

The smallest such constant α is known as the Lipschitz constant of f. It is easy to see that if f is differentiable and $\beta = \sup_{x \in D} |f'(x)| < \infty$, then f is Lipschitz in D with Lipschitz constant $\alpha \leqslant \beta$. We now extend this definition to a vector-valued function. Let $D \subset \mathbb{R}^n$ be a domain.

Definition 1.4 (Lipschitz Continuity). A function $\mathbf{f}(t, \mathbf{x}) : (a, b) \times D \to \mathbb{R}^n$ is said to be Lipschitz continuous (globally in D) with respect to $\mathbf{x}$ if there exists $\alpha > 0$ such that $|\mathbf{f}(t, \mathbf{x_1}) - \mathbf{f}(t, \mathbf{x_2})| \leq \alpha |\mathbf{x_1} - \mathbf{x_2}|$, for all $(t, \mathbf{x_1})$ and $(t, \mathbf{x_2})$ in $(a, b) \times D$.

The smallest such constant α is known as the Lipschitz constant of $\mathbf{f}$. We can also define local Lipschitzness analogously.

Theorem 1.5 (Uniqueness). Let a and b be positive real numbers. Suppose that $f = f(t, x)$ is continuous in the rectangle $\mathcal{R} = \{(t, x) \in \mathbb{R}^2 : |t - t_0| \leq a, |x - x_0| \leq b\}$ and Lipschitz continuous with respect to x in $\mathcal{R}$. Then, the IVP (1.1.1) has at most one solution.

Regarding existence, we have the following couple of theorems.

Theorem 1.6 (Picard's Existence Theorem). Let D be an open connected set in $\mathbb{R}^2$. Assume $f : D \to \mathbb{R}$ satisfies the following conditions:

(i) f is continuous in D.

(ii) f is Lipschitz continuous with respect to x in D with Lipschitz constant $\alpha > 0$.

Let $(t_0, y_0) \in D$ and a and b are positive constants such that the rectangle $\mathcal{R}$ is a subset of D and put $M = \max_{(t,y) \in \mathcal{R}} |f(t, y)|$ and $h = \min(a, b/M)$. Then, the IVP (1.1.1) has a unique solution in the interval $|t - t_0| \leq h$.

The proof is based on the convergence of the sequence of iterates known as *Picard's iterates*, which are recursively defined by

$$x_k(t) = x_0 + \int_{t_0}^{t} f(\tau, x_{k-1}(\tau)) d\tau, \ t \in (a, b), \tag{1.1.5}$$

for $k = 2, 3, \ldots$, and the first iterate is $x_1(t) = x_0$, the initial value. Under the assumptions stated in the theorem, we can show that the sequence $\{x_k\}$ converges uniformly to some x in $C(I)$, the space of continuous functions defined on I and equipped with sup norm. It is then shown that this x is the unique solution of the integral equation.

The assumption of Lipschitz continuity is needed only to prove the uniqueness. However, we do have the existence result with only continuity assumption on f. Indeed, the proof of Picard's theorem uses Lipschitz continuity even for existence.

Theorem 1.7 (Cauchy–Peano Existence Theorem). Let a, b and $\mathcal{R}$ be as in Theorem 1.6. Assume $f(t, y)$ is continuous on the rectangle $\mathcal{R}$. Then, there exists a solution to the IVP (1.1.1) in the interval $|t - t_0| \leq h$, where $h = \min\left(a, \dfrac{b}{M}\right)$ and $M = \max_{\mathcal{R}} |f|$.

We remark that the interval of existence given by the previous theorems need not be the best possible and the solution may exist on larger intervals leading to the concept of the *maximal interval of existence*. This is nothing but the union of all intervals of existence of the IVP, which is an open interval. We now move on to the discussion of the continuous dependence of the solution to the IVP on the data: the initial condition and the function f. This concept is crucial in applications; it essentially says that if the error in input data (initial value x_0, dynamics f) is small in appropriate norms, then the error in solution is also small. We consider a more general IVP with an input function u in the dynamics as:

$$x'(t) = f(t, x(t), u(t)),\ t \in I,\ x(t_0) = x_0. \tag{1.1.6}$$

Such a problem has wide applications in many areas. For example, in control theory, such input functions are known as *controls* that can be manipulated to obtain a desired trajectory to reach a desired state. For a given function u, the standard theory will give the existence and uniqueness with appropriate assumptions on f, and there will be a unique solution x starting from x_0. However, a control problem is that whether an input function (control) u can be chosen so that the trajectory x reaches a pre-designed state x_1 at a prescribed time T. In general, this is not true, and we may see examples in Chapter 2 on the linear systems. We now state the continuous dependence result.

Theorem 1.8 (Continuous Dependence). Let $\mathcal{R}$ be as in Theorem 1.5. Suppose $f, \tilde{f} \in C(\mathcal{R})$ are Lipschitz continuous with respect to x variable, with Lipschitz constants $\alpha, \tilde{\alpha}$ respectively. Let x and $\tilde{x}$ be, respectively, the solutions of the IVP $x' = f(t, x),\ x(t_0) = x_0$ and $\tilde{x}' = \tilde{f}(t, \tilde{x}),\ \tilde{x}(\tilde{t}_0) = \tilde{x}_0$ in some closed intervals I_1, I_2 containing t_0 and $\tilde{t}_0$. For small $|t_0 - \tilde{t}_0|$, let I be any finite interval containing t_0 and $\tilde{t}_0$, where both x and $\tilde{x}$ are defined. Then,

$$\max_{t \in I} |x(t) - \tilde{x}(t)| \leq \left(|x_0 - \tilde{x}_0| + |I| \max_{\mathcal{R}} |f(t,x) - \tilde{f}(t,x)| + M|t_0 - \tilde{t}_0| \right) e^{\alpha_0 |I|},$$

where $|I|$ is the length of the interval I, $M = \max\left(\max_{\mathcal{R}} |f|, \max_{\mathcal{R}} |\tilde{f}|\right)$ and $\alpha_0 = \min(\alpha, \tilde{\alpha})$.

1.2 Some Applications

We now discuss applications of the first-order equations to some two-dimensional geometric problems. As is customary, we use (x, y) to denote a point in the plane, and x is considered an independent variable and y a dependent variable.

1.2.1 The Envelope of a Family of Curves

Consider a one-parameter family of smooth two-dimensional curves represented by

$$F(x, y, \alpha) = 0. \tag{1.2.1}$$

Here, α denotes the parameter which is a real number; each α gives rise to a curve in the family. By smoothness, we mean that the partial derivatives $\dfrac{\partial F}{\partial x}, \dfrac{\partial F}{\partial y}$ and $\dfrac{\partial F}{\partial \alpha}$ exist and are continuous. For example, for the family of straight lines passing through the origin, the slope of the line plays the role of the parameter; for the family of circles with same radius and different centres, all lying on a straight line, the parameter comes through the centres of the circles. On the other hand, the family of circles with distinct radii and distinct centres, all lying on a straight line, is an example of more than one-parameter family of curves.

Definition 1.9. A curve L is said to be an *envelope* of a one-parameter family of curves if at each point the curve L touches a curve of the family and different curves in the family touch the curve L at different points. That is, at each point on L, the curve L is tangential to exactly one curve of the family and vice versa.

Given a family of curves as in equation (1.2.1), we will now determine an envelope of it using the following procedure. Suppose the curve given by $y = \varphi(x)$, where φ is a C^1 function, is an envelope. Since at each point (x, y), the envelope touches a curve in the family, which in turn determines a value of the parameter α, obviously depending on the point (x, y). Denote this by $\alpha = \alpha(x, y)$. Thus, $F(x, y, \alpha(x, y)) = 0$ for each point on the envelope. Assuming that $\alpha(x, y)$ is a differentiable function, we obtain after differentiation with respect to x the following relation:

$$\frac{\partial F}{\partial x} + \frac{\partial F}{\partial y} y' + \frac{\partial F}{\partial \alpha} \left(\frac{\partial \alpha}{\partial x} + \frac{\partial \alpha}{\partial y} y' \right) = 0, \tag{1.2.2}$$

at each point of the envelope. The slope of the tangent to the curve of the family (1.2.1) at the point (x, y) is obtained from the equation

$$\frac{\partial F}{\partial x} + \frac{\partial F}{\partial y} y' = 0, \tag{1.2.3}$$

as α is a constant as far as the family of curves is concerned. Assume $\dfrac{\partial F}{\partial y} \neq 0$; we reverse the roles of x and y if $\dfrac{\partial F}{\partial y} = 0$ but $\dfrac{\partial F}{\partial x} \neq 0$. Since the envelope touches every curve of the family, by definition, the tangent of the envelope is the same as that of the tangent of the curve of the family, where the envelope touches the curve. It therefore follows from equations (1.2.2) and (1.2.3) that $\frac{\partial F}{\partial \alpha}\left(\frac{\partial \alpha}{\partial x}+\frac{\partial \alpha}{\partial y}y'\right)=0$. On the envelope, $\frac{\partial \alpha}{\partial x}+\frac{\partial \alpha}{\partial y}y' \neq 0$ as α is not a constant on the envelope. Thus, on the envelope, the following two relations must hold:

$$F(x,y,\alpha) = 0 \text{ and } \frac{\partial F}{\partial \alpha}(x,y,\alpha) = 0. \tag{1.2.4}$$

Conversely, if, by eliminating α from the two equations in (1.2.4), we can obtain a function $y = \varphi(x)$, where φ is a C^1 function and α is not a constant on this curve, then $y = \varphi(x)$ determines an envelope of the family in equation (1.2.1).

The points (x,y) where both $\frac{\partial F}{\partial x} = 0$ and $\frac{\partial F}{\partial y} = 0$ vanish are termed as *singular points*. It is not difficult to show that the singular points also satisfy equation (1.2.4). Thus, equation (1.2.4) either determines an envelope or the locus of singular points or a combination of both.

Numerical Range of a Matrix: We now take a short digression into linear algebra [17], [21]. If A is a real or complex $n \times n$ matrix, its *numerical range*, denoted by $W(A)$, is the set of complex numbers $\{(Az,z) : \|z\| = 1\}$, where $(\cdot,\cdot)$ and $\|\cdot\|$ denote the standard scalar product and the norm in $\mathbb{C}^n$, respectively. It is not difficult to show that the spectrum (the set of eigenvalues) of A is a subset of $W(A)$. Clearly, $W(A)$ is a bounded subset as $|(Az,z)| \leqslant \|A\|$ for all $\|z\| = 1$. It is also a convex set. Furthermore, $W(U^*AU) = W(A)$ for any unitary matrix U. We use this to determine $W(A)$ when $n = 2$. By a well-known result, there is a unitary matrix $\mathbf{U}$ such that

$U^*AU = \begin{bmatrix} \lambda_1 & m \\ 0 & \lambda_2 \end{bmatrix}$, where λ_1 and λ_2 are the eigenvalues of A and $m = 0$ if and only if A is a normal matrix. Therefore, $W(A) = \{\lambda_1|z_1|^2 + \lambda_2|z_2|^2 + m\bar{z}_1 z_2\}$, where z_1 and z_2 are arbitrary complex numbers such that $|z_1|^2 + |z_2|^2 = 1$. Put $w = \lambda_1|z_1|^2 + \lambda_2|z_2|^2 + m\bar{z}_1 z_2$ and $t = |z_1|^2$. Then, $w = t\lambda_1 + (1-t)\lambda_2 + |m|\sqrt{t(1-t)}\, e^{i\gamma}$, where t varies in $[0,1]$ and γ is an arbitrary real number. From this, we conclude that $W(A)$ is the line segment joining λ_1 and λ_2, if $m = 0$; if $\lambda_1 = \lambda_2 = \lambda$, say, then $W(A)$ is the circular disk centred at λ and radius $|m|/2$. Now, assume that λ_1 and λ_2 are distinct and $m \neq 0$. Observe that $\frac{w - \lambda_2}{\lambda_1 - \lambda_2} = t + \frac{|m|\sqrt{t(1-t)}}{\lambda_1 - \lambda_2} e^{i\gamma}$. Thus, if we put $z = \frac{w - \lambda_2}{\lambda_1 - \lambda_2} = x + iy$, where x and y are real, it follows $(x-t)^2 + y^2 = a^2 t(1-t)$, where $a = \frac{|m|}{|\lambda_1 - \lambda_2|}$. This is a family of circles with $t \in [0.1]$ playing the role of a parameter. Therefore, if we find the envelope of this family of circles, that in turn will determine the required numerical range of the 2×2 matrix A. This will be done in an exercise.

1.2.2 Orthogonal and Isogonal Trajectories

Consider a one-parameter family of smooth curves in the plane:
$$F(x, y, \alpha) = 0. \tag{1.2.5}$$

Definition 1.10. The curve that intersects every curve of the family in equation (1.2.5) at a *constant* angle is called an *isogonal trajectory*. If the constant angle is a right angle, it is called an *orthogonal trajectory*.

An isogonal trajectory is also called an *oblique trajectory*. If two straight lines with slopes m_1 and m_2 intersect at a right angle, then $m_1 m_2 = -1$; an appropriate statement may be made while considering the lines parallel to axes. Using this simple fact, it is now straightforward to find the orthogonal trajectories.

The slope of the tangent line to any curve in the family (1.2.5) at a point (x, y) is determined by the equation
$$\frac{\partial F}{\partial x} + \frac{\partial F}{\partial y} y' = 0.$$

Eliminating α from equation (1.2.5), we obtain the differential equation
$$G\left(x, y, \frac{dy}{dx}\right) = 0. \tag{1.2.6}$$

Therefore, at (x, y), the differential equation satisfied by the orthogonal trajectories is
$$G\left[x, y, -\left(\frac{dy}{dx}\right)^{-1}\right] = 0. \tag{1.2.7}$$

Suppose $\tilde{F}(x, y, \beta) = 0$ describes the general solution of equation (1.2.7) with an arbitrary parameter β. This yields a family of orthogonal trajectories of the given family of curves (1.2.5).

A family of isogonal trajectories to the family of curves in equation (1.2.5) can be found using the following elementary observation. If two straight lines with slopes m_1 and m_2 intersect at an (acute) angle γ, then $\tan \gamma = \dfrac{|m_2 - m_1|}{|m_1 m_2 + 1|}$ or $m_1 = \dfrac{m_2 - \kappa}{\kappa m_2 + 1}$ with $\kappa = \tan \gamma$. The differential equation satisfied by the family of curves (1.2.5) is the same as equation (1.2.6), namely
$$G\left(x, y, \frac{dy}{dx}\right) = 0.$$

Therefore, the differential equation satisfied by an isogonal trajectory is given by
$$G\left[x, y, \frac{p - \kappa}{1 + \kappa p}\right] = 0, \tag{1.2.8}$$

where $p = \dfrac{dy}{dx}$ and κ is as above. Solving this equation for a general solution with one arbitrary constant gives a family of isogonal trajectories.

The general references are [46], [10], [11] and [29].

1.3 Exercises

Exercise 1.1

Discuss the existence and uniqueness of the following IVP:

(1) $x_1' = x_2,\ x_2' = -\dfrac{g}{l}\sin(x_1) - \dfrac{k}{m}x_2,\ x_1(0) = 0,\ x_2(0) = 0$.

(2) $x_1' = x_2 - x_1(x_1^2 + x_2^2),\ x_2' = -x_1 - x_2(x_1^2 + x_2^2),\ x_1(0) = x_2(0) = 0$.

Exercise 1.2

Discuss the solvability of the IVP $y' = -\operatorname{sgn}(y)$, $y(0) = y_0 > 0$. Here,

$$\operatorname{sgn}(y) = \begin{cases} 1 & \text{if } y \geqslant 0, \\ -1 & \text{if } y < 0. \end{cases}$$

Exercise 1.3

State the conditions under which the following differential equations will have a unique solution:

(i) The nth-order non-linear equation $y^{(n)}(t) = g(t, y(t), y^{(1)}(t), \ldots, y^{(n-1)}(t))$.

(ii) The nth-order linear non-homogeneous equation

$$y^{(n)}(t) + a_1(t)y^{(n-1)}(t) + \cdots + a_{n-1}(t)y^{(1)}(t) + a_n(t)y(t) = b(t).$$

Exercise 1.4

In the following exercises, describe the domain of definition of the given function and corresponding initial condition, find the maximal interval (α, β) of existence of the solution and find its limits as t approaches α, β:

(1) $y' = \dfrac{1}{1+y^2}$,   (2) $y' = \dfrac{1}{ty}$,   (3) $y' = \dfrac{1}{2y}$.

Exercise 1.5

Assume the IVP $y' = ay(t) - by^2(t)$, $y(t_0) = y_0$, where a and b are positive real numbers and $t_0, y_0 \in \mathbb{R}$, has a unique solution $y = y(t)$ in an interval (t_1, t_2) containing t_0.

(1) Without attempting an explicit representation of the solution, show that y satisfies the relation $\operatorname{sgn}(y(t)(y(t) - (a/b))) = \operatorname{sgn}(y_0(y_0 - (a/b)))$.

(2) Solve the IVP and obtain the solution in implicit form:

$$\log\left(\frac{|y|}{|y_0|}\frac{|a-by_0|}{|a-by|}\right) = t - t_0.$$

(3) Use (1) to obtain the solution y in the explicit form

$$y(t) = \frac{ay_0}{by_0 + (a-by_0)e^{-a(t-t_0)}}.$$

(4) For each of the cases $y_0 < 0$, $0 < y_0 < a/b$ and $y_0 > a/b$, describe the maximal interval (t_*, t^*) of existence, where the solution y is defined. Furthermore, compute the limits $\lim_{t\uparrow t^*} y(t)$ and $\lim_{t\downarrow t_*} y(t)$.

(5) In each of the cases mentioned in (4) above, find $\dfrac{dy}{dt}$ and $\dfrac{d^2y}{dt^2}$ and analyse the shape of the solution curve.

(6) Find the conditions on y_0 so that $t_* = -\infty$ and/or $t^* = +\infty$.

(7) Plot the graphs of the solutions y in the $t-y$ plane for y_0 lying in the regions described in (4).

(8) Let $z = z(t)$ be the solution of IVP: $\dfrac{dz}{dt} = az - bz^2$, $z(t_1) = y_0$. Represent z in terms of y. Analyse further by taking different initial times. Do you observe any property? Describe the observed properties for the general problem: $\dfrac{dy}{dt} = f(y)$, $y(t_0) = y_0$.

Exercise 1.6

Consider the modified population model with a real parameter λ, namely the IVP $y' = ay - by^2 - \lambda$, $y(t_0) = y_0$. Do a similar analysis for various values of λ as in Exercise 1.5. More precisely, show that there is a critical value λ_{cr} such that for $\lambda < \lambda_{cr}$, the behaviour of the solution is exactly similar to the one in Exercise 1.5, but for $\lambda \geqslant \lambda_{cr}$, the behaviour of the solution is completely different.

Exercise 1.7

Consider the linear model of the atomic waste disposal problem:

$$\frac{dV}{dt} + \frac{cg}{W}V(t) = \frac{g}{W}(W - B),\ V(0) = 0,$$

where $V = V(t)$ is the velocity of the container of the atomic waste at time t.

(1) Find the solution V, and find the limit $\lim_{t\to\infty} V(t)$.

(2) Derive the non-linear model by considering the velocity as a function of the distance:

$$\frac{v}{W - B - cv}\frac{dv(y)}{dy} = \frac{g}{W},\ v(0) = 0,$$

where $v = v(y)$ is the velocity at the distance y, and solve the same to obtain the solution in the implicit form:

$$\frac{gy}{W} = -\frac{v}{c} - \frac{W-B}{c^2}\log\left(\frac{W-B-cv}{W-B}\right).$$

Exercise 1.8

Using appropriate dilation of the variables x and t, convert the unforced Duffing equation $x'' - \alpha x + \beta x^3 + \delta x' = 0$, where α and β are non-zero constants and $\delta \geqslant 0$, to the standard forms

$$x'' - x + x^3 + \tilde{\delta} x' = 0, \text{ if } \alpha\beta > 0 \text{ and}$$
$$x'' + x + x^3 + \tilde{\delta} x' = 0, \text{ if } \alpha\beta < 0,$$

for some appropriate constant $\tilde{\delta} \geqslant 0$.

Exercise 1.9

Analyse and find the solution(s) of the IVP $x' = |x|^{1/2}$ and $x(t_0) = x_0$.

Exercise 1.10

Prove that every separable equation is exact.

Exercise 1.11

Solve the following linear equations with the prescribed initial conditions:

(i) $\dfrac{dy}{dt} = \dfrac{2y}{t}$, $y(t_0) = y_0$, where $t_0 \neq 0$.

(ii) $y' + (\sin t)y = 0$, $y(0) = 3/2$.

(iii) $\dfrac{dy}{dt} + e^{t^2} y = 0$, $y(1) = 2$.

Exercise 1.12

Classify the following equations into linear and non-linear equations:

(1) $y' = ay - by^2$, $\quad y' = -t/y$, $\quad y' = -y/t$, $\quad y' = \sin t$,
$\sin y + x\cos(y') = 0$, $\quad y' = |y|$, $\quad yy' = y$, $\quad y' = \sin y$, $\quad yy' = \dfrac{g}{W}(W - B - cy)$.

(2) (Duffing equation): $y'' + \delta y' + \alpha y + \beta x^3 = 0$.

(3) (van der Pol equation): $y'' - \mu(y^2 - 1)y' + y = 0$.

(4) (Prey–predator system) $x' = ax - bxy$, $y' = -cy + dxy$.
(5) (Epidemiology): $S' = -\beta SI$, $I' = \beta SI - \gamma I$.
(6) (Bernoulli equation): $y' + \phi(t)y = \psi(t)y^n$.
(7) (Reduced Bernoulli equation): $y' + (1-n)\phi(t)y = (1-n)\psi(t)$.
(8) (Generalized Riccati equation): $y' + \psi(t)y^2 + \phi(t)y + \chi(t) = 0$.

Exercise 1.13

Consider the Bernoulli equation $x' + \phi x = \psi x^n$, where ϕ and ψ are continuous functions and n is a real number. Show that it can be reduced to a linear equation by a suitable substitution. Hence, find the general solution of the equation.

Exercise 1.14

Find the general solution of (i) $x' + e^t x = e^t x^2$, (ii) $x' + t^n x = x^n$.

Exercise 1.15

Consider the Jacobi equation:

$$(a_1 + b_1 t + c_1 x)(tdx - xdt) - (a_3 + b_3 t + c_3 x)dx + (a_3 + b_3 t + c_3 x)dt = 0,$$

where a_i, b_i and c_i's are constants. Using a change of variables $t = \tau + \alpha$ and $x = y + \beta$ and choosing α and β appropriately, obtain the equation $\tau dy - y d\tau + \phi(y/\tau)dy + \psi(y/\tau)d\tau = 0$. Finally, making the substitution $y = \tau u$, show that τ as a function of u satisfies the Bernoulli equation: $\dfrac{d\tau}{du} + h(u)\tau + g(u)\tau^2 = 0$.

Exercise 1.16

Consider the generalized Riccati equation $x' + \psi(t)x^2 + \phi(t)x + \chi(t) = 0$, where ψ, χ and ϕ are functions of t. In general, it is not possible to obtain the solutions in the explicit form. However, if we know one solution, then we can obtain all other solutions of this equation. Assume that $x = x_1$ is one known solution, and let x be any other solution. Write $x = x_1 + y$. Show that y satisfies the Bernoulli equation: $y' + (2x_1\psi + \phi)y + \psi y^2 = 0$.

Exercise 1.17

Find the general solution of the following equations:

(1) $x' + x^2 + x - (1 + t + t^2) = 0$, given that $x(t) = t$ is a solution of this equation.
(2) $x' = (1 - 2t - t^3) + 2(1 + t^2)x - tx^2$, given that $y(t) = t$ is a solution of this equation.
(3) $x' = (t + t^{-1})x - (t^2 + x^2)/2$, given that this equation has a solution $x(t) = t$.

Exercise 1.18

(a) Assume $\psi(t) \neq 0$ for all t in the Riccati equation: $y' + \psi(t)y^2 + \phi(t)y + \chi(t) = 0$. Show that the transformation $y = \dfrac{1}{\psi}\dfrac{z'}{z}$ reduces the Riccati equation to a linear second-order equation for z.

(b) Reduce the original Riccati equation, namely $y' + ay^2 = bt^m$, where a and b are constants, to a second-order linear equation $z'' - abt^m z = 0$ using the transformation mentioned in (a) above.

Exercise 1.19

Assume that the functions p, p_1, p_2, q, q_1 and q_2 are continuous real-valued functions defined on an interval $[a, b]$.

(a) Consider the linear equation $y' + py = q$, and assume $q \geqslant 0$ in $[a, b]$. Show that $y \geqslant 0$ in $[a, b]$, if $y(a) \geqslant 0$. Use this result to compare the solutions of the linear equations $x' + px = q_1$ and $y' + py = q_2$, assuming that $q_1 \geqslant q_2$ in $[a, b]$.

(b) Suppose x and y are solutions of the linear equations $x' + p_1 x = q$ and $y' + p_2 y = q$ in the interval $[a, b]$ and $y \geqslant 0$ in $[a, b]$. If $p_2 \geqslant p_1$ in $[a, b]$ and $x(a) \geqslant y(a)$, show that $x \geqslant y$ in $[a, b]$.

(c) Consider the differential inequality $y' + py \leqslant q$ in the interval $[a, b]$. Derive the inequality

$$y(t) \leqslant \exp\left(-\int_a^t p(s)\,ds\right)\left[y(a) + \int_0^t q(s)\exp\left(\int_a^s p(\tau)\,d\tau\right)ds\right].$$

(d) Derive the Gronwall's inequality: Assume f and g are continuous real-valued functions defined on the interval $[a, b]$ and $g \geqslant 0$ in $[a, b]$. Assume[1]

$$f(t) \leqslant c + k\int_a^t f(s)g(s)\,ds,$$

for all $t \in [a, b]$, where c and k are constants and $k \geqslant 0$. Then,

$$f(t) \leqslant c\exp\left(k\int_a^t g(s)\,ds\right) \quad \text{for all } t \in [a, b].$$

(e) (Uniqueness) Show that the linear IVP $x' + p(t)x = q(t)$, $x(a) = x_0$ has at most one solution in $[a, b]$.

---

[1] To obtain a similar estimate for $t \leqslant a$, interchange the limits in the integral.

Exercise 1.20

Find the general solution of the following equations:

(1) $y' - a\dfrac{y}{t} = \dfrac{t+1}{t}$.

(2) $(t - t^3)y' + 2(t^2 - 1)y - at^3 = 0$.

(3) $y' + ty = t^3 y^3$.

(4) $y - y' \cos t = y^2 \cos t(1 - \sin t)$.

(5) $\left(\dfrac{1}{t^2} + \dfrac{3y^2}{t^4}\right) dt = \dfrac{2y}{t^3} dy$.

(6) $\dfrac{t^2 dy - y^2 dt}{(t-y)^2} = 0$.

Exercise 1.21

Verify whether the following pairs of functions are linearly independent or not, in any interval of $\mathbb{R}$:

(i) $y_1(t) = e^{r_1 t}$, $\quad y_2(t) = e^{r_2 t}$, where $r_1 \neq r_2$.

(ii) $y_1(t) = e^{rt}$, $\quad y_2(t) = te^{rt}$.

(iii) $y_1(t) = e^{\alpha t} \cos \beta t$, $\quad y_2(t) = e^{\alpha t} \sin \beta t$, where $\beta \neq 0$.

(iv) $y_1(t) = t^3$, $\quad y_2(t) = |t|^3$.

Exercise 1.22

Find the general solution of the following equations:

(1) $ty'' - (t+n)y' + ny = 0$.

(2) $y'' - f(t)y' + (f(t) - 1)y = 0$.

(3) $y'' + 2ty' + (1 + t^2)y = 0$.

(4) $t^2 y'' + ty' - y = 0$.

(5) $t^2 y'' + 5ty' - 2y = 0$.

(6) $t^2 y'' - ty' - 2y = 0$.

(7) $t^2 y'' - 3ty' + 4y = 0$.

Exercise 1.23

Three solutions of a certain second-order non-homogeneous linear equation in $\mathbb{R}$ are $\varphi_1(t) = t^2$, $\varphi_2(t) = t^2 + e^{2t}$ and $\varphi_3(t) = 1 + t^2 + 2e^{2t}$. Find the general solution of this equation.

Exercise 1.24

Assume the unique existence of a solution to the nth-order IVP:

$$Ly := y^{(n)} + p_1(t)y^{(n-1)} + \cdots + p_{n-1}(t)y^{(1)} + p_n(t)y = 0$$

$$y(t_0) = y_0, y'(t_0) = y_1, \ldots, y^{(n-1)}(t_0) = y_{n-1},$$

for arbitrary real numbers $y_0, \ldots, y_{n-1}$, where $p_1, \ldots, p_n$ are continuous functions. Let S be the set of all solutions to $Ly = 0$. Show that S is a linear (vector) space of dimension n.

Exercise 1.25

Let $f(t, y, y') = h(t)$ be the general form of the first-order equation, where $h = h(t)$ is all the combined non-homogeneous terms. Consider $L(r, s) = f(t, r, s)$, where t is fixed. If L is linear in (r, s), show that there exist functions $p_0 = p_0(t)$ and $p_1 = p_1(t)$ so that f takes of the form

$$f(t, y, y') = p_0(t)y'(t) + p_1(t)y(t).$$

More generally, an nth-order linear equation has the general form

$$p_0(t)y^{(n)}(t) + p_1(t)y^{(n-1)}(t) + \cdots + p_n(t)y(t) = h(t).$$

Exercise 1.26

Find the envelope and/or locus of singular points of the following families of two-dimensional curves:

(i) Let $0 < r < R$ be fixed constants and $a = (R+r)/2$, $b = (R-r)/2$. The family of curves is given by $(x - a\cos\alpha)^2 + (y - a\sin\alpha)^2 = b^2$, where the parameter $\alpha \in [0, 2\pi]$.

(ii) The family of curves is given by $y^{2n+1} = (x - \alpha)^{2n}$, where the parameter α varies over all the real numbers and n is a fixed positive integer.

(iii) The family of curves is given by $(y - \alpha)^{2n} = k(x - \alpha)^{2n+1}$, where the parameter α varies over all the real numbers, n is a fixed positive integer and k is a fixed non-zero real number.

Exercise 1.27

Find the envelope of the family of circles $(x - t)^2 + y^2 = a^2 t(1 - t)$ where the parameter t varies in the interval $[0, 1]$ and $a > 0$ is a fixed real number. Hence, determine the numerical range of a 2×2 real or complex matrix A, as discussed in Section 1.2.1.

Exercise 1.28

A projectile is launched at a constant velocity v at an angle α. Assume that the trajectory of the projectile is confined to a two-dimensional plane. Describe the trajectories of the projectile for a range of values of α. Find the envelope of these trajectories.

Exercise 1.29

(i) Find the orthogonal trajectories of the family of parabolas $y = Cx^2$, where C is a real parameter.

(ii) Find the isogonal trajectories of the family of concentric circles $x^2 + y^2 = r^2$, where r is a positive real parameter.

Exercise 1.30

Prove the continuity of the solution in appropriate norm with respect to the initial data $\mathbf{x}_0$ and $\mathbf{f}$ of the IVP $\mathbf{x}' = \mathbf{f}(t, \mathbf{x})$, $\mathbf{x}(t_0) = \mathbf{x}_0$, assuming that $\mathbf{f}$ is continuous and Lipschitz continuous with respect to $\mathbf{x}$ variables, in its domain of definition.

Exercise 1.31

Consider the IVP: $\mathbf{x}' = \mathbf{f}(t, \mathbf{x}(t), \mathbf{u}(t))$, $\mathbf{x}(t_0) = \mathbf{x}_0$, where the function $\mathbf{f} : \mathbb{R} \times \mathbb{R}^n \times \mathbb{R}^m \to \mathbb{R}^n$ is continuous and is Lipschitz continuous with respect to $\mathbf{x}$ and $\mathbf{u}$ variables, and the continuous function $\mathbf{u}(t)$ is an external control input applied to the system. Prove that the system has a unique solution for a given initial condition $\mathbf{x}_0$ and a given control function $\mathbf{u}(t)$. Also, prove the following:

(a) Let $\mathbf{x}$ be the unique solution with initial state $\mathbf{x}_0$ and $\tilde{\mathbf{x}}$ be the unique solution with initial condition $\tilde{\mathbf{x}}_0$, for a fixed control input $\mathbf{u}$. Then, there exists a constant $K_1 > 0$ such that $\|\mathbf{x} - \tilde{\mathbf{x}}\| \leqslant K_1 \|\mathbf{x}_0 - \tilde{\mathbf{x}}_0\|$.

(b) Let $\mathbf{x}_\mathbf{u}$ be the unique solution with a control $\mathbf{u}$ and $\mathbf{x}_{\tilde{\mathbf{u}}}$ be the unique solution with a control $\tilde{\mathbf{u}}$ for a fixed initial state $\mathbf{x}_0$. Then, there exists a constant $K_2 > 0$ such that $\|\mathbf{x}_\mathbf{u} - \mathbf{x}_{\tilde{\mathbf{u}}}\| \leqslant K_2 \|\mathbf{u} - \tilde{\mathbf{u}}\|$.

Exercise 1.32

Escape Velocity Problem: Determine the minimum velocity with which a body must be thrown vertically upward so that it will not return to the earth; assume that the air resistance is negligible.

Let M denote the mass of the earth and m denote the mass of the body to be thrown. By Newton's law of gravitation, the force of attraction f acting on the body is $f = k\dfrac{Mm}{r^2}$, where r is the distance between the centre of the earth and the centre of gravity of the body and k is the gravitational constant. The second law of motion, then, implies that the motion of the body is governed by the differential equation

$$m\frac{d^2 r}{dt^2} = -k\frac{Mm}{r^2} \quad \text{or} \quad \frac{d^2 r}{dt^2} = -k\frac{M}{r^2}.$$

The minus sign indicates that the acceleration is negative; air resistance has been omitted. Solve the above equation with initial conditions: $r = R$, $\dfrac{dr}{dt} = v_0$ at time $t = 0$, where R is the radius of the earth, and determine v_0 so that the body in question will not return to earth.

1.4 Solutions

Exercise 1.1

(1) The given system is non-linear. However, the non-linear part, namely $\sin x_1$, is bounded and Lipschitz continuous in $\mathbb{R}$. Hence, global solution exists for any initial condition. This system will be later considered for stability analysis. Since $x_1 = x_2 \equiv 0$ is a solution for the prescribed initial conditions, this is the only solution of the IVP.

(2) Here, the non-linear terms $-x_1(x_1^2 + x_2^2)$ and $-x_2(x_1^2 + x_2^2)$ are cubic polynomials in x_1 and x_2. They are thus Lipschitz continuous in any finite rectangle in $x_1 - x_2$ plane. Hence, local unique solution exists for any initial conditions $x_1(t_0) = x_0^1$ and $x_2(t_0) = x_0^2$. For zero initial conditions, the solution, is the trivial solution, which is global, but for non-zero initial conditions, the solution need not be global. Recall the single equation $y' = y^2$. We may encounter similar equations later in Chapter 5 on qualitative theory.

Exercise 1.2

Note that this problem is non-linear and the function sgn is discontinuous at 0. It can be shown that (see [46]) the given ODE has no C^1 solution when $y(0) = 0$. If $y_0 > 0$, then $y(t) > 0$ for t near 0, by continuity. Thus, we need to solve $y' = -1$ near 0 whose solution is given by $y(t) = -t + c$. Using the initial condition, we get $y(t) = y_0 - t$. Observe that the above solution is valid for $t \in (-\infty, y_0)$ as $y(t)$ changes sign when t crosses y_0.

Exercise 1.3

(i) The equation, in general, is non-linear in all the variables $y, y^{(1)}, \ldots$ and $y^{(n-1)}$ and is linear in the highest derivative term $y^{(n)}$. The equation can be written as
$$\mathbf{x}' = \mathbf{f}(t, \mathbf{x}(t)), \quad \text{where} \quad \mathbf{x} = \begin{bmatrix} x_1 & x_2 & \cdots & x_n \end{bmatrix}^{\mathrm{T}} \quad \text{with} \quad x_i = y^{(i-1)} \quad \text{for} \quad i = 1, \ldots, n$$
$(y^{(0)} = y)$ and $\mathbf{f} = \begin{bmatrix} f_1 & f_2 & \cdots & f_n \end{bmatrix}^{\mathrm{T}}$ with $f_i(t, \mathbf{x}) = x_{i+1}(t) = y^{(i)}(t)$ for $i = 1, \ldots, n-1$ and $f_n(t, \mathbf{x}) = g(t, \mathbf{x})$. Here, the superscript T denotes the transpose. For $i = 1, \ldots, n-1$, the function f_i is linear and hence trivially Lipschitz continuous. Thus, we additionally require that the function f_n is also a locally Lipschitz continuous function. Hence, if we assume that there are neighbourhoods I of t_0 in $\mathbb{R}$, U of $\mathbf{x}_0$ in $\mathbb{R}^n$ such that
$$|g(t, \mathbf{x}) - g(t, \mathbf{y})| \leqslant L\|\mathbf{x} - \mathbf{y}\|,$$
for all $t \in I$, $\mathbf{x}, \mathbf{y} \in U$ and for some constant $L > 0$, then we obtain the existence and uniqueness of a solution to the IVP with $\mathbf{x}(t_0) = \mathbf{x}_0$.

(ii) The given IVP is a linear nth-order equation or equivalently a system of n first-order equations: $\mathbf{x}' = \mathbf{A}(t)\mathbf{x}(t) + \mathbf{b}(t)$, where $\mathbf{x} = \begin{bmatrix} x_1 & x_2 & \cdots & x_n \end{bmatrix}^T$ with $x_i = y^{(i-1)}$ for $i = 1, \ldots, n$ $(y^{(0)} = y)$, $\mathbf{b} = \begin{bmatrix} 0 & 0 & \cdots & b \end{bmatrix}^T$ and

$$\mathbf{A}(t) = \begin{bmatrix} 0 & 1 & \cdots & \cdots & 0 \\ 0 & 0 & 1 & \cdots & 0 \\ \cdots & \cdots & \cdots & \cdots & \cdots \\ -a_1(t) & -a_2(t) & \cdots & \cdots & -a_n(t) \end{bmatrix}.$$

Hence, the continuity of elements of the matrix function $\mathbf{A}(t)$ and the function b in a closed neighbourhood I of t_0 will automatically validate the uniform Lipschitz condition in I. More precisely, assume that a_i, $i = 1, \ldots, n$, and b are continuous in I. Then, there exists a unique solution to the IVP in I. ∎

Exercise 1.4

(1) Here, $f(t, y) = (1 + y^2)^{-1}$, which is defined for all t and y and is uniformly Lipschitz, as $|f(t, y_1) - f(t, y_2)| \leq |y_1 - y_2|$ for all y_1 and y_2. Thus, there is a unique solution $y(t)$ defined for all t, satisfying the initial condition $y(t_0) = y_0$ for arbitrary t_0 and y_0. Since $0 \leq y' \leq 1$, it follows that $y_0 \leq y(t) \leq y_0 + (t - t_0)$ for all $t \geq t_0$. Hence, the solution has linear growth. To find the solution in the explicit form, integrate the equation once to obtain $y + \dfrac{y^3}{3} = t - t_0 + y_0 + \dfrac{y_0^3}{3}$. As the function $y \mapsto y + \dfrac{y^3}{3}$ is a strictly increasing function from $\mathbb{R}$ onto $\mathbb{R}$, we can obtain a unique real solution of the above cubic equation for any y_0 and t_0, but it is cumbersome.[2] Furthermore, we see that $\alpha = -\infty$ and $\beta = \infty$ and $y(t) \to \pm\infty$ as $t \to \pm\infty$.

(2) Here, $f(t, y) = (ty)^{-1}$, which is defined for all $t \neq 0$ and $y \neq 0$. Hence, the initial condition $y(t_0) = y_0$ can be prescribed only for $t_0 \neq 0$ and $y_0 \neq 0$. Since f being the inverse of a quadratic is certainly uniformly Lipschitz in the domain $\Omega = \{(t, y) \in \mathbb{R}^2 : |y| \geq \delta \text{ and } |t| \geq \delta\}$ for any $\delta > 0$ as $|f(t, y_1) - f(t, y_2)| \leq (1/\delta^3)|y_1 - y_2|$ for all $(t, y_1), (t, y_2) \in \Omega$. If $t_0 \neq 0$ and $y_0 \neq 0$, we can choose $\delta > 0$ such that $\delta < (1/2)\min(|t_0|, |y_0|)$. Thus, there is a unique solution $y(t)$ defined for all t in a neighbourhood of t_0. To find the solution in the explicit form, rewrite the given equation as $(y^2/2)' = 1/t$ and integrate. Let $t_0 > 0$ and $y_0 > 0$; the other cases are similar. In this case, we obtain the solution as $y(t) = \sqrt{y_0^2 + 2\log(t/t_0)}$ for $t > 0$. We further observe that $\alpha = t_0 \exp(-y_0^2/2)$ and $\beta = \infty$ and $y(t) \to 0$ as $t \to \alpha+$, $y(t) \to \infty$ as $t \to \beta$.

---

[2] The unique real solution of the cubic equation $z^3 + 3z = k$, $k \in \mathbb{R}$ is given by $z = 2^{-1/3}[(k + \sqrt{k^2 + 4})^{1/3} + (k - \sqrt{k^2 + 4})^{1/3}]$.

(3) Here, the function $f(t,y) = 1/(2y)$ is defined for $y \neq 0$. Hence, the initial condition should be prescribed as $y(t_0) = y_0 \neq 0$ for any t_0. However, f (which is also referred to as the dynamics when the given differential equation is viewed as a dynamical system) is uniformly Lipschitz in $\Omega = \{(t,y) \in \mathbb{R}^2 : |y| \geq \delta\}$ for any $\delta > 0$ as $|f(y_1) - f(y_2)| \leq (1/(2\delta^2))|y_1 - y_2|$ for all $y_1, y_2 \in \Omega$.

Thus, the given ODE with $y(t_0) = y_0 \neq 0$ has a unique solution in a neighbourhood of t_0 as long as $y(t) \neq 0$. Rewriting the equation as $\frac{d}{dt}(y^2) = 1$, we find that the general solution is $y^2 = t + c$. If $y_0 > 0$, then the unique solution is $y(t) = +\sqrt{t - t_0 + y_0^2}$, which is valid for all $t > t_0 - y_0^2$. On the other hand, if $y_0 < 0$, then the unique solution is $y(t) = -\sqrt{t - t_0 + y_0^2}$, which again is valid for all $t > t_0 - y_0^2$. In either case, the maximum interval of existence is $(t_0 - y_0^2, \infty)$ and $y(t) \to 0$ as $t \to (t_0 - y_0^2)+$, which lies on the boundary of the domain where f is defined, and $y(t) \to \pm\infty$ as $t \to \infty$ for $y_0 > 0$ and $y_0 < 0$, respectively. In particular, if $y(1) = 1$, then $y(t) = \sqrt{t}$, $t > 0$ is the solution.

Exercise 1.5

(1) It is easy to see that $y_1(t) = 0$ and $y_2(t) = a/b$ for all t are two trivial constant solutions of the IVP. In fact, these solutions are equilibrium solutions, obtained by solving the algebraic equation $ay - by^2 = 0$. The graphs of the above two solutions, which are nothing but the lines $y = 0$ and $y = a/b$, divide the $t - y$ plane into three disjoint regions: $y > 0$, $0 < y < a/b$ and $y < 0$. Thus, by uniqueness assumption, any solution to the ODE, with an initial condition in any of the three regions, will remain in the same region for all $t \in (t_1, t_2)$. To see this, suppose $0 < y_0 < a/b$ and the corresponding solution y meets the t-axis at $\tilde{t}$, that is, $y(\tilde{t}) = 0$. Then, the given ODE has two different solutions y_1 and y having the same initial condition at $\tilde{t}$, which is a contradiction to the uniqueness assumption. Similar arguments hold for other cases as well. Thus, the sign of $y(t)(y(t) - (a/b))$ is retained for all $t \in (t_1, t_2)$. This proves (1).

(2) Integrating the ODE with respect to t, we get

$$\int_{t_0}^{t} \frac{y'(t)}{ay(t) - by^2(t)} \, dt = \int_{t_0}^{t} dt = t - t_0.$$

Now, making a change of the variable in the integral on the left-hand side (i.e., applying the change of variable formula), we get

$$\int_{t_0}^{t} \frac{y'(t)}{ay(t) - by^2(t)} \, dt = \int_{y_0}^{y(t)} \frac{dy}{ay - by^2}.$$

This integral can be evaluated using the partial fractions of the integrand, and we obtain the solution in the implicit form as

$$\log\left(\frac{|y|}{|y_0|}\frac{|a-by_0|}{|a-by|}\right) = a(t-t_0).$$

(3) We can remove the modulus in the expression obtained above using the relation on signs derived in (1). A bit of computation leads to the explicit form of the solution:

$$y(t) = \frac{ay_0}{by_0 + (a-by_0)e^{-a(t-t_0)}}.$$

(4) If the initial value satisfies $0 < y_0 < a/b$, then $y(t)$ exists for all the time as the denominator never vanishes. Furthermore, $0 < y(t) < a/b$ for all $t \in \mathbb{R}$ and as $t \to \infty$, $y(t) \to a/b$ and as $t \to -\infty$, $y(t) \to 0$. Hence, $t_* = -\infty$ and $t^* = \infty$ in this case.

On the other hand, if $y_0 > a/b$, it follows that $y(t) > a/b > 0$ by (1) for t in the interval of existence. In this case, the term $(a-by_0)e^{-a(t-t_0)} < 0$, and hence, the denominator vanishes only at $\bar{t}$, where

$$\bar{t} = t_0 - \frac{1}{a}\log\left(\frac{by_0}{by_0-a}\right).$$

Since $by_0 > 0$ and $by_0 - a > 0$, we see that $\frac{by_0}{by_0-a} > 1$ and hence $\log\left(\frac{by_0}{by_0-a}\right) > 0$. Thus, $\bar{t} < t_0$ and the solution exists for all $t > \bar{t}$. Furthermore, it is easy to see that $\lim_{t\downarrow\bar{t}} y(t) = \infty$ and $\lim_{t\uparrow\infty} y(t) = a/b$. We thus have $t_* = \bar{t}$ and $t^* = \infty$.

A similar analysis can be carried out when $y_0 < 0$. The formula for $\bar{t}$ is the same as above. In this case, $by_0 - a < by_0 < 0$ and $0 < \frac{by_0}{by_0-a} = \frac{b|y_0|}{b|y_0|+a} < 1$ Thus, $\log\left(\frac{by_0}{by_0-a}\right) < 0$ which gives $\bar{t} > t_0$, and the solution exists for all $t < \bar{t}$ which is a neighbourhood of t_0. In this case, $\lim_{t\uparrow\bar{t}} y(t) = -\infty$ and $\lim_{t\downarrow-\infty} y(t) = 0$ and $t^* = \bar{t}$ and $t_* = -\infty$.

(5) First, consider the case $0 < y_0 < a/b$. Thus, $0 < y(t) < a/b$ for all t, and we have $\frac{dy}{dt} = y(a-by) > 0$. Therefore, $y(t)$ is a strictly increasing function on $\mathbb{R}$ whose range is $(0, a/b)$. From the equation, we see that $\frac{d^2y}{dt^2} = (a-2by)\frac{dy}{dt}$ which is positive if $0 < y(t) < a/(2b)$ and $\frac{d^2y}{dt^2} < 0$ if $a/(2b) < y(t) < a/b$. Thus, there is an accelerated growth of $y(t)$ till it reaches half of the limiting population, which is $a/(2b)$, after which it still grows but in a decelerated way and eventually moves towards the limiting population a/b. In the second case $y_0 > a/b$, it is easy to see that $\frac{dy}{dt} < 0$, whereas $\frac{d^2y}{dt^2} > 0$ for all

$t \in (\bar{t}, \infty)$, the maximal interval of existence. Thus, the solution is a strictly decreasing function. In the third case $y_0 < 0$, the solution remains negative for all t in $(-\infty, \bar{t})$, as we observed above, and both $\dfrac{dy}{dt}$ and $\dfrac{d^2y}{dt^2}$ are negative in this interval, and hence, it decreases from 0 to $-\infty$.

(6) From the above analysis, we conclude that in the first case $0 < y_0 < a/b$, we have $t_* = -\infty$, $t^* = \infty$; in the second case $y_0 > a/b$, $t_* = \bar{t}$, $t^* = \infty$ and in the third case $y_0 < 0$, $t_* = -\infty$, $t^* = \bar{t}$, where $\bar{t}$ is as mentioned in (4).

(7) We take $t_0 = 0$. A typical solution graph when $0 < y_0 < a/b$ is shown in Figure 1.1(i). In this case, the maximal interval of existence is $(-\infty, \infty)$. If $y_0 > a/b$, the maximum interval of existence is (t_*, ∞), where $t_* < 0$. A typical solution graph of this case is shown in Figure 1.1(ii). Finally, if $y_0 < 0$, the maximum interval of existence is $(-\infty, t^*)$, where $t^* > 0$. A typical solution graph of this case is shown in Figure 1.1(iii).

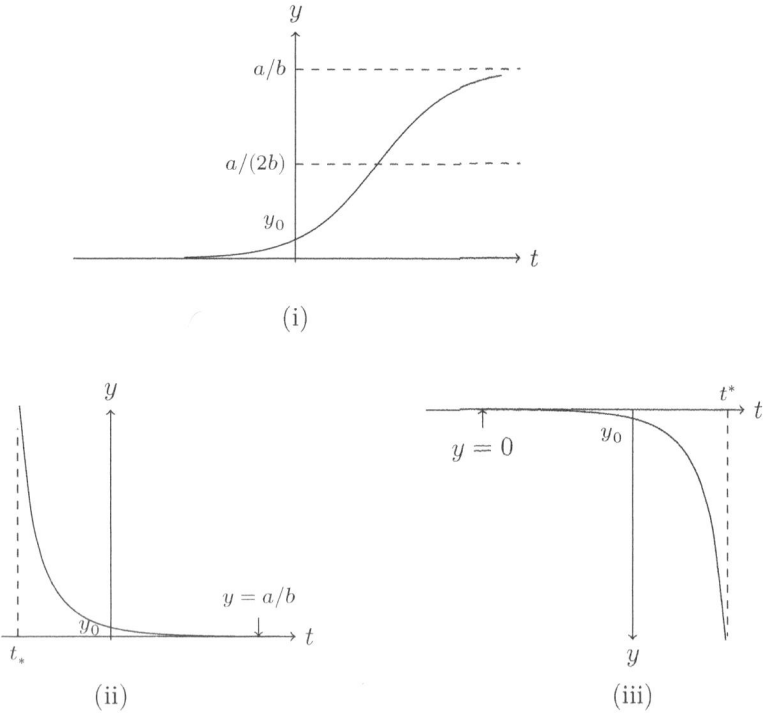

Figure 1.1 | Solution graphs of Exercise 1.5: (i) $0 < y_0 < a/b$, (ii) $y_0 > a/b$, (iii) $y_0 < 0$.

(8) Define $w(t) = y(t + t_0 - t_1)$, and then it is easy to verify that w satisfies the same differential equation and $w(t_1) = y(t_0) = y_0$. Thus, z and w satisfy the same IVP, and by uniqueness, $z(t) = w(t) = y(t + t_0 - t_1)$. This is a very general property enjoyed by an autonomous ODE or even a system. This essentially means that we can take any point on the graph of y and reset as the initial value and it will give the same solution.

In fact, for any constant c, the translated function $w(t) = y(t+c)$ is a solution to the same ODE, and w passes through the initial value y_0 of y at time $t_0 - c$. The same is true with the general autonomous system given in the problem; care should be taken with regard to the interval of existence as the solution w is defined in a different interval. Note that the solution may not exist for $t_0 - c$.

Exercise 1.6

Consider the algebraic equation $ay - by^2 - \lambda = 0$ to get the equilibrium (trivial or constant solutions). The roots of the equation are given by $y = (a \pm \sqrt{a^2 - 4b\lambda})/(2b)$. The roots are real if $a^2 - 4b\lambda \geqslant 0$, that is, $\lambda \leqslant \dfrac{a^2}{4b}$. Thus, let $\lambda_{\text{cr}} = a^2/(4b)$. The case $\lambda < \lambda_{\text{cr}}$ can be analysed exactly as in Exercise 1.5. We have two real and distinct roots $y_1 = (a - \sqrt{a^2 - 4b\lambda})/(2b)$ and $y_2 = (a + \sqrt{a^2 - 4b\lambda})/(2b)$.

Observe that $y_2 > 0$ and $y(t) = y_1$, $y(t) = y_2$ are two constant solutions. The analysis of IVP is exactly similar to the standard logistic equation, by considering three different cases, namely: (i) $y_1 < y_0 < y_2$; (ii) $y_0 > y_2$ and (iii) $y_0 < y_1$. A solution in the implicit form is given by

$$\frac{1}{y_1 - y_2} \log \left(\left|\frac{y - y_1}{y - y_2}\right| \left|\frac{y_0 - y_2}{y_0 - y_1}\right| \right) = a(t - t_0).$$

We will not pursue the analysis further. In fact, if we put $z = y - y_1$, then z satisfies the equation

$$z' = z\left(\sqrt{a^2 - 4b\lambda} - bz\right),$$

which is the same as the standard logistic equation, with a different constant. We conclude that the solution y exists for all the time if the initial value satisfies $y_1 < y_0 < y_2$ and has the limiting values y_1 and y_2 as $t \to \mp\infty$, respectively.

For $\lambda = \lambda_{\text{cr}}$, the quadratic equation $ay - by^2 - \lambda = 0$ has the double real root $y = a/(2b)$. An integration of the given equation in this case gives the solution as $y(t) = a/(2b) + (b(t - t_0 + \gamma))^{-1}$, where the constant of integration γ is expressed in terms of the initial value $y(t_0) = y_0$ as $\gamma = (by_0 - a/2)^{-1}$. Observe that $\gamma = \infty$ when $y_0 = a/(2b)$, and we obtain the constant solution $y(t) = a/(2b)$ for all t. In general, we see that the solution becomes infinite when $t = t_0 - \gamma$ and $y(t) \to a/(2b)$ as $t \to -\infty$ for $\gamma < \infty$.

When $\lambda > \lambda_{\text{cr}}$, there are no real roots of the quadratic equation, and hence, the ODE does not have any constant (real) solutions. However, since $ay(t) - by^2(t) - \lambda$ is locally Lipschitz, the IVP has a local unique solution. When $\lambda > \lambda_{\text{cr}}$, rewrite the quadratic as

$$ay - by^2 - \lambda = -b\left[\left(y - \frac{a}{2b}\right)^2 + \frac{1}{b}(\lambda - \lambda_{\text{cr}})\right],$$

and put $\mu = \sqrt{b(\lambda - \lambda_{\text{cr}})}$. Thus, y' is negative for all the values of y, and y is a strictly decreasing function. An integration of the equation gives the solution in the implicit form as

$$\arctan\left(\frac{by - a/2}{\mu}\right) = -\mu(t + \gamma),$$

where γ is a constant of integration. The left-hand side term has a finite range as y varies over $\mathbb{R}$, whereas the right-hand side term has infinite range as t varies over $\mathbb{R}$. This suggests that the solution y approaches infinity as t approaches a finite value. More precisely, we have $\int_{-\infty}^{\infty} \frac{dy}{by^2 - ay + \lambda} < \infty$. Therefore,[3] there are finite $t_1 < t_2$ such that $\lim_{t \downarrow t_1} y(t) = \infty$ and $\lim_{t \uparrow t_2} y(t) = -\infty$; here, we have used the strictly decreasing property of y. With these observations, we can now write the solution in the explicit form:

$$y(t) = \frac{a}{2b} - \frac{\mu}{b} \tan(\mu(t - t_0 + \gamma)),$$

which is valid in the interval (t_1, t_2). The constant γ can be expressed in terms of initial condition $y(t_0) = y_0$. We have $t_1 = -\frac{\pi}{2\mu} + t_0 - \gamma$ and $t_2 = \frac{\pi}{2\mu} + t_0 - \gamma$. Thus, $t_2 - t_1 = \frac{\pi}{\mu}$.

From the above analysis, we conclude that the given equation is suitable to model the population growth of a species only if $\lambda < \lambda_{\text{cr}}$.

Exercise 1.7

(1) The given equation can be written as

$$\frac{d}{dt}\left(\exp\left(\frac{cg}{W}t\right) V(t)\right) = \frac{g}{W}(W - B) \exp\left(\frac{cg}{W}t\right).$$

Hence, $\exp\left(\frac{cg}{W}t\right)$ is an IF. Using a direct integration and the initial value $V(0) = 0$, we get the solution as

$$V(t) = \frac{W - B}{c}\left(1 - \exp\left(-\frac{cg}{W}t\right)\right).$$

The solution exists for all $t \in \mathbb{R}$, and clearly, $\lim_{t \to \infty} V(t) = \frac{W - B}{c}$ and $\lim_{t \to -\infty} V(t) = -\infty$.

---

[3] Let $f : (c, \infty) \to \mathbb{R}$ be either a positive or a negative function, and assume $\int^{\infty} \frac{dy}{f(y)} < \infty$. Then, any solution of $y' = f(y)$ blows up in finite time. That is, there is a finite τ such that $y(t) \to \infty$ or $-\infty$ as $t \to \tau$. The conclusion is similar if we replace (c, ∞) by $(-\infty, c)$ or $(-\infty, \infty)$.

(2) The velocity $v = v(y)$ satisfies the relation $V(t) = v(y(t))$. Thus, $\dfrac{dV}{dt} = \dfrac{dv}{dy}\dfrac{dy}{dt}$, and substituting for $\dfrac{dV}{dt}$, we arrive at an equation satisfied by v as

$$\frac{v}{W - B - cv}\frac{dv}{dy} = \frac{g}{w}, \quad v(0) = 0.$$

This is a non-linear model, whereas the equation for V was a linear model. Nevertheless, the equation can be solved in the implicit form as follows. We write

$$\frac{v}{W - B - cv} = -\frac{1}{c} + \frac{W - B}{c}\frac{1}{W - B - cv}.$$

Now, the ODE can be directly integrated (using a change of variable) to get

$$-\frac{1}{c}\int_0^v dv + \frac{W-B}{c}\int_0^v \frac{1}{W-B-cv}dv = \frac{g}{w}\int_0^y dy,$$

which gives

$$-\frac{v}{c} - \frac{W-B}{c^2}\log\left(\frac{|W-B-cv|}{W-B}\right) = \frac{gy}{w}.$$

Since the limiting velocity is $\dfrac{W-B}{c}$ and v is an increasing function, we have $v < \dfrac{W-B}{c}$ and thus $W - B - cv > 0$. Hence, we get the required expression. ∎

Exercise 1.8

Make a change of variables $y = ax$ and $\tau = bt$, where a and b are constants to be chosen. Then, $y = y(\tau) = ax(\tau) = ax(bt)$, and thus $\dfrac{dy}{d\tau} = a\dfrac{dx}{d\tau} = \dfrac{a}{b}\dfrac{dx}{dt}$. Similarly, $\dfrac{d^2 y}{d\tau^2} = \dfrac{a^2}{b^2}\dfrac{dx}{dt}$. Hence, the equation becomes

$$\frac{b^2}{a^2}\frac{d^2 y}{d\tau^2} - \frac{\alpha}{a}y + \frac{\beta}{a^3}y^3 + \frac{\delta b}{a}\frac{dy}{d\tau} = 0,$$

or after some simplification,

$$\frac{d^2 y}{d\tau^2} - \frac{\alpha a}{b^2}y + \frac{\beta}{ab^2}y^3 + \frac{\delta a}{b}\frac{dy}{d\tau} = 0.$$

Now, let $ab^2 = \beta$. Then, $\dfrac{\alpha a}{b^2} = \dfrac{\alpha a^2}{\beta}$. Thus, if α and β have the same sign, take $a = \pm\sqrt{\alpha/\beta}$ and choose the sign of a as that of β so that $b = \pm\sqrt{\beta/a}$. Again, choose the sign of b as that of a. If α and β have the opposite signs, take $a = \pm\sqrt{-\alpha/\beta}$ and $b = \pm\sqrt{\beta/a}$. Again, a and b have the same sign as β. This gives the required standard forms of the equation. ∎

Exercise 1.9

First, we make some observations. The non-linear function $f(x) = |x|^{1/2}$ is indeed continuous in the whole line, but it is not Lipschitz continuous in any neighbourhood of 0. It is Lipschitz continuous in any deleted neighbourhood of 0, that is, in $\mathbb{R} \setminus (-\delta, \delta)$ for any $\delta > 0$. This follows from $||x_1|^{1/2} - |x_2|^{1/2}| \leqslant (4\delta)^{-1/2}|x_1 - x_2|$ for all $x_1, x_2 \in \mathbb{R} \setminus (-\delta, \delta)$. Hence, we cannot expect uniqueness in the neighbourhood of the origin with initial values $x(t_0) = 0$, whereas local uniqueness holds when $x(t_0) = x_0 \neq 0$. Thus, we need to consider the cases of zero and non-zero initial conditions separately. Clearly, $x \equiv 0$ is a (trivial) solution to the ODE.

First, consider the case $x(0) = 0$. Since $|x|^{1/2} \geqslant 0$, we see that $x(t)$ is a non-decreasing function whenever it exists. Thus, we have $x(t) \geqslant 0$ for $t \geqslant 0$ and $x(t) \leqslant 0$ for $t \leqslant 0$. Hence, we solve the equation $\dfrac{dx}{dt} = x^{1/2}$ for $t \geqslant 0$ and the equation $\dfrac{dx}{dt} = (-x)^{1/2}$ for $t \leqslant 0$. These equations can be integrated to get the solution as $x(t) = t^2/4$ for $t \geqslant 0$ and $x(t) = -t^2/4$ for $t \leqslant 0$, which can be written as $x(t) = t|t|/4$ for $t \in \mathbb{R}$. This is a global solution to the IVP in the present case in addition to the trivial solution.

In fact, we can construct infinitely many solutions as follows. Let $\alpha, \beta \geqslant 0$, and define x_α, y_β by

$$x_\alpha(t) = \frac{1}{4}(t-\alpha)|t-\alpha|\chi_{[\alpha,\infty)} = \begin{cases} \dfrac{1}{4}(t-\alpha)^2 & \text{if } t \geqslant \alpha, \\ 0 & \text{if } t \leqslant \alpha. \end{cases}$$

$$y_\beta(t) = \frac{1}{4}(t+\beta)|t+\beta|\chi_{(-\infty,\beta]} = \begin{cases} -\dfrac{1}{4}(t+\beta)^2 & \text{if } t \leqslant -\beta, \\ 0 & \text{if } t \geqslant -\beta. \end{cases}$$

Note that $y_\beta(t) = -x_\beta(-t)$. Furthermore, define $z_{\alpha\beta}(t) = x_\alpha(t) + y_\beta(t)$. Then, x_α, y_β and $z_{\alpha\beta}$ are all global solutions to the IVP in the present case. The verification that these functions are differentiable, especially at $t = \alpha$ and $t = \beta$, is left to the reader. Typical graphs of x_α and y_β are depicted in Figure 1.2. Next, consider the case $x(t_0) = 0$ for $t_0 \neq 0$.

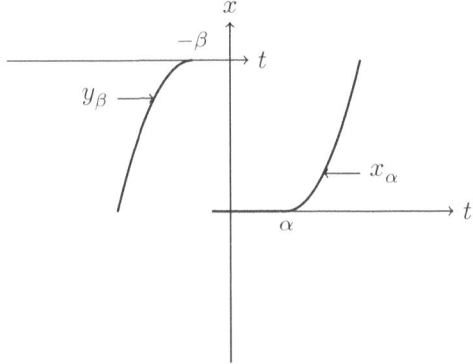

Figure 1.2 | Solution graphs of Exercise 1.9.

First- and Second-Order ODE | 29

The solutions can be obtained by translating the above solutions. More precisely, $x_\alpha(t-t_0)$, $y_\beta(t-t_0)$ and $z_{\alpha\beta}(t-t_0)$ are all global solutions. In fact, it is easily checked that this procedure gives all the solutions, including the trivial one, by using the local uniqueness when $x(t_0) = 0$.

Finally, consider the case $x(t_0) = x_0 \neq 0$. In this case, we will get unique local solution. Assume $x_0 > 0$, the other case being similar. The IVP can be solved to obtain $x(t) = \frac{1}{4}(t - t_0 + 2\sqrt{x_0})^2$. This is the unique solution for $t \geq t_0$. On the left side of t_0, it exists uniquely till it meets the t-axis at the point $t_* = t_0 - 2\sqrt{x_0}$. Thus, the solution exists uniquely in the interval (t_*, ∞). The solution $x(t)$ can then be extended for $t < t_*$ in many ways, using the functions x_α or y_β constructed above. Thus, there is no global uniqueness of the solution.

Exercise 1.10

The equation in separable variable form is given by $\frac{dy}{dt} = h(t)g(y)$. Let $H(t)$ and $G(y)$ be the primitives of $h(t)$ and $\frac{1}{g(y)}$, respectively. Define $\phi(t,y) = G(y) - H(t)$, and then

$$\frac{d}{dt}(\phi(t,y(t))) = \frac{1}{g(y(t))}\frac{dy}{dt}(t) - h(t) = 0.$$

Hence, an ODE in separable variable form is exact.

Exercise 1.11

(i) First, note that the function $f(t,y) = \frac{2y}{t}$ on the right-hand side is not defined on the line $t = 0$. It is therefore necessary to restrict the analysis to the region $t \neq 0$ in the $t-y$ plane. As $\frac{\partial f}{\partial y} = \frac{2}{t}$ is bounded in the region $|t| \geq \delta$ for any $\delta > 0$, the function f is uniformly Lipschitz continuous in this region. Hence, there is a unique solution to the stated IVP. As the given equation is a linear first-order equation, let us rewrite it as $(y/t^2)' = 0$, using an IF. Upon an integration, we see that the general solution to the ODE is given by $y = Ct^2$, for an arbitrary constant C. In fact, this gives us a one-parameter family of solutions. Strictly speaking, the ODE is not defined at $t = 0$. However, the general solution obtained may be interpreted as a solution for all t, with appropriate meaning given at $t = 0$, namely $y(t)/t = 0$ when $t = 0$. However, all these solutions to the IVP satisfy the initial condition $y(0) = 0$, and thus uniqueness is lost.

When $t_0 \neq 0$, we get $C = \frac{y_0}{t_0^2}$, and the solution is given by $y(t) = \frac{y_0}{t_0^2}t^2$, which exists for all $t \in \mathbb{R}$, with appropriate interpretation at $t = 0$. Even in this case, we have uniqueness of the solution only in the interval $(0, \infty)$ or $(-\infty, 0)$ depending on the sign of t_0. For, if $t_0 > 0$, the function y defined by

$$y(t) = \begin{cases} \frac{y_0}{t_0^2} t^2 & \text{if } t \geq 0 \\ Ct^2 & \text{if } t < 0, \end{cases}$$

where C is an arbitrary real number, is also a solution of the IVP. In fact, the function y defined by

$$y(t) = \begin{cases} c_1 t^2 & \text{if } t \geq 0 \\ c_2 t^2 & \text{if } t < 0, \end{cases}$$

where c_1 and c_2 are arbitrary real numbers, is also a solution to the given equation. Geometrically, such a solution is the result of joining a branch of parabola, passing through the origin in the $t-y$ plane, from the positive t-axis to a branch of another parabola from the negative t-axis, at $t = 0$.

Remark: The nature of the general solution enables us to interpret the ODE at $t = 0$ in a suitable way. The reason for this is that the limit $\lim_{t \to 0} \frac{y(t)}{t}$ is 0, for any solution y. The same analysis holds good for the equation $\frac{dy}{dt} = \frac{\mu y}{t}$ for any real $\mu \geq 1$. If $\mu > 1$ and is not an integer, then the general solution is given by $y(t) = C|t|^\mu$. If μ is an odd integer > 1, then there are two families of general solutions: $y(t) = Ct^\mu$ and $y(t) = C|t|^\mu$. However, the same cannot be expected in other situations. As an example, the reader should analyse the equation $y' = \frac{ky}{t^n}$, where $k > 0$ is a constant and $n \geq 2$, an integer.

(ii) The function $f(t, y) = (\sin t)y$ is uniformly Lipschitz in $\mathbb{R} \times \mathbb{R}$. Thus, there is a unique global solution. An IF is given by $\exp\left(\int \sin t \, dt\right) = e^{-\cos t}$, and the given equation can be rewritten as $\frac{d}{dt}(e^{-\cos t} y(t)) = 0$. An integration therefore gives the general solution as $y(t) = Ce^{\cos t}$. Now, applying the initial condition $y(0) = 3/2$, we get the required solution as $y(t) = (3/2) \exp(\cos t - 1)$.

(iii) Since the anti-derivative of e^{t^2} cannot be represented in terms of elementary functions, we do definite integration from 1 to t in the process of obtaining an IF. Then, the unique global solution to the IVP with the given initial condition $y(1) = 2$ is given by

$$y(t) = 2 \exp\left(-\int_1^t e^{s^2} \, ds\right).$$

First- and Second-Order ODE | 31

Exercise 1.12

(1) The given equations consecutively are non-linear, non-linear, linear, linear, non-linear, non-linear, linear, non-linear and non-linear.

(2) Linear if $\beta = 0$ and non-linear if $\beta \neq 0$.

(3) Non-linear.

(4) Linear if $b = d = 0$ and non-linear if either b or d is non-zero.

(5) Non-linear if $\beta \neq 0$ and linear if $\beta = 0$.

(6) Non-linear if $n \neq 0, 1$ and linear if $n = 0, 1$.

(7) Linear.

(8) Non-linear if ψ is not identically a zero function.

Exercise 1.13

For $n = 0, 1$, the equation is linear. So, assume $n \neq 0, 1$. Multiplying[4] the equation by x^{-n}, we get $x^{-n} x' + \phi x^{1-n} = \psi$. This suggests to look for a change of variable $y = x^{1-n}$. Then, y satisfies the linear equation $y' + (1-n)\phi y = (1-n)\psi$, as can easily be verified. An IF for the above equation is given by $Q(t) = \exp\left((1-n)\int \phi\, dt\right)$. Thus, the above equation can be written as $(Q(t)y(t))' = (1-n)Q(t)\psi(t)$. Upon integration, we get the general solution as

$$y(t) = \frac{1}{Q(t)}\left((1-n)\int Q(t)\psi(t)\, dt + C\right)$$

and $x(t) = (y(t))^{\frac{1}{1-n}}$. While determining a particular solution with a prescribed initial value, care should be exercised to insert appropriate limits in the integral.

Exercise 1.14

(i) This is a Bernoulli equation with $n = 2$, $\phi(t) = e^t$ and $\psi(t) = e^t$. Make the substitution $y = x^{-1}$, and see that y satisfies the linear equation $\dfrac{dy}{dt} - e^t y = -e^t$. An IF is $Q(t) = \exp(-e^t)$. Thus, the general solution of this linear equation is given by

$$y(t) = \exp(e^t)\left(-\int e^t \exp(-e^t)\, dt + C\right) = 1 + C\exp(e^t).$$

Therefore, $x(t) = (1 + C\exp(e^t))^{-1}$.

(ii) This is a linear equation if $n = 0$ or $n = 1$ and a Bernoulli equation otherwise with $\phi(t) = (1-n)t^n$ and $\psi(t) = 1$. If $n = 1$, the equation is linear homogeneous equation

---

[4] As n is a real number, the powers make sense only if $x > 0$. However, this is no restriction as any solution to the given equation with a positive initial value would remain positive, at least locally.

and can readily be integrated to obtain the solution as $x(t) = C \exp(t - t^2/2)$. If $n = 0$, the equation is a linear inhomogeneous equation $x' + x = 1$ with a general solution given by $x(t) = 1 + Ce^{-t}$. If $n \neq 1$, make the substitution $y = x^{1-n}$. Then, y satisfies the equation $y' + (1-n)t^n y = 1 - n$. An IF is given by $Q(t) = \exp\left(\dfrac{(1-n)t^{n+1}}{1+n}\right)$. Thus, the general solution is given by

$$y(t) = \exp\left(-\frac{(1-n)t^{n+1}}{1+n}\right)\left((1-n)\int Q(t)dt + C\right)$$

and $x(t) = (y(t))^{\frac{1}{1-n}}$.

Exercise 1.15

The idea is to choose α and β so that the coefficients of $\tau dy - y d\tau$, dy and $d\tau$ are homogeneous of degree 1, that is, of the form $A\,dy + B\,d\tau$ for suitable constants A and B. Rearranging the coefficients, the transformed equation can be written as

$$(b_1\tau + c_1 y)(\tau dy - y d\tau) - [A_2 + b_2\tau + c_2 y - \alpha(A_1 + b_1\tau + c_1 y) - A_1\tau]dy$$

$$+ [A_3 + b_3\tau + c_3 y - \beta(A_1 + b_1\tau + c_1 y) - A_1 y]d\tau = 0,$$

where $A_i = a_i + b_i\alpha + c_i\beta$ for $i = 1, 2, 3$. Choose $A_1 = \eta$, $A_2 = \alpha\eta = \alpha A_1$ and $A_3 = \beta\eta = \beta A_1$. The coefficient of dy in the second term can be written as

$$(b_2 - \alpha b_1 - \alpha\eta)\tau + (c_2 - \alpha c_1 - \beta\eta)y,$$

which is of the form $\tilde{B}_1\tau + \tilde{B}_2 y$. Similarly, the coefficient of $d\tau$ in the third term can be written as

$$(b_3 - \beta b_1 - \beta\eta)\tau + (c_3 - \beta c_1 - \alpha\eta)y,$$

which is again of the form $\tilde{C}_1\tau + \tilde{C}_2 y$. Dividing the equation by $b_1\tau + c_1 y$, we get the equation of the form as required, where

$$\phi(y/\tau) = \frac{\tilde{B}_1\tau + \tilde{B}_2 y}{b_1\tau + c_1 y} = \frac{\tilde{B}_1 + \tilde{B}_2(y/\tau)}{b_1 + c_1(y/\tau)}$$

and

$$\psi(y/\tau) = \frac{\tilde{C}_1\tau + \tilde{C}_2 y}{b_1\tau + c_1 y} = \frac{\tilde{C}_1 + \tilde{C}_2(y/\tau)}{b_1 + c_1(y/\tau)}.$$

Thus, we need to solve for α and β from $A_1 = \eta$ and $A_2 = \alpha\eta = \alpha A_1$, that is, from the equations

$$(a_1 - \eta) + b_1\alpha + c_1\beta = 0, \quad a_2 + (b_2 - \eta)\alpha + c_2\beta = 0, \quad a_3 + b_2\alpha + (c_3 - \eta)\beta = 0.$$

If η is determined by the consistent condition

$$\begin{vmatrix} a_1 - \lambda & b_1 & c_1 \\ a_2 & b_2 - \lambda & c_2 \\ a_3 & b_3 & c_3 - \lambda \end{vmatrix} = 0,$$

then α and β are the solutions of any two of the three equations mentioned above.

Now, make the substitution $y = \tau u$, that is, $dy = u\, d\tau + \tau\, du$, the transformed equation becomes

$$\tau(u\, d\tau + \tau\, du) - \tau u\, d\tau + \phi(u)(u\, d\tau + \tau\, du) + \psi(u) d\tau = 0.$$

Now, rearranging the terms, we get the required Bernoulli equation $\dfrac{d\tau}{du} + h(u)\tau + g(u)\tau^2 = 0$, with appropriate h and g.

Exercise 1.16

Since x_1 and $x = x_1 + y$ are solutions to the given Riccati equation, we have $x_1' + \psi(t)x_1^2 + \phi(t)x_1 + \chi(t) = 0$ and

$$x_1' + y' + \psi(t)(x_1^2 + 2x_1 y + y^2) + \phi(t)(x_1 + y) + \chi(t) = 0.$$

Subtracting the first equation from the second, we arrive at the Bernoulli equation $y' + (2x_1\psi + \phi)y + \psi y^2 = 0$.

Exercise 1.17

(1) Since the given Riccati equation has a solution $x_1(t) = t$, we look for a solution of the form $x(t) = t + y(t)$ as described in Exercise 1.16. Then, y satisfies the Bernoulli's equation $y' + (2t + 1)y + y^2 = 0$, with $n = 2$. Now, make a substitution $z(t) = 1/y(t)$ to obtain a linear equation for z: $z' - (2t+1)z - 1 = 0$. An IF can be determined as $\exp(-(t^2 + t))$, and the solution is given by

$$z(t) = \exp(t^2 + t)\left(\int^t e^{-(t^2+t)}\, dt + C\right),$$

from which y can be determined, and the general solution of the Riccati equation is given by $x(t) = t + y(t)$.

(2) As in (1) above, we look for the general solution in the form $x(t) = x_1(t) + y(t)$ as $x_1(t) = t$ is a solution. Therefore, y satisfies the Riccati equation $y' - 2y + ty^2 = 0$. Making the substitution $y = 1/z$, we see that z satisfies the linear equation $z' + 2z -$

$t = 0$ and its solution is given by $z(t) = Ce^{-2t} + t/2 - 1/4$, where C is a constant of integration. Thus, the general solution to the given problem is

$$x(t) = t + y(t) = t + \frac{1}{z(t)} = t + \frac{4}{4Ce^{-2t} + 2t - 1}.$$

(3) This is a Riccati equation with $\psi(t) = 1/2$, $\phi(t) = -(t + t^{-1})$ and $\chi(t) = t^2/2$. It is given that $x_1(t) = t$ is a solution. Thus, we look for another solution of the form $x(t) = x_1(t) + y(t) = t + y(t)$. Then, y satisfies the equation $y' - (1/t)y + (1/2)y^2 = 0$. The substitution $y = 1/z$ leads to the linear equation $z' + (1/t)z - 1/2 = 0$, whose general solution can be easily obtained as $z(t) = (t^2 + 4C)/(4t)$, where C is a constant of integration. Therefore, $y(t) = 4t/(t^2 + 4C)$ and $x(t) = t + 4t/(t^2 + 4C)$.

Exercise 1.18

(a) First, consider the substitution $v = \psi y$. After a bit of computation, we see that v satisfies the equation $v' + v^2 + \Phi(t)v + \psi(t)\chi(t) = 0$, where $\Phi(t) = \phi(t)\psi(t) - \frac{\psi'}{\psi}$. One more change of variable $v = \frac{z'}{z}$ and some more computation show that z satisfies the equation $z'' + \Phi(t)z' + \psi(t)\chi(t)z = 0$, which is a linear second-order equation.

(b) Here, ψ is the constant function a. Thus, using the transformation $y = \frac{z'}{az}$, a direct computation gives the required equation.

Exercise 1.19

(a) These are comparison results and quite useful in applications, and it is easy to establish them. In fact, the first one follows from the solution representation. We have

$$y(t) = \exp\left(-\int_a^t p(s)\,ds\right)\left(y(a) + \int_a^t q(s)\exp\left(\int_0^s p(\tau)\,d\tau\right)ds\right).$$

Hence, $y(t) \geq 0$ whenever $y(a) \geq 0$, as $q \geq 0$. We cannot conclude similar result when $y(a) < 0$. However, we remark that the solution cannot cross the t-axis more than once. This is the effect of uniqueness. By uniqueness, we can replace $y(a)$ by an initial value at any time t_1, that is, by $y(t_1)$. Thus, if $y(t_1) \geq 0$ at some time t_1, then $y(t)$ will remain non-negative for all $t \geq t_1$.

In the second comparison case, subtract the second equation from the first equation, and then apply the first part to conclude that $x \geq y$ in $[a, b]$.

(b) Let $z(t) = x(t) - y(t)$. Then, z satisfies the equation $z' + p_1 z = (p_2 - p_1)y$. By hypothesis, the term on the right-hand side of this equation is non-negative. Thus, applying the result of (a), we see that $x \geq y$ if $x(a) \geq y(a)$. In particular, the stated conclusion holds true if $x(a) \geq y(a) \geq 0$ and $q \geq 0$ in $[a, b]$.

(c) This is a linear integral inequality. We know that $\exp\left(\int_a^t p(s)\,ds\right)$ is a non-negative IF for the linear operator $\dfrac{d}{dt}+p$. Hence, we can multiply the inequality by the IF without affecting the inequality sign. Thus, we get

$$\frac{d}{dt}\left[y\exp\left(\int_a^t p(s)\,ds\right)\right] \leqslant q\exp\left(\int_a^t p(s)\,ds\right).$$

Now, an integration over $[a,t]$ will give the required inequality.

(d) Let $F(t)=c+k\displaystyle\int_a^t f(s)g(s)\,ds$ for $t\in[a,b]$. Then, $f(t)\leqslant F(t)$, by hypothesis. Furthermore, by differentiating F, we get $F'(t)=kg(t)f(t)\leqslant kg(t)F(t)$, as $g\geqslant 0$ in $[a,b]$. This is a linear inequality for F as in (c) above with $p(t)=-kg(t)$ and $q(t)=0$. Hence, we have the Gronwall's inequality as required.

(e) Let $z=x-y$, where x and y are two solutions to the given IVP. Then, z satisfies $z'+pz=0$, $z(a)=0$. Representing the solution using an IF, we see that

$$|z(t)|\leqslant k\exp\left(\int_a^t p(s)\,ds\right),$$

where k is non-negative constant. Here, we have $c=0$ in (d). Hence, we conclude using Gronwall's inequality that $|z|\leqslant 0$ and hence $x(t)=y(t)$ for all $t\in[a,b]$.

Exercise 1.20

(1) Observe that the line $t=0$ should be avoided in the $t-y$ plane. An initial condition should be of the form $y(t_0)=y_0$ with $t_0\neq 0$. Then, we have a unique local solution for the IVP.

When $a=0$, the solution is given by $y(t)=t+\log|t|+C$, and when $a=1$, the solution is $y(t)=t\left(\log|t|-(1/t)+C\right)$, for suitable constant C. For $a\neq 0,1$, the IF can be computed as t^{-a} for $t>0$ and $(-t)^{-a}$ for $t<0$. For $t>0$, the equation can be rewritten as $(yt^{-a})'=t^{-a}+t^{-a-1}$. Therefore, upon an integration, we obtain the general solution as $y(t)=\dfrac{t}{1-a}-\dfrac{1}{a}+Ct^a$. For $t<0$, we just have to replace t by $|t|$ in the above expression. Hence, $y(t)=\dfrac{|t|}{1-a}-\dfrac{1}{a}+C|t|^a$ for all $t\neq 0$.

Note that the expressions for the general solution indicate the necessity to consider the cases $t>0$ and $t<0$ separately.

(2) Here, the values of $t=0,\pm 1$ should be avoided, and the intervals for the initial should be one of the intervals $(-\infty,-1)$, $(-1,0)$, $(0,1)$ and $(1,\infty)$. The equation can be written

in the standard form as $y' - (2/t)y = (at^2)/(1-t^2)$, and the corresponding IF is t^{-2}. Thus, rewriting the equation as $(t^{-2}y)' = a/(1-t^2)$ and performing an integration, the solution is given by

$$y(t)t^{-2} = a \int^t \frac{1}{(1-t)(1+t)} \, dt + C.$$

From an integration of the integral on the right-hand side term, we get the general solution as

$$y(t) = \frac{at^2}{2} \left(\log|1-t| + \log|1+t| + C \right).$$

(3) This is a Bernoulli equation with $n = 3$. Thus, we make the substitution $z = y^{-2}$. Then, z satisfies the linear equation $z' - 2tz = t^3$. An IF is $\exp(-t^2)$, and the equation is rewritten as $[\exp(-t^2)y]' = t^3 \exp(-t^2)$. Hence, upon an integration, the solution z is given by

$$z(t)\exp(-t^2) = \int^t t^3 \exp(-t^2) \, dt + C,$$

where C is a constant of integration. It is not difficult to evaluate the integral on the right-hand side by a change of variable. The end result is $z(t) = -\frac{1}{2}(1+t^2) + C\exp(t^2)$.

Thus, we have $y(t) = \dfrac{\sqrt{2}}{\sqrt{2C\exp(t^2) - 1 - t^2}}$.

(4) This is also a Bernoulli equation with $n = 2$. If we make the substitution $z = y^{-1}$, we get the linear equation $z' + \dfrac{z}{\cos t} = 1 - \sin t$. Before proceeding further, observe that the term $\cos t$ sits in the denominator, and as such, we should be concerned with the points where it vanishes. We do the analysis in the interval $(-\pi/2, \pi/2)$, and the reader should do the analysis in different intervals. In this interval, we have $\cos t > 0$ and $\sec t + \tan t > 0$. An IF can be found to be

$$\exp\left(\int \frac{dt}{\cos t} \right) = \exp(\log|\tan t + \sec t|) = \tan t + \sec t,$$

by the above observation. Thus, from an integration of the linear equation, we get

$$z(t)(\tan t + \sec t) = \int (1 - \sin t)(\tan t + \sec t) \, dt.$$

Now,

$$(1 - \sin t)(\tan t + \sec t) = (1 - \sin t)\left(\frac{\sin t}{\cos t} + \frac{1}{\cos t} \right) = \frac{1 - \sin^2 t}{\cos t} = \cos t.$$

Thus, evaluating the integral on the right-hand side, the solution is given by

$$z(t) = \frac{\sin t + C}{\tan t + \sec t} \quad \text{and} \quad y(t) = \frac{\tan t + \sec t}{\sin t + C}.$$

Remark: Consider the IVP $z' + z/\cos t = 1 - \sin t$, $z(0) = 1$. What is the maximal interval of existence? What happens if the initial value is changed to $z(0) = 2$?

(5) Writing the equation as $y' = (3/(2t))y + t/(2y)$, we see that it is a Bernoulli equation with $n = -1$ (and $t = 0$ to be avoided). Making a substitution $z = y^2$, we see that z satisfies the linear equation $z' - (3/t)y = t$, which can be solved by noting that t^{-3} is an IF:
$$z(t) = Ct^3 - t^2, \quad y(t) = |t|\sqrt{Ct - 1}.$$

(6) Note that the line $y = t$ in the $t - y$ plane should be avoided. If the initial condition is $y(t_0) = y_0 \neq t_0$, then by continuity the solution satisfies the condition $y(t) \neq t$ for $|t - t_0|$ small. This observation helps us rewrite the equation as $t^2 y' = y^2$, which is in the variable's separable form. Integrating this equation, we get $y(t) = \dfrac{t}{1 - Ct}$ as the general solution, where C is a constant of integration.

Exercise 1.21

Linear independence can be proved either using the Wronskian or directly; the argument using the Wronskian may not work in some cases. However, the latter remark does not apply to the solutions of linear second-order equations.

(i) Here, the Wronskian is $W(t) = y_1(t)y_2'(t) - y_1'(t)y_2(t) = (r_2 - r_1)e^{(r_1+r_2)t} \neq 0$. Hence, y_1 and y_2 are linearly independent. To see the result directly, suppose $e^{r_1 t} = Ce^{r_2 t}$ for some constant C. Then, $C = e^{(r_1 - r_2)t}$ which is not a constant as $r_1 \neq r_2$.

(ii) Here, the Wronskian $W(t) = -2e^{2rt} \neq 0$. Hence, y_1 and y_2 are linearly independent.

(iii) Again, the linear independence can be established using the Wronskian. This can also be seen directly. If $y_2 = Cy_1$, for some constant C, then we see that $C = \tan \beta t$, which is not a constant as $\beta \neq 0$.

(iv) In any interval on the positive (respectively, negative) t-axis, we see that $y_2 = y_1$ (respectively, $y_2 = -y_1$). However, if we consider any interval containing 0 in its interior, then we see that y_1 and y_2 are linearly independent, as the relations $y_2 = \pm y_1$ cannot simultaneously hold. However, it is easy to see that the Wronskian of y_1 and y_2 is 0 everywhere.

Exercise 1.22

(1) Here, $y_1 = e^t$ is a solution. Again, we look for another independent solution $y_2(t) = C(t)y_1(t) = C(t)e^t$. Substituting y_2 and its derivatives in the equation, we see that $C(t)$ satisfies the equation $C'' = ((n/t) - 1)C'$. Integrate this equation twice, without adding constants, to get $C(t) = \int^t t^n e^{-t}$, and the second solution is given by $y_2(t) = e^t \int^t t^n e^{-t}$. Again, it is not difficult to verify the linear independence of y_1 and y_2. Hence, the general solution is $y(t) = e^t (A + BI_n)$, where A and B are

arbitrary constants and $I_n = \int t^n e^{-t} dt$. This integral can be written down explicitly by deriving a recurrence formula. An integration by parts gives $I_n = -t^n e^{-t} + nI_{n-1}$, for $n = 1, 2, \ldots$ with $I_0 = -e^{-t}$. In particular, $I_1 = -e^{-t}(1+t)$ and $I_2 = -e^{-t}(2 + 2t + t^2)$. In fact, we can observe a pattern in the recursion relation, and it is possible to write down an explicit formula for I_n, $n \geq 0$.

(2) In this case, observe that $y_1 = e^t$ is a solution. Again, we look for another independent solution $y_2(t) = C(t)y_1(t) = C(t)e^t$. We see that C is given by

$$C(t) = \int^t e^{2t} \exp\left(-\int^t f(s)\, ds\right) dt,$$

and the general solution to the given equation is

$$y(t) = e^t \left(A + B \int^t e^{2t} \exp\left(-\int^t f(s)\, ds\right) dt\right).$$

Appropriate limits in the integration should be provided while solving an IVP.

(3) We look for a solution of the form $y = uv$. Substituting this into the given equation, we get

$$(uv'' + 2u'v' + u''v) + 2t(uv' + u'v) + (1 + t^2)uv = 0.$$

We choose u and v so that the coefficient of v' term is absent, in the above equation. This gives $2u' + 2tu = 0$. Thus, we choose $u(t) = \exp(-t^2/2)$. Substituting this in the above equation, we get an equation for v as $v'' = 0$. Thus, $v(t) = At + B$, where A and B are arbitrary constants. Hence, the general solution to the given equation is $y(t) = (At + B) \exp(-t^2/2)$. Observe that the two solutions $\exp(-t^2/2)$ and $t \exp(-t^2/2)$ of the given equation are linearly independent.

The equations in (4)–(7) are Euler equations. An Euler equation of second order[5] is the equation $t^2 y'' + \alpha t y' + \beta y = 0$, where α and β are constants. This equation is invariant under the transformation $t \mapsto -t$. Hence, we may assume that $t > 0$. This equation is transformed into a linear constant coefficient equation by the change of variable $\tau = \log t$:

$$\frac{d^2 y}{d\tau^2} + (\alpha - 1)\frac{dy}{d\tau} + \beta y = 0.$$

The characteristic roots of this equation then determine the general solution. By the substitution $\tau = \log t$ will give the general solution in t variable. For example, for the equation in (4), we have $\alpha = 1$ and $\beta = -1$. Therefore, the characteristic roots are the roots of the quadratic equation $m^2 - 1 = 0$. Thus, the general solution of the

---

[5] An nth-order Euler equation is a linear equation of the form

$$t^n y^{(n)} + a_{n-1} t^{n-1} y^{(n-1)} + \cdots + a_0 y = 0,$$

where $a_0, \ldots, a_{n-1}$ are constants.

given equation is $y(\tau) = Ae^\tau + Be^{-\tau}$ or $y(t) = At + B/t$, where A and B are arbitrary constants. The other equations can be worked out in a similar fashion.

Remark: On passing, we make a remark on the Euler equation $t^2 y'' + y/4 = 0$, whose general solution is given by $y(t) = (A + B \log t)t^{1/2}$, $t > 0$, where A and B are arbitrary constants. Thus, y has exactly one zero (root) in $t > 0$ if $AB \neq 0$. On the other hand, the general solution of the equation $t^2 y'' + \lambda y = 0$ has infinitely many zeroes in $t > 0$ if $\lambda > 1/4$ and has at most one zero if $\lambda < 1/4$. We say that the equation $t^2 y'' + \lambda y = 0$ is *oscillatory* if $\lambda > 1/4$ and *non-oscillatory* if $\lambda < 1/4$. Finding out sufficient conditions on the function $q(t)$ so that the second-order equation $y'' + q(t)y = 0$ is oscillatory is required in the study of the eigenvalue problem of second-order equations.

Exercise 1.23

Since ϕ_1, ϕ_2 and ϕ_3 are solutions to the linear second-order non-homogeneous equation, we see that their pairwise differences would be solutions to the corresponding homogeneous equation. Thus, in particular, $z = \phi_2 - \phi_1 = e^{2t}$ and $w = \phi_3 - \phi_1 = 1 + 2e^{2t}$ are solutions to the corresponding homogeneous equation. Since the Wronskian $zw' - z'w = 4e^{4t} - 2e^{2t}(1 + 2e^{2t}) = -2e^{2t}$, of z and w, is non-zero, they are linearly independent. Thus, a general solution to the homogeneous equation is $Ae^{2t} + B(1 + 2e^{2t})$ which can also be rewritten as $Ae^{2t} + B$ by combining the last term with the first term. Thus, the solution to the non-homogeneous equation can be obtained by adding any of the particular solutions given. Thus, $y(t) = Ae^{2t} + t^2 + B$ is the general solution.

In particular, the solution of the non-homogeneous linear equation $\mathcal{L}y = g$, satisfying the initial conditions $y(0) = 1$ and $y'(0) = 2$, can be found to be $y(t) = e^{2t} + t^2$, using the expression for the general solution.

Exercise 1.24

Let z_i, $i = 1, \ldots, n$ be the unique solution to the IVP with initial data $z_i^{(j-1)}(t_0) = \mathbf{e}_{ji}$, for $j = 1, \ldots, n$, where $\{\mathbf{e}_i, 1 \leqslant i \leqslant n\}$ is the canonical basis in $\mathbb{R}^n$. Now, let $y \in S$ be any solution. Then, it is easy to see by uniqueness that $y = \sum_{i=1}^{n} y^{(i-1)}(t_0) z_i$. This means that the set $\{z_1, \ldots, z_n\}$ spans S and hence $\dim(S) \leqslant n$. To get the equality, it suffices to prove that the vectors $z_1, \ldots, z_n$ are linearly independent.

Assume $c_1 z_1(t) + \cdots + c_n z_n(t) = 0$ for all t and some constants $c_1, \ldots, c_n$. Differentiating this relation $n - 1$ times, we obtain the equation

$$\sum_{i=1}^{n} c_i z_i^{(j-1)}(t) = 0, \quad \text{for} \quad j = 1, 2, \ldots, n.$$

In particular, taking $t = t_0$ and using the initial conditions, we get

$$\sum_{i=1}^{n} c_i \mathbf{e}_{ji} = 0, \quad \text{for} \quad j = 1, 2, \ldots, n.$$

This immediately gives that $c_1 = \cdots = c_n = 0$, proving the assertion that $\dim(S) = n$.

Exercise 1.25

The proof follows from a familiar result from linear algebra. Since L is a continuous linear functional, by Riesz representation theorem, there are constants p_0 and p_1 such that $L(r,s) = p_0 r + p_1 s$. Since L depends on t, we see that p_0 and p_1 also depend on t and we have the representation as required. In the general case, keeping t fixed, we have $L(r_1, \ldots, r_{n+1}) = f(t, r_1, \ldots, r_{n+1})$, where $f(t, y, y^{(1)}, \ldots, y^{(n)}) = h(t)$ be the general nth-order equation. Again, L is a continuous linear functional on $\mathbb{R}^{n+1}$ and hence the existence of $p_0, p_1, \ldots, p_n$ satisfying the given representation for f. ∎

Exercise 1.26

(i) The given family of curves is

$$(x - a\cos\alpha)^2 + (y - a\sin\alpha)^2 = b^2. \tag{1.3.1}$$

This equation can be rewritten as

$$2a(x\cos\alpha + y\sin\alpha) = x^2 + y^2 + a^2 - b^2 = x^2 + y^2 + Rr. \tag{$*$}$$

These are circles situated in the annulus $r^2 \leqslant x^2 + y^2 \leqslant R^2$. Differentiating equation (1.3.1) with respect to α, we get

$$x\sin\alpha - y\cos\alpha = 0. \tag{1.3.2}$$

To eliminate α from equations $(*)$ and (1.3.2), square both the equations and add. Some simplification leads to

$$4a^2(x^2 + y^2) = (x^2 + y^2)^2 + 2Rr(x^2 + y^2) + R^2 r^2,$$

which is the same as

$$(x^2 + y^2 - r^2)(x^2 + y^2 - R^2) = 0.$$

Each factor represents a circle. We therefore conclude that the circles $x^2 + y^2 = r^2$ and $x^2 + y^2 = R^2$ are the envelopes of the given family of circles. Furthermore, it is straightforward to check that the given family of circles has no singular point.

(ii) The given family of curves is $y^{2n+1} = (x - \alpha)^{2n}$. When $n = 1$, this curve is called a *semi-cubic parabola*. Differentiation with respect to α gives $\alpha = x$. Substituting this into the given equation gives $y = 0$. However, the points on the x-axis are easily seen to be singular points. Thus, the x-axis is a locus of singular points and the given family of curves has no envelope.

(iii) The family of curves is given by $(y - \alpha)^{2n} = k(x - \alpha)^{2n+1}$. Differentiation with respect to α gives the relation

$$2n(y - \alpha)^{2n-1} = k(2n+1)(x - \alpha)^{2n}.$$

Substituting this into the given equation, we get

$$k(x-\alpha)^{2n+1} = (y-\alpha)^{2n} = (y-\alpha)^{2n-1}(y-\alpha)$$
$$= \frac{k(2n+1)}{2n}(x-\alpha)^{2n}(y-\alpha).$$

The last relation implies $\alpha = x$ or $\alpha = (2n+1)y - 2nx$. The first choice gives, using the given equation, $y = x$ and the second choice gives (after some computations) $y = x - \frac{(2n)^{2n}}{k(2n+1)^{2n+1}}$ and also $y = x$. It is straightforward to observe that the points on the line $y = x$ are all singular points of the given family of curves. Thus, the line $y = x$ is the locus of singular points. On the other hand, the line $y = x - \frac{(2n)^{2n}}{k(2n+1)^{2n+1}}$ is an envelope of the given family of curves.

Exercise 1.27

The given family of curves is $(x-t)^2 + y^2 = a^2 t(1-t)$. Differentiating this equation with respect to t, we get $-2(x-t) = a^2(1-2t)$ or $t = \frac{a^2+x}{2(1+a^2)}$. Substituting this value of t in the given equation and performing some simple computations give

$$\frac{(x-1/2)^2}{(1+a^2)/4} + \frac{y^2}{a^2/4} = 1,$$

which represents an ellipse with centre at $(1/2, 0)$, with semi-major axis equal to $\sqrt{1+a^2}/2$ and semi-minor axis equal $a/2$. Writing $z = x + iy$ and using a well-known property of an ellipse, the above equation of the ellipse can be written as $|z| + |z-1| = \sqrt{1+a^2}$.

Now, return to the problem of determining the numerical range of a 2×2 real or complex matrix A with eigenvalues λ_1 and λ_2. Using the notations in Section 1.2.1, we rewrite the above equation of the ellipse as $|w - \lambda_1| + |w - \lambda_2| = \sqrt{|\lambda_1 - \lambda_2|^2 + |m|^2}$, where $z = \frac{w - \lambda_2}{\lambda_1 - \lambda_2}$. This represents the ellipse with foci at λ_1 and λ_2 with semi-major axis equal to $(1/2)\sqrt{|\lambda_1 - \lambda_2|^2 + |m|^2}$. The numerical range $W(A)$ is then the elliptic disk whose boundary is this ellipse. We conclude that

$$W(A) = \left\{ w \in \mathbb{C} : |w - \lambda_1| + |w - \lambda_2| \leqslant \sqrt{|\lambda_1 - \lambda_2|^2 + |m|^2} \right\}.$$

Exercise 1.28

Choose the co-ordinate system so that the projectile is launched from the origin and that its trajectory is confined to the two-dimensional $x - y$ plane. The x and y co-ordinates of a point on the trajectory at any time t are $x = vt \cos \alpha$ and $y = vt \sin \alpha - (1/2)gt^2$, assuming that the air resistance is negligible. Here, g denotes the acceleration due to gravity in appropriate units. Only the time range that is necessary is the time when the projectile hits back the ground. Eliminating the time variable from these equations, we find that $y = x \tan \alpha - \frac{g}{2v^2}(\sec^2 \alpha)x^2$.

This is a family of parabolas with α as parameter. Differentiating this equation with respect to α, we get $x \sec^2 \alpha - (g/v^2) \sec^2 \alpha \tan \alpha = 0$ or $\tan \alpha = v^2/(gx)$. Substituting this into the equation of the parabola (and thus eliminating α), we obtain $y = \frac{v^2}{2g} - \frac{g}{2v^2}x^2$, which also represents a parabola. This is the envelope of the family of the projectiles under consideration and is called the *safety parabola*, as there is no impact of the projectile outside this zone.

Exercise 1.29

(i) The differential equation satisfied by the given family of parabolas is $\frac{dy}{dx} = 2Cx = \frac{2y}{x}$. Therefore, the differential equation satisfied by an orthogonal trajectory is given by $\frac{dy}{dx} = -\frac{x}{2y}$. Upon integration, we get $\frac{x^2}{2} + y^2 = C^2$. This represents the family of ellipses with the same centre and the ratio of the major axis and minor axis being $\sqrt{2}$.

(ii) The differential equation satisfied by the given family of circles is $\frac{dy}{dx} = -\frac{x}{y}$. Therefore, the differential equation satisfied by an isogonal trajectory is given by $\frac{\frac{dy}{dx} - \kappa}{1 + \kappa \frac{dy}{dx}} = -\frac{y}{x}$, where $\kappa = \tan \alpha$, with α being the given constant (acute) angle. Upon simplification, we get $\frac{dy}{dx} = \frac{\kappa y - x}{y + \kappa x}$. This is a homogeneous equation. By introducing a new dependent variable $v = y/x$, we obtain $x \frac{dv}{dx} = -\frac{v^2 + 1}{v + \kappa}$, which is an equation in separable variable form. Solving this equation and substituting $v = y/x$, we get

$$\log(\sqrt{x^2 + y^2}) = C - \kappa \arctan(y/x),$$

where C is an arbitrary constant. It is convenient to introduce the polar co-ordinates: $x = r \cos \theta$ and $y = r \sin \theta$. The above equation may now be written as $r = Ce^{-\kappa \theta}$, where C is now a positive real parameter. A member of this family of curves is called an *equiangular spiral*. It is also called a *logarithmic spiral*.

Exercise 1.30

The equivalent integral formulation is given by

$$\mathbf{x}(t) = \mathbf{x}_0 + \int_{t_0}^{t} \mathbf{f}(s, \mathbf{x}(s))\, ds.$$

If **y** is a solution of the IVP $\mathbf{y}' = \mathbf{g}(t, \mathbf{y}(t))$, $\mathbf{y}(t_0) = \mathbf{y}_0$, then by subtraction, we get

$$\|\mathbf{x}(t) - \mathbf{y}(t)\| \leq \|\mathbf{x}_0 - \mathbf{y}_0\| + \int_{t_0}^{t} \|\mathbf{f}(s, \mathbf{x}(s)) - \mathbf{g}(s, \mathbf{y}(s))\| \, ds$$

$$\leq \|\mathbf{x}_0 - \mathbf{y}_0\| + \int_{t_0}^{t} \|\mathbf{f}(s, \mathbf{x}(s)) - \mathbf{f}(s, \mathbf{y}(s))\| \, ds$$

$$+ \int_{t_0}^{t} \|\mathbf{f}(s, \mathbf{y}(s)) - \mathbf{g}(s, \mathbf{y}(s))\| \, ds$$

$$\leq \|\mathbf{x}_0 - \mathbf{y}_0\| + L_{\mathbf{f}} \int_{t_0}^{t} \|\mathbf{x}(s) - \mathbf{y}(s)\| \, ds + \|\mathbf{f} - \mathbf{g}\|_{\infty} (T - t_0),$$

where $L_{\mathbf{f}}$ is the Lipschitz constant of f, and we are considering the interval $[t_0 - T, t_0 + T]$. By Gronwall's inequality, we get

$$\|\mathbf{x} - \mathbf{y}\| \leq (\|\mathbf{x}_0 - \mathbf{y}_0\| + \|\mathbf{f} - \mathbf{g}\|_{\infty} (T - t_0)) \exp(L_{\mathbf{f}}(T - t_0)).$$

In fact, by changing the role of f and g, we get

$$\|\mathbf{x} - \mathbf{y}\| \leq (\|\mathbf{x}_0 - \mathbf{y}_0\| + \|\mathbf{f} - \mathbf{g}\|_{\infty} (T - t_0)) \exp(L(T - t_0)),$$

where $L = \min(L_{\mathbf{f}}, L_{\mathbf{g}})$, and $L_{\mathbf{g}}$ is the Lipschitz constant of **g**. This proves the continuity with respect to the initial data $\mathbf{x}_0$ and **f**.

Exercise 1.31

The existence and uniqueness can be proved in a similar fashion as in equation (1.1.5) using Picard's iterates:

$$\mathbf{x}_{n+1}(t) = \mathbf{x}_0 + \int_{t_0}^{t} \mathbf{f}(s, \mathbf{x}_n(s), \mathbf{u}(s)) \, ds.$$

A similar procedure can be repeated to show that the series

$$S_n(t) = \mathbf{x}_0 + \sum_{i=1}^{n} [\mathbf{x}_i(t) - \mathbf{x}_{i-1}(t)] = \mathbf{x}_n(t)$$

converges uniformly to some **x** in some interval $[t_0, t_0 + h]$ for h small. It is easy to show that **x** satisfies the integral equation. We prove the uniqueness in the course of proving the statements (a) and (b).

Suppose $\mathbf{y}$ is another solution to the same ODE with the initial condition $\mathbf{y}(t_0) = \mathbf{y}_0$ and a control function $\mathbf{v}$. Then, using the Lipschitz property of f, we get

$$\|\mathbf{x}(t) - \mathbf{y}(t)\| \leqslant \|\mathbf{x}_0 - \mathbf{y}_0\| + \int_{t_0}^{t} \|\mathbf{f}(s, \mathbf{x}(s), \mathbf{u}(s)) - \mathbf{f}(s, \mathbf{y}(s), \mathbf{v}(s))\| \, ds$$

$$\leqslant \|\mathbf{x}_0 - \mathbf{y}_0\| + \int_{t_0}^{t} \|\mathbf{f}(s, \mathbf{x}(s), \mathbf{u}(s)) - \mathbf{f}(s, \mathbf{y}(s), \mathbf{u}(s))\| \, ds$$

$$+ \int_{t_0}^{t} \|\mathbf{f}(s, \mathbf{y}(s), \mathbf{u}(s)) - \mathbf{f}(s, \mathbf{y}(s), \mathbf{v}(s))\| \, ds$$

$$\leqslant \|\mathbf{x}_0 - \mathbf{y}_0\| + L \int_{t_0}^{t} \|\mathbf{x}(s) - \mathbf{y}(s)\| \, ds + L \int_{t_0}^{t} \|\mathbf{u}(s) - \mathbf{v}(s)\| \, ds$$

$$\leqslant (\|\mathbf{x}_0 - \mathbf{y}_0\| + L\|\mathbf{u} - \mathbf{v}\|(T - t_0)) + L \int_{t_0}^{t} \|\mathbf{x}(s) - \mathbf{y}(s)\| \, ds,$$

where L is the Lipschitz constant. The Gronwall's inequality now implies the estimate

$$\|\mathbf{x} - \mathbf{y}\| \leqslant K(\|\mathbf{x}_0 - \mathbf{y}_0\| + \|\mathbf{u} - \mathbf{v}\|),$$

for some constant K. Thus, we have the continuity with respect to the data including the control function. Clearly, if $\mathbf{x}_0 = \mathbf{y}_0$ and $\mathbf{u} = \mathbf{v}$, we obtain the uniqueness. On the other hand, if $\mathbf{u} = \mathbf{v}$, we get the estimate in (a) and the estimate in (b) is obtained by taking $\mathbf{x}_0 = \mathbf{y}_0$.

Exercise 1.32

Multiplying the second-order equation by $\dfrac{dr}{dt}$, we get

$$\frac{dr}{dt} \frac{d^2r}{dt^2} = -k \frac{M}{r^2} \frac{dr}{dt} \quad \text{or} \quad \frac{1}{2} \frac{d}{dt}\left(\frac{dr}{dt}\right)^2 = k \frac{d}{dt}\left(\frac{M}{r}\right).$$

An integration of this equation gives

$$\frac{v^2}{2} = \frac{kM}{r} + \left(\frac{v_0^2}{2} - \frac{kM}{R}\right),$$

where $v = \dfrac{dr}{dt}$. Here, we have used the prescribed initial conditions. This is referred to as the conservation of total energy, which is the sum of kinetic energy and potential energy.

We notice that the first term on the right-hand side of this conservation equation is always positive. Hence, if the second term therein is also non-negative, then v can never become 0 and we would have achieved the task of making the body not to return to earth. Therefore, we conclude that v_0 should satisfy the condition $v_0^2 \geqslant \dfrac{2kM}{R}$.

In the usual CGS units, $k = 6.66 \times 10^{-8} \text{cm}^3/\text{g.sec}^2$ and $R = 63 \times 10^7 \text{cm}$. At the earth's surface $r = R$, the acceleration due to gravity is $g = 981 \text{cm/sec}^2$. Therefore, $g = \dfrac{kM}{R^2}$. Hence, we require $v_0^2 \geqslant 2gR$, which gives an approximate value of v_0 as $11.2 \times 10^5 \text{cm/sec}^2$. This particular value of the velocity of the body is termed as the *escape velocity*. Notice that it does not depend on the mass of the body.

2

Linear Systems

2.1 Introduction

Consider the two-dimensional linear system

$$x' = ax + by, \quad y' = cx + dy, \tag{2.1.1}$$

where a, b, c, d are real numbers. Depending on the nature of eigenvalues of the coefficient matrix $\mathbf{A} = \begin{bmatrix} a & b \\ c & d \end{bmatrix}$, we can construct a non-singular real matrix $\mathbf{C}$ such that the substitution $\begin{bmatrix} u \\ v \end{bmatrix} = \mathbf{C} \begin{bmatrix} x \\ y \end{bmatrix}$ reduces the system (2.1.1) into one of the following systems:

(i) $u' = \lambda u, \; v' = \mu v$

(ii) $u' = \lambda u + v, \; v' = \lambda v$

(iii) $u' = \alpha u + \beta v, \; v' = -\beta u + \alpha v.$

Here λ and μ are the eigenvalues of $\mathbf{A}$, in case $\mathbf{A}$ has real eigenvalues; they may be equal. Whereas α and $\beta (\neq 0)$ are respectively the real and the imaginary parts of the eigenvalues, in case $\mathbf{A}$ has non-real eigenvalues (they occur in conjugate pair) $\alpha \pm i\beta$. It is easy to write down the solutions of the reduced systems. We have $u(t) = u_0 e^{\lambda t}$, $v(t) = v_0 e^{\mu t}$ for the case (i). The solutions are $u(t) = (u_0 + v_0 t)e^{\lambda t}$, $v(t) = v_0 e^{\lambda t}$ for the case (ii). Finally, for the case (iii), we have $u(t) = e^{\alpha t}(u_0 \cos(\beta t) + v_0 \sin(\beta t))$, $v(t) = e^{\alpha t}(v_0 \cos(\beta t) - u_0 \sin(\beta t))$. Using the matrix $\mathbf{C}$, we can then write down the expressions for the solutions x, y.

The same procedure works for higher-dimensional linear systems. However, it is not easy to construct the matrix $\mathbf{C}$ as it involves deeper aspects of the eigenvalues and the corresponding eigenvectors of the coefficient matrix $\mathbf{A}$ and involves the Jordan canonical form. We only remark that each component of the solution of the linear system $\mathbf{x}' = \mathbf{A}\mathbf{x}$ is a linear combination of the products of the functions of t, and these functions are polynomials, exponentials and (in the presence of complex eigenvalues) the trigonometric functions–sine and cosine functions.

Initial Value Problem (IVP): The IVP of the general non-homogeneous, non-autonomous linear system can be written as

$$\mathbf{x}'(t) = \mathbf{A}(t)\mathbf{x}(t) + \mathbf{g}(t), \quad \mathbf{x}(t_0) = \mathbf{x}_0, \quad t > 0. \qquad (2.1.2)$$

Here $\mathbf{x}(t) = \begin{bmatrix} x_1(t) & \cdots & x_n(t) \end{bmatrix}^T$ is an unknown trajectory vector, $\mathbf{A}(t) = [a_{ij}(t)]$ is a given matrix function, and $\mathbf{g}(t)$ is a given non-homogeneous vector function. If the elements $a_{ij}(t)$ and $g_i(t)$ are continuous, then the general theorem gives the existence of a unique solution. Of course, for the linear system, the solution can be written explicitly via the transition matrix. If $\mathbf{A}(t) = \mathbf{A}$ is a constant matrix, then the system (2.1.2) is an autonomous, non-homogeneous system and the solution can be written down in terms of the exponential matrix. For example, when $\mathbf{g} \equiv 0$ in equation (2.1.2) is an autonomous homogeneous system and the solution is given by

$$\mathbf{x}(t) = e^{(t-t_0)\mathbf{A}} \mathbf{x}_0. \qquad (2.1.3)$$

Here, for any matrix $\mathbf{A}$, the exponential $e^{\mathbf{A}} = \exp(\mathbf{A})$ is defined by

$$e^{\mathbf{A}} = \sum_{k=0}^{\infty} \frac{\mathbf{A}^k}{k!}$$

and the series converges in $L(\mathbb{R}^n)$, the space of linear operators in $\mathbb{R}^n$. Recall that a matrix $\mathbf{A}$ can be viewed as a linear operator from $\mathbb{R}^n$ to $\mathbb{R}^n$.

It is worthwhile to recall at this stage that any nth-order linear ordinary differential equation (ODE) can be converted into a first-order system. The computation of $e^{t\mathbf{A}}$ in general is difficult, as mentioned, for example, in the book [46]. Even though the solution can be explicitly written as $\mathbf{x}(t) = e^{(t-t_0)\mathbf{A}} \mathbf{x}_0$, this expression itself does not reveal the behaviour of $\mathbf{x}(t)$ as t changes over time. This is the most important issue in the analysis of dynamical systems, namely the stability or instability properties of the trajectories $\mathbf{x}(t)$. This is achieved via diagonalization or Jordan decomposition of the matrix $\mathbf{A}$.

There are special cases, however, where we can get a better understanding of $e^{t\mathbf{A}}$. For example, when $\mathbf{A} = \text{diag}\begin{bmatrix} \lambda_1 & \cdots & \lambda_n \end{bmatrix}$ is a *diagonal matrix*, then $e^{\mathbf{A}} = \text{diag}\begin{bmatrix} e^{\lambda_1} & \cdots & e^{\lambda_n} \end{bmatrix}$. Consequently, $\mathbf{x}(t) = \begin{bmatrix} e^{t\lambda_1}x_{01} & \cdots & e^{t\lambda_n}x_{0n} \end{bmatrix}^T$. The sign of the eigenvalues gives us the behaviour of the components $x_i(t) = e^{t\lambda_i} x_{0i}$. For, $\lim_{t \to \infty} e^{t\lambda_i} x_{0i} = 0$ if $\lambda_i < 0$. Other cases can similarly be analysed. If $\mathbf{A}$ is not a diagonal matrix, we look for a similarity matrix $\mathbf{B}$; that is, look for an invertible (real) matrix $\mathbf{P}$ so that $\mathbf{A} = \mathbf{P}\mathbf{B}\mathbf{P}^{-1}$ with $\mathbf{B}$ as a diagonal matrix. In this situation, it is easy to see that $e^{\mathbf{A}} = \mathbf{P} e^{\mathbf{B}} \mathbf{P}^{-1}$ and $\mathbf{B} = \text{diag}\begin{bmatrix} \lambda_1 & \cdots & \lambda_n \end{bmatrix}$. Recall that $\mathbf{A}$ and $\mathbf{B}$ have the same eigenvalues $\lambda_1, \cdots, \lambda_n$. Further, we arrive at the solution $\mathbf{x}(t) = \mathbf{P} \, \text{diag} \begin{bmatrix} e^{t\lambda_1} & \cdots & e^{t\lambda_n} \end{bmatrix} \mathbf{P}^{-1} \mathbf{x}_0$. In this scenario, $\mathbf{A}$ is said to be a diagonalizable matrix. Unfortunately, all matrices are not diagonalizable. A matrix $\mathbf{A}$ is diagonalizable if and only if the algebraic and geometric multiplicities coincide for all the eigenvalues of $\mathbf{A}$; equivalently, a set of linear independent eigenvectors forms a basis for $\mathbb{R}^n$. If $\mathbf{A}$ is not diagonalizable, the next step is to find an equivalent block diagonal matrix, possibly with simple diagonal blocks. This is the content of the Jordan decomposition theorem. See Ref. [46] for more details.

Definition 2.1. A linear system $\mathbf{x}' = \mathbf{A}\mathbf{x}$ is said to be (linearly) equivalent to another linear system $\mathbf{y}' = \mathbf{B}\mathbf{y}$ if there is an invertible matrix $\mathbf{P}$ so that $\mathbf{A} = \mathbf{P}\mathbf{B}\mathbf{P}^{-1}$.

Since $\mathbf{A}$ and $\mathbf{B}$ have the same eigenvalues, the behaviour of the trajectories will not change. Under the equivalence, we can move from one system to the other and vice versa. We have $\mathbf{A}\mathbf{x} = \mathbf{P}\mathbf{B}\mathbf{P}^{-1}\mathbf{x}$. Thus, the substitution $\mathbf{y} = \mathbf{P}^{-1}\mathbf{x}$ gives $\mathbf{y}' = \mathbf{P}^{-1}\mathbf{x}' = \mathbf{P}^{-1}\mathbf{A}\mathbf{x} = \mathbf{B}\mathbf{y}$; that is, the variables $\mathbf{x}$ and $\mathbf{y}$ are connected by $\mathbf{y} = \mathbf{P}^{-1}\mathbf{x}$ or $\mathbf{x} = \mathbf{P}\mathbf{y}$. In the case of a 2×2 system $\mathbf{x}' = \mathbf{A}\mathbf{x}$, all the equivalent linear systems are given by $\mathbf{y}' = \mathbf{B_i}\mathbf{y}$, $i = 1, 2, 3$, where

$$\mathbf{B}_1 = \begin{bmatrix} \lambda & 0 \\ 0 & \mu \end{bmatrix}, \quad \mathbf{B}_2 = \begin{bmatrix} \lambda & 1 \\ 0 & \lambda \end{bmatrix}, \quad \mathbf{B}_3 = \begin{bmatrix} a & -b \\ b & a \end{bmatrix}.$$

Here λ, μ, a, b are real numbers with $b \neq 0$. We obtain $\mathbf{B}_1$ if $\mathbf{A}$ has two linearly independent real eigenvectors; $\mathbf{B}_2$ if $\mathbf{A}$ has a single eigenvalue with algebraic multiplicity 2 and geometric multiplicity 1; $\mathbf{B}_3$ if $\mathbf{A}$ has non-real eigenvalues $a \pm ib$, $b \neq 0$.

The fact is that the exponential $e^{t\mathbf{B}_i}$ can be computed easily, and hence the solution $\mathbf{y}$ and thus $\mathbf{x}$ can be written down explicitly. We have

$$e^{t\mathbf{B}_1} = \begin{bmatrix} e^{\lambda t} & 0 \\ 0 & e^{\mu t} \end{bmatrix}, \quad e^{t\mathbf{B}_2} = e^{\lambda t} \begin{bmatrix} 1 & t \\ 0 & 1 \end{bmatrix}, \quad e^{t\mathbf{B}_3} = e^{at} \begin{bmatrix} \cos(bt) & -\sin(bt) \\ \sin(bt) & \cos(bt) \end{bmatrix}.$$

The solutions $\mathbf{y}$ are given by $\mathbf{y}(t) = \text{diag}\begin{bmatrix} e^{\lambda t} & e^{\mu t} \end{bmatrix} \mathbf{y}_0$,

$$\mathbf{y}(t) = e^{\lambda t}\begin{bmatrix} 1 & t \\ 0 & 1 \end{bmatrix}\mathbf{y}_0 \text{ and } \mathbf{y}(t) = e^{at}\begin{bmatrix} \cos(bt) & -\sin(bt) \\ \sin(bt) & \cos(bt) \end{bmatrix}\mathbf{y}_0 \text{ for } i=1,2,3,$$

respectively; in the $\mathbf{x}$ variables, the corresponding solutions are thus written as

$$\mathbf{x}(t) = \mathbf{P}\operatorname{diag}\begin{bmatrix} e^{\lambda t} & e^{\mu t} \end{bmatrix}\mathbf{P}^{-1}\mathbf{x}_0, \; \mathbf{x}(t) = e^{\lambda t}\mathbf{P}\begin{bmatrix} 1 & t \\ 0 & 1 \end{bmatrix}\mathbf{P}^{-1}\mathbf{x}_0 \text{ and}$$

$$\mathbf{x}(t) = e^{at}\mathbf{P}\begin{bmatrix} \cos(bt) & -\sin(bt) \\ \sin(bt) & \cos(bt) \end{bmatrix}\mathbf{P}^{-1}\mathbf{x}_0.$$ We can also analyse higher-dimensional systems, which involve many more combinations and computations will become hard.

For the n-dimensional system $\mathbf{x}' = \mathbf{A}\mathbf{x}$, the elements of the null space $N(\mathbf{A}) = \{\mathbf{x} \in \mathbb{R}^n : \mathbf{A}\mathbf{x} = \mathbf{0}\}$ are called equilibrium points. If $\mathbf{A}$ is invertible, then $\mathbf{0}$ is the only equilibrium point. If $\bar{\mathbf{x}}$ is an equilibrium point, then $\mathbf{x}(t) \equiv \bar{\mathbf{x}}$ is a solution of the IVP $\mathbf{x}' = \mathbf{A}\mathbf{x}$, $\mathbf{x}(0) = \bar{\mathbf{x}}$. The stability analysis is concerned with the situation when the initial value (data), say $\mathbf{x}_0$ is close, but not equal to $\bar{\mathbf{x}}$, whether the corresponding solution $\mathbf{x}(t)$ will approach $\bar{\mathbf{x}}$ (asymptotic stability) or remain close to $\bar{\mathbf{x}}$ (stability) or move away from $\bar{\mathbf{x}}$ (instability) as $t \to \infty$. According to the various cases, the equilibrium points are classified as saddle point, node, focus and centre. See Ref. [46] for a complete discussion.

Definition 2.2. Let $\mathbf{A}$ be an $n \times n$ matrix. A subspace E of $\mathbb{R}^n$ is said to be *invariant* with respect to the flow $e^{t\mathbf{A}}$ if $e^{t\mathbf{A}}(E) \subset E$, for all t, that is, if $\mathbf{x}_0 \in E$, then the entire trajectory $\mathbf{x}(t)$ through $\mathbf{x}_0$ lies in E.

We can decompose the state space $\mathbb{R}^n$ into three invariant subspaces E^s, E^u, E^c as $\mathbb{R}^n = E^s \oplus E^u \oplus E^c$, which are, respectively, the stable, unstable and centre subspaces; one or two of these subspaces may contain only the zero vector. These subspaces are respectively spanned by the generalized eigenvectors of $\mathbf{A}$, corresponding to the eigenvalues with negative real parts, positive real parts and zero real parts (purely imaginary). Thus, we have

$$e^{t\mathbf{A}}(E^s) \subset E^s, \quad e^{t\mathbf{A}}(E^u) \subset E^u \quad e^{t\mathbf{A}}(E^c) \subset E^c.$$

We remark that the above properties remain unchanged under a linear equivalence. Indeed, the spaces E^s, E^u, E^c may change, but not their dimensions and hence the stability properties. In other words, the stability properties of a system (perhaps complicated) can be derive from that of a corresponding linear equivalent system (possibly simpler).

Some Facts: When the matrix $\mathbf{A}$ has either only real eigenvalues or only non-real eigenvalues (this requires the order of the matrix to be even), we can write it as a sum of a diagonalizable matrix and a nilpotent matrix. A matrix $\mathbf{N}$ is said to be a *nilpotent matrix* of order k if k is the smallest integer such that $\mathbf{N}^k$ is the zero matrix. By Cayley–Hamilton theorem, $k \leqslant n$. For a nilpotent matrix $\mathbf{N}$ of order k, we have

$$e^{\mathbf{N}} = I + \mathbf{N} + \cdots + \frac{\mathbf{N}^{k-1}}{(k-1)!}.$$

Theorem 2.3. Suppose the $n \times n$ matrix $\mathbf{A}$ has only real eigenvalues $\lambda_1, \lambda_2, \ldots \lambda_n$, counted according to their (algebraic) multiplicities. Then, there exists an invertible matrix $\mathbf{P} = \begin{bmatrix} \mathbf{u}_1 & \mathbf{u}_2 & \cdots & \mathbf{u}_n \end{bmatrix}$ consisting of generalized eigenvectors of $\mathbf{A}$ such that $\mathbf{A} = \mathbf{S} + \mathbf{N}$, where $\mathbf{S}$ is diagonalizable using $\mathbf{P}$, that is, $\mathbf{P}^{-1}\mathbf{S}\mathbf{P} = \mathrm{diag}\begin{bmatrix} \lambda_1 & \cdots & \lambda_n \end{bmatrix}$ and $\mathbf{N} = \mathbf{A} - \mathbf{S}$ is nilpotent of order k. Furthermore, $\mathbf{SN} = \mathbf{NS}$.

Since $\mathbf{S}$ and $\mathbf{N}$ commute, the solution to the homogeneous IVP is given by $\mathbf{x}(t) = e^{t(\mathbf{S}+\mathbf{N})}\mathbf{x}_0 = e^{t\mathbf{S}}e^{t\mathbf{N}}\mathbf{x}_0$ so that

$$\mathbf{x}(t) = \mathbf{P}\,\mathbf{D}\,\mathbf{P}^{-1}\left[I + t\mathbf{N} + \cdots + \frac{t^{k-1}\mathbf{N}^{k-1}}{(k-1)!}\right]\mathbf{x}_0, \qquad (2.1.4)$$

where $\mathbf{D} = \mathrm{diag}\begin{bmatrix} e^{\lambda_1 t} & \cdots & e^{\lambda_n t} \end{bmatrix}$. A similar theorem can be stated when the eigenvalues are all non-real. Indeed, n is even $= 2m$, say. In this case, $\mathbf{S}$ is block diagonalizable, where each block is a 2×2 matrix of the form $\mathbf{B}_j = \begin{bmatrix} a_j & -b_j \\ b_j & a_j \end{bmatrix}$, where $a_j \pm ib_j$, $1 \leqslant j \leqslant m$ are the non-real eigenvalues of $\mathbf{A}$. Further, the invertible matrix $\mathbf{P} = \begin{bmatrix} \mathbf{v}_1 & \mathbf{u}_1 & \mathbf{v}_2 & \mathbf{u}_2 & \cdots & \cdots & \mathbf{v}_m & \mathbf{u}_m \end{bmatrix}$, where $\mathbf{w}_j = \mathbf{u}_j + i\mathbf{v}_j$'s are the generalized eigenvectors. The final solution is given by $\mathbf{x}(t) = e^{t(\mathbf{S}+\mathbf{N})}\mathbf{x}_0 = e^{t\mathbf{S}}e^{t\mathbf{N}}\mathbf{x}_0$ or

$$\mathbf{x}(t) = \mathbf{P}\,\tilde{\mathbf{D}}\,\mathbf{P}^{-1}\left[I + t\mathbf{N} + \cdots + \frac{t^{k-1}\mathbf{N}^{k-1}}{(k-1)!}\right]\mathbf{x}_0, \qquad (2.1.5)$$

where $\tilde{\mathbf{D}} = \mathrm{diag}\begin{bmatrix} e^{t\mathbf{B}_1} & \cdots & e^{t\mathbf{B}_m} \end{bmatrix}$.

Non-homogeneous System: The solution of an autonomous, non-homogeneous IVP is given explicitly as

$$\mathbf{x}(t) = e^{(t-t_0)\mathbf{A}}\mathbf{x}_0 + \int_{t_0}^{t} e^{(t-s)\mathbf{A}}\mathbf{g}(s)\,ds.$$

This is obtained via the variation of parameters (constants).

Non-autonomous Systems: For the general system (2.1.2), the representation of the unique solution can be given in terms of the fundamental matrix . Let $\boldsymbol{\phi}_i(t,t_0)$ be the unique solution of the IVP $\mathbf{x}'(t) = \mathbf{A}(t)\mathbf{x}(t)$, $\mathbf{x}(t_0) = \mathbf{e}_i$, where $\{\mathbf{e}_1, \ldots, \mathbf{e}_n\}$ is the canonical basis in $\mathbb{R}^n$. Since $\mathbf{x}_0 = \sum_{i=1}^{n} x_{0i}\mathbf{e}_i$, we see that $\mathbf{x}(t) = \sum_{i=1}^{n} x_{0i}\boldsymbol{\phi}_i(t,t_0)$ is the unique solution to the system $\mathbf{x}'(t) = \mathbf{A}(t)\mathbf{x}(t)$, $\mathbf{x}(t_0) = \mathbf{x}_0$. Define the matrix

$$\boldsymbol{\Phi}(t,t_0) = \begin{bmatrix} \boldsymbol{\phi}_1(t,t_0) & \cdots & \boldsymbol{\phi}_n(t,t_0) \end{bmatrix}.$$

Then, $\boldsymbol{\Phi}$ satisfies the matrix system $\boldsymbol{\Phi}' = \mathbf{A}\boldsymbol{\Phi}$, $\boldsymbol{\Phi}(t_0,t_0) = I$. In the case of an autonomous system, we have $\boldsymbol{\Phi}(t,t_0) = e^{(t-t_0)\mathbf{A}}$. Indeed, the matrix $\boldsymbol{\Phi}(t,t_0)$ is invertible, and it also satisfies

$$\boldsymbol{\Phi}^{-1}(t,t_0) = \boldsymbol{\Phi}(t_0,t), \quad \boldsymbol{\Phi}(t,s)\boldsymbol{\Phi}(s,t_0) = \boldsymbol{\Phi}(t,t_0), \quad \boldsymbol{\Phi}(t_0,t_0) = I.$$

The last two properties are known as semigroup properties. The matrix $\boldsymbol{\Phi}(t,t_0)$ is known as a *transition matrix*. Finally, the unique solution of the non-homogeneous, non-autonomous system (2.1.2) is given by

$$\mathbf{x}(t) = \boldsymbol{\Phi}(t,t_0)\mathbf{x}_0 + \int_{t_0}^{t} \boldsymbol{\Phi}(t,s)\mathbf{g}(s)ds.$$

As a final remark, any non-singular $\boldsymbol{\Psi}$ satisfying the matrix system $\boldsymbol{\Psi}' = \mathbf{A}\boldsymbol{\Psi}$ is known as a *fundamental matrix*. The transition matrix constructed above is a special fundamental matrix. If $\mathbf{C}$ is any non-singular matrix, it is readily seen that $\boldsymbol{\Psi}(t) = \boldsymbol{\Phi}(t)\mathbf{C}$ is also a fundamental matrix. Conversely, if $\boldsymbol{\Psi}(t)$ is a fundamental matrix, then $\boldsymbol{\Psi}(t) = \boldsymbol{\Phi}(t)\mathbf{C}$, where $\mathbf{C} = \boldsymbol{\Psi}(t_0)$.

2.2 Some Applications

We now discuss a couple of applications of linear systems. The first one is from kinematics, and the second one is finite-dimensional linear control systems. As these topics need some knowledge of advanced topics in linear algebra, the reader is advised to acquire the same.

2.2.1 Kinematics

Here we discuss small oscillations of a mechanical system with n degrees of freedom ([26], [37], [6]). Consider the free oscillations of a conservative mechanical system with n degrees of freedom near a stable position of an equilibrium. For $n = 1$, think of a simple pendulum equation or a simple harmonic motion of a single spring. We now perturb this system from the position of its equilibrium by means of a position vector $\mathbf{x}(t) = (x_1(t), \ldots, x_n(t))$, which is a function of the time variable t; the position of equilibrium corresponds to $x_1 = \cdots = x_n = 0$. The kinetic energy T of the system is represented as a quadratic form in the components of the velocity vector $\mathbf{x}'(t) = (x_1'(t), \ldots, x_n'(t))$: $T = \sum_{i,j=1}^{n} b_{ij} x_i' x_j'$. The coefficients b_{ij} are symmetric and in general are functions of the position vector. Since we are considering only small oscillations, we assume that b_{ij} are constants. The kinetic energy is always positive and is zero only when $x_1' = \cdots = x_n' = 0$. Thus, the matrix $\mathbf{B} = [b_{ij}]$ is a real symmetric positive definite matrix.

The potential energy P is a function of the position vector. We may assume that $P(0, \ldots, 0) = 0$ as adding a constant will not change the equations of motion. Again, at equilibrium, the potential energy always has stationary value: $\left.\frac{\partial P}{\partial x_i}\right|_{\mathbf{x}=\mathbf{0}} = 0$ for all i. We therefore assume that $P(x_1, \cdots, x_n) = \sum_{i,j=1}^{n} a_{ij} x_i x_j$, where the matrix $\mathbf{A} = [a_{ij}]$ is a real symmetric matrix. We now write down the equations of motion in the form of Lagrange equations:

$$\frac{d}{dt}\frac{\partial T}{\partial x_i'} - \frac{\partial T}{\partial x_i} = -\frac{\partial P}{\partial x_i}, \; i = 1, 2, \ldots, n.$$

The equations of motion are thus written as

$$\sum_{j=1}^{n} b_{ij} x_j'' + \sum_{j=1}^{n} a_{ij} x_j = 0, \; i = 1, 2, \ldots, n. \quad (2.2.1)$$

This is a linear system of n equations of second order. Using the tools of linear algebra, solutions can be written down in explicit form. This will be done in an exercise.

2.2.2 Controllability and Observability of Linear Systems

Recall that in the linear system (2.1.1), the source term $\mathbf{g}$ is a given input function. For the given data $\mathbf{A}$ and $\mathbf{g}$ with suitable conditions (e.g., $\mathbf{g}$ is continuous), there is a unique solution $\mathbf{x}(t)$ starting from the initial value $\mathbf{x}(t_0) = \mathbf{x}_0$. Quite often in applications (for instance, in the missile and satellite trajectories), we would like a trajectory to reach a desired position $\mathbf{x}_1$ at a time $T > t_0$, that is, $\mathbf{x}(T) = \mathbf{x}_1$. To achieve this, we may need to choose $\mathbf{g}$ suitably. Such problems are known as *control problems*, where **g** acts as a control to steer the solution **x** to a desired state at a desired time. The basic question is whether such a control exists, of course, in some appropriate space. It is possible to formulate very general non-linear control problems both for the dynamics given by ODEs and partial differential equations (PDEs) as well as for deterministic and non-deterministic cases. In this short note, however, we discuss only a bare minimum about the control problems given by linear systems with controls also acting linearly. More precisely, we consider

$$\mathbf{x}'(t) = \mathbf{A}(t)\mathbf{x}(t) + \mathbf{B}(t)\mathbf{u}(t), \quad t > 0. \tag{2.2.2}$$

Here $\mathbf{B}(t)$ is an $n \times m$ matrix-valued function and $\mathbf{u}(t)$ is an m vector. The idea is to control all the n components of the trajectory $\mathbf{x}(t)$, possibly with a less number $m < n$ of controls. Indeed, in general, controllability cannot be achieved even for an autonomous system with n controls. The examples will be discussed in the exercises.

Definition 2.4. The linear system (2.2.2) is said to be *exactly controllable* in time $T > t_0$ if for any $\mathbf{x}_0, \mathbf{x}_1 \in \mathbb{R}^n$, there exists a control function $t \mapsto \mathbf{u}(t) \in \mathbb{R}^m$ such that the corresponding solution $\mathbf{x}(t)$ of equation (2.2.2) satisfies the conditions $\mathbf{x}(t_0) = \mathbf{x}_0$ and $\mathbf{x}(T) = \mathbf{x}_1$.

For the two-dimensional system $x_1' = x_1 + u$, $x_2' = 0$, the controllability cannot be achieved since the control u does not influence the component x_2 and $x_2(t) = x_{02}$ is a constant, which cannot be moved to any other point. This does not mean that a single control is not enough to achieve controllability of systems. On the other hand, consider a single equation $x' = ax + bu$, with $b \neq 0$. Then $u = (1/b)(x' - ax)$. This suggests to look for a control of the form $u(t) = (1/b)(z'(t) - az(t))$ for some z to be chosen later. Putting $y = x - z$, it is trivial to see that $y' = ay$. Now choose any z satisfying the boundary conditions $z(t_0) = x_0$ and $z(T) = x_1$. There are many such functions z to choose; for example, the straight line connecting the two points x_0 and x_1. Thus, $y(t_0) = 0$ and, by uniqueness, $y \equiv 0$ and hence $x(t) = z(t)$. Thus, a more general trajectory controllability is achieved with the control u given above.

This example also suggests the controllability of the autonomous system in equation (2.2.2), that is, $\mathbf{A}(t) = \mathbf{A}$ and $\mathbf{B}(t) = \mathbf{B}$ are constant matrices. We also assume that $\mathbf{B}$ is invertible. In this case, we are therefore using n controls. Choose any vector function $\mathbf{z}(t)$ such that $\mathbf{z}(t_0) = \mathbf{x}_0$ and $\mathbf{z}(T) = \mathbf{x}_1$. Then $\mathbf{u} = \mathbf{B}^{-1}(\mathbf{z}' - \mathbf{A}\mathbf{z})$ will act as the control, and $\mathbf{x} = \mathbf{z}$ is the controlled solution.

Now, we consider the general case. The solution to the IVP corresponding to equation (2.2.2) is given by

$$\mathbf{x}(t) = \mathbf{\Phi}(t, t_0)\mathbf{x}_0 + \int_{t_0}^{t} \mathbf{\Phi}(t, s)\mathbf{B}(s)\mathbf{u}(s)\,ds,$$

where $\boldsymbol{\Phi}$ is the transition matrix. Thus, the system is controllable if there is a $\mathbf{u}$ (normally it is taken to be an element in $L^2(0,T;\mathbb{R}^m)$) such that

$$\mathbf{x}_1 = \boldsymbol{\Phi}(T,t_0)\mathbf{x}_0 + \int_{t_0}^{T} \boldsymbol{\Phi}(T,s)\mathbf{B}(s)\mathbf{u}(s)\,ds. \tag{2.2.3}$$

The linear system is said to be *null controllable* if it is controllable for arbitrary $\mathbf{x}_0$ and $\mathbf{x}_1 = \mathbf{0}$. For a linear system, the null controllability is equivalent to the controllability. But this is not true in general for non-linear systems. From equation (2.2.3), we see that the problem of controllability reduces to that of understanding the range of the operator $C \colon X := L^2(0,T;\mathbb{R}^m) \to \mathbb{R}^n$, given by

$$C\mathbf{u} := \int_{t_0}^{T} \boldsymbol{\Phi}(T,s)\mathbf{B}(s)\mathbf{u}(s)\,ds.$$

In other words, the control problem is equivalent to C being onto, that is, $\mathrm{im}(C) = \mathbb{R}^n$. This in turn is equivalent to the fact that C^*, the adjoint of C, is one-one, which is a bounded linear operator from $\mathbb{R}^n$ to $L^2(0,T;\mathbb{R}^m)$. This in turn is equivalent to the operator CC^* being one-one. It is straightforward to compute the operator $CC^* \colon \mathbb{R}^n \to \mathbb{R}^n$ and is given by

$$CC^* = \int_{t_0}^{T} \boldsymbol{\Phi}(T,s)\mathbf{B}(s)\mathbf{B}^*(s)\boldsymbol{\Phi}^*(T,s)\,ds. \tag{2.2.4}$$

Observe that $CC^* = \mathbf{W}(t_0,T)$ is an $n \times n$ matrix and is known as the *controllability Gramian matrix*. Thus, controllability is equivalent to the invertibility of the matrix $\mathbf{W}(t_0,T)$, with the control given by

$$\mathbf{u}(t) = \mathbf{B}^*(t)\boldsymbol{\Phi}^*(T,t)\mathbf{W}^{-1}(\mathbf{x}_1 - \boldsymbol{\Phi}(T,t_0)\mathbf{x}_0). \tag{2.2.5}$$

The above control actually has a minimum L^2 norm among all the admissible controls. For the autonomous control system, the controllability can be reduced to a rank condition of an $n \times nm$ block matrix, namely

$$\mathrm{rank}\begin{bmatrix} \mathbf{B} & \mathbf{AB} & \cdots & \mathbf{A}^{n-1}\mathbf{B} \end{bmatrix} = n. \tag{2.2.6}$$

Thus, the controllability is equivalent to the above-mentioned block matrix having the full rank. This criterion is known as the *Kalman's rank condition* for controllability of autonomous linear systems. We once again remark this is not an elaborate discussion on various topics in control theory but only a brief discussion, which fits into a discussion on linear systems. Some of the above-stated results will be discussed in the exercises.

Observability: There is another companion (dual) notion of observability of linear systems, without control, but with an (partial) observation related to the state for sufficiently large enough time. The objective is to recover or understand the state completely. This is formulated as follows: Consider the uncontrolled system

$$\mathbf{y}'(t) = \mathbf{G}(t)\mathbf{y}(t), \quad t > 0, \qquad (2.2.7)$$

with an observation

$$\boldsymbol{\omega}(t) = \mathbf{H}(t)\mathbf{y}(t), \quad t > 0. \qquad (2.2.8)$$

Here $\mathbf{G}(t)$ and $\mathbf{H}(t)$ are, respectively $n \times n$ and $n \times m$ given matrices. Note that $\boldsymbol{\omega}(t)$ is a partial information related to the state $\mathbf{y}(t)$.

Definition 2.5. We say the system (2.2.7) is observable with the observation (2.2.8) in $[t_0, T]$ if $\boldsymbol{\omega}(t) = 0$ for all $t \in [t_0, T]$ implies $\mathbf{y} = \mathbf{0}$ in $[t_0, T]$.

We have $\boldsymbol{\omega}(t) = \mathbf{H}(t)\boldsymbol{\Phi}(t, t_0)\mathbf{y}_0$, where $\boldsymbol{\Phi}(t, t_0)$ is the transition matrix corresponding to the homogeneous linear system (2.2.7). We remark that the observability is equivalent to the fact that the initial state $\mathbf{y}(t_0) = \mathbf{y}_0$ is recovered from the knowledge of the observation $\boldsymbol{\omega}(t)$ over the interval $[t_0, T]$ (it is a partial information of the state over time), which in turn gives the state $\mathbf{y}(t)$. See the discussion below. Thus, we define the operator $L:\mathbb{R}^n \to L^2(0, T; \mathbb{R}^m)$ by $(L\mathbf{y}_0)(t) = \boldsymbol{\omega}(t)$. Hence, the system is observable if and only if L is invertible. Analogous to controllability results, we also deduce that observability $\iff L$ is $1-1 \iff$ the adjoint operator L^* is onto $\iff L^*L$ is onto. The adjoint operator L^* can be computed as

$$L^*\mathbf{v} = \int_{t_0}^{T} \boldsymbol{\Phi}^*(t, t_0)\mathbf{H}^*(t)\mathbf{v}(t)\, dt.$$

The related observability Gramian matrix denoted by $\mathbf{M}(t_0, T) = L^*L$, is given by

$$\mathbf{M}(t_0, T) = \int_{t_0}^{T} \boldsymbol{\Phi}^*(t, t_0)\mathbf{H}^*(t)\mathbf{H}(s)\boldsymbol{\Phi}(t, t_0)\, dt.$$

Again, observability is equivalent to the invertibility of $\mathbf{M}$. For the autonomous system, the observability can be reduced to the rank condition given by

$$\operatorname{rank} \begin{bmatrix} \mathbf{H} & \mathbf{H}\mathbf{A} & \cdots & \mathbf{H}\mathbf{A}^{n-1} \end{bmatrix}^T = n.$$

We can reconstruct the initial state $\mathbf{y}_0$ from the observation $\boldsymbol{\omega}$ as follows: We have $\boldsymbol{\omega} = L\mathbf{y}_0$, which implies $L^*\boldsymbol{\omega} = L^*L\mathbf{y}_0$. Thus, we get

$$\mathbf{y}_0 = (L^*L)^{-1}L^*\boldsymbol{\omega} = \mathbf{M}^{-1}\int_{t_0}^{T} \boldsymbol{\Phi}^*(t,t_0)\mathbf{H}^*(t)\boldsymbol{\omega}(t)\,dt.$$

Connection between Controllability and Observability (A Duality Process): We just remark that the controllability of the system (2.2.2) is equivalent to the observability of the adjoint system

$$\mathbf{y}'(t) = -\mathbf{A}^*(t)\mathbf{y}(t), \quad \text{with observation} \quad \boldsymbol{\omega}(t) = \mathbf{B}^*(t)\mathbf{y}(t),\ t > 0.$$

See exercises for details.

Control via Minimizers: Consider the adjoint system

$$\mathbf{y}'(t) = -\mathbf{A}^*(t)\mathbf{y}(t), \quad t > 0 \quad \mathbf{y}(T) = \mathbf{y}_T \tag{2.2.9}$$

with the terminal condition $\mathbf{y}_T$. For any solution $\mathbf{x}$ of the linear system (2.2.2), we then get

$$(\mathbf{x}', \mathbf{y}) = (\mathbf{A}\mathbf{x}, \mathbf{y}) + (\mathbf{B}\mathbf{u}, \mathbf{y}) = (\mathbf{x}, \mathbf{A}^*\mathbf{y}) + (\mathbf{B}\mathbf{u}, \mathbf{y}) = -(\mathbf{x}, \mathbf{y}') + (\mathbf{B}\mathbf{u}, \mathbf{y}).$$

It follows that $\dfrac{d}{dt}(\mathbf{x},\mathbf{y}) = (\mathbf{B}\mathbf{u},\mathbf{y})$. Upon integration, we get

$$(\mathbf{x}(T), \mathbf{y}_T) = (\mathbf{x}_0, \mathbf{y}(t_0)) + \int_{t_0}^{T} (\mathbf{u}(t), \mathbf{B}^*(t)\mathbf{y}(t))\,dt.$$

Thus, the system (2.2.2) is null controllable, that is, $\mathbf{x}(T) = 0$, if and only if

$$(\mathbf{x}_0, \mathbf{y}(t_0)) + \int_{t_0}^{T} (\mathbf{u}(t), \mathbf{B}^*(t)\mathbf{y}(t))\,dt = 0 \tag{2.2.10}$$

for all the choices of terminal condition $\mathbf{y}_T \in \mathbb{R}^n$ and the solution $\mathbf{y}$ of the adjoint system (2.2.9). It is a simple observation that the equation (2.2.10) is an optimality condition for the critical points of the quadratic functional $J:\mathbb{R}^n \to \mathbb{R}$ defined by

$$J(\mathbf{y}_T) = \frac{1}{2}\int_{t_0}^{T} |\mathbf{B}^*(t)\mathbf{y}(t)|^2\,dt + (\mathbf{x}_0, \mathbf{y}(t_0)). \tag{2.2.11}$$

More precisely, if $\hat{\mathbf{y}}_T$ is a solution of the minimization problem $J(\hat{\mathbf{y}}_T) = \min\limits_{\mathbf{y}_T} J(\mathbf{y}_T)$, then $\mathbf{u} = \mathbf{B}^*\hat{\mathbf{y}}$ satisfies (2.2.10), where $\hat{\mathbf{y}}$ is the solution of (2.2.9) with the terminal condition $\hat{\mathbf{y}}_T$. Thus, $\mathbf{u} = \mathbf{B}^*\hat{\mathbf{y}}$ is a control driving the system from $\mathbf{x}_0$ to $\mathbf{0}$.

We conclude this section by making some comments. The previous discussion is a variational approach for obtaining a control if J has a minimum. Further, by varying J, it may be possible to obtain different types of controls that are more practical like *bang-bang* control. In applications, quite often we may not need to apply controls continuously and apply controls as and when it is necessary, getting information from the trajectory till that time. The variational method is very powerful in the analysis of control systems governed by partial differential equations. The above method tells us that if we begin from $\mathbf{y}(T) = \mathbf{y}_T$, which evolves (reversely) according to the adjoint system, and observe the quantity $\mathbf{B}^*\mathbf{y}(t)$ for all $t_0 < t < T$. If we can recover $\mathbf{y}(t_0)$ uniquely (that is observability), then controllability is achieved. Indeed, such a thing is possible if we have the so-called *observability inequality*

$$\int_{t_0}^{T} |\mathbf{B}^*(t)\mathbf{y}(t)|^2 \, dt \geqslant \tilde{c}\,|\mathbf{y}(t_0)|^2 \qquad (2.2.12)$$

for some constant $\tilde{c}$ and all $\mathbf{y}_T \in \mathbb{R}^n$. This inequality is equivalent to the *unique continuation principle* (**u.c.p.**):

$$\mathbf{B}^*(t)\mathbf{y}(t) = \mathbf{0} \text{ for all } t \in [t_0, T], \text{ then } \mathbf{y}(t_0) = \mathbf{0}. \qquad (2.2.13)$$

Consequently, $\mathbf{y} \equiv \mathbf{0}$. In general, in infinite dimensions, observability inequality is a stronger concept than **u.c.p.** In PDEs, the controllability results are proved by establishing appropriate observability estimates.

Thus, in conclusion, a control problem is reduced to that of studying an uncontrolled problem together with an observation and an estimate (inequality). This is an ideal situation for the PDE analysis. The literature for control, stabilizability and optimal control problems, including the notions not discussed here, in the settings of both ODE and PDE, is abundant. Here we mention a few for the finite-dimensional control problems: [12], [13], [14], [53], [51], [28].

2.3 Exercises

Exercise 2.1

Write down the general nth-order linear equation in regular form and convert it into a first-order system.

Exercise 2.2

Discuss the stability of the zero solution in the following linear systems:
(a) $x' = -x - 2y, \quad y' = 4x - 5y$.
(b) $x' = -3x + 4y, \quad y' = -2x + 3y$.
(c) $x' = 4x - 2y, \quad y' = 5x + 2y$.

Exercise 2.3

Find a matrix $\mathbf{A}$ of order n that has an eigenvalue λ of algebraic multiplicity n and geometric multiplicity k for each $k = 1, \ldots, n$.

Exercise 2.4

Write the system $x_1' = -x_1 + x_2, \quad x_2' = x_1 - x_2$ in the matrix form as $\mathbf{x}' = \mathbf{A}\mathbf{x}$ and find the eigenvalues, eigenvectors and equivalent diagonal system of the matrix $\mathbf{A}$. Further analyse the orbits of this system.

Exercise 2.5

Consider the system $x_1' = -x_1 + 3x_2, \quad x_2' = 2x_2$. Do an analysis of this system similar to Exercise 2.4.

Exercise 2.6

For the system $x_1' = 2x_1 + x_2, \quad x_2' = -4x_1 - 2x_2$, show that 0 is a double eigenvalue with geometric multiplicity 1 of the coefficient matrix $\mathbf{A}$ and find a corresponding eigenvector and a generalized eigenvector. Further, find a matrix $\mathbf{P}$ so that $\mathbf{P}^{-1}\mathbf{A}\mathbf{P} = \mathbf{B} = \begin{bmatrix} 0 & 2 \\ 0 & 0 \end{bmatrix}$; solve the systems $\mathbf{x}' = \mathbf{A}\mathbf{x}$ and $\mathbf{y}' = \mathbf{B}\mathbf{y}$ and do the stability analysis as well.

Exercise 2.7

Consider the three-dimensional homogeneous linear system given by the coefficient matrix
$\mathbf{A}_1 = \begin{bmatrix} -3 & 0 & 0 \\ 0 & 3 & -2 \\ 0 & 1 & 1 \end{bmatrix}$. Analyse the phase portrait and determine the stable and unstable spaces. Workout a similar problem with the coefficient matrix $\mathbf{A}_2 = \begin{bmatrix} -2 & -1 & 0 \\ 1 & -2 & 0 \\ 0 & 0 & 3 \end{bmatrix}$. Sketch a few trajectories of these systems and observe the differences in their behaviour.

Exercise 2.8

Consider the 4×4 system given by the coefficient matrix

$$\mathbf{A} = \begin{bmatrix} 1 & -1 & 0 & 0 \\ 1 & 1 & 0 & 0 \\ 0 & 0 & 3 & -2 \\ 0 & 0 & 1 & 1 \end{bmatrix}.$$

Find the eigenvalues, eigenvectors and the matrix $\mathbf{P}$. Write down the equivalent system and solve it.

Exercise 2.9

Obtain the decomposition $\mathbf{A} = \mathbf{S} + \mathbf{N}$ for the matrix $\mathbf{A} = \begin{bmatrix} 1 & 0 & 0 \\ -1 & 2 & 0 \\ 1 & 1 & 2 \end{bmatrix}$.

Exercise 2.10

Show that the solutions are confined to the centre subspace of the matrix $\mathbf{A} = \begin{bmatrix} 0 & -1 & 0 \\ 1 & 0 & 0 \\ 0 & 0 & 2 \end{bmatrix}$ are all bounded. Give an example of a two-dimensional system to show that this need not be the case always.

Exercise 2.11

List all the possible (upper)Jordan canonical forms of a real 4×4 matrix with a real eigenvalue λ of algebraic multiplicity 4 and find the corresponding deficiency index in each case.

Exercise 2.12

Suppose the matrix function $t \mapsto \mathbf{A}(t)$ is continuous from a compact interval $[a, b]$ into the space of $n \times n$ real matrices and assume that the matrices $\mathbf{A}(t)$ and $\mathbf{A}(s)$ commute for all $t, s \in [a, b]$. Verify that the vector function $\mathbf{x}(t) \equiv \exp\left(\int_{t_0}^{t} \mathbf{A}(s)\, ds\right) \mathbf{x}_0$ is the solution of the IVP:

$$\mathbf{x}'(t) = \mathbf{A}(t)\mathbf{x}(t), \quad \mathbf{x}(t_0) = \mathbf{x}_0.$$

Here $t_0 \in [a, b]$ is the initial time and $\mathbf{x}_0$ is the initial value.

Exercise 2.13

Consider the matrix valued function $t \mapsto \mathbf{A}(t) = \begin{bmatrix} 1 & 1+t \\ 0 & t \end{bmatrix}$, from $\mathbb{R}$ into the space of 2×2 real matrices. Analyse the two-dimensional non-autonomous system $\mathbf{x}'(t) = \mathbf{A}(t)\mathbf{x}(t)$ and compare it with the result of Exercise 2.12.

Exercise 2.14

Solve the following system of second-order equations derived in the discussion on kinematics:

$$\sum_{j=1}^{n} b_{ij} x_j'' + \sum_{j=1}^{n} a_{ij} x_j = 0, \ i = 1, 2, \ldots, n.$$

Exercise 2.15

Consider the following two control problems:

(1) $x_1' = -ax_1, \quad x_2' = ax_1 - bx_2 + u,$
(2) $x_1' = -ax_1 + u, \quad x_2' = ax_1 - bx_2,$

where a, b are given positive real numbers. Analyse the controllability in each case and present a physical interpretation.

Exercise 2.16

Consider the control problem given by the nth order equation as in equation (2.3.1), where $u(t) = g(t)$ is the control function. Show that with only this control, the trajectories of $x(t)$ and its derivatives $x^{(j)}(t)$ can be controlled for all $j = 1, \ldots, n-1$.

Exercise 2.17

Show the following statements are equivalent for the linear system (2.2.2):

(i) The system is controllable for any arbitrary $\mathbf{x}_0, \mathbf{x}_1 \in \mathbb{R}^n$.
(ii) The system is controllable for any arbitrary $\mathbf{x}_0 \in \mathbb{R}^n$ and the final state $\mathbf{x}_1 = \mathbf{0} \in \mathbb{R}^n$. This is known as null controllability.
(iii) The system is controllable for any arbitrary final state $\mathbf{x}_1 \in \mathbb{R}^n$ and initial state $\mathbf{x}_0 = \mathbf{0} \in \mathbb{R}^n$.

Exercise 2.18

Compute the adjoint operators C^* and L^* corresponding to the operators C and L mentioned in Section 2.2.2.

Exercise 2.19

Show that for an autonomous linear system, the controllability and observability are equivalent to the respective rank conditions mentioned in Section 2.2.2.

Exercise 2.20

Study the controllability of the spring-mass system $y'' + y = u$. Compute the transition matrix, controllability Gramian matrix and a steering control for a choice of initial and terminal points.

Exercise 2.21

(a) Derive the optimality condition for the minimization problem given by the functional as in equation (2.2.11).

(b) Show that the observability inequality equation (2.2.12) is equivalent to the inequality

$$\int_{t_0}^{T} \|\mathbf{B}^*(s)\mathbf{y}(s)\|^2 \, ds \geq \tilde{c}|\mathbf{y}_T|^2, \tag{2.3.2}$$

for some constant $\tilde{c}$.

(c) Show that the **u.c.p.** in equation (2.2.13) is equivalent to the observability inequality.

Exercise 2.22

Using the estimate in equation (2.3.2), show that J has a minimizer and hence the linear system is controllable. Conversely, if the linear system is controllable, then the observability estimate holds.

Exercise 2.23

Give examples of linear non-autonomous systems $\mathbf{x}'(t) = \mathbf{A}(t)\mathbf{x}(t)$, to show that the nature of the eigenvalues of $\mathbf{A}(t)$ may or may not imply the stability of an equilibrium point, in contrast with the linear autonomous systems.

2.4 Solutions

Exercise 2.1

A general nth-order regular linear ODE can be written in the form

$$\frac{d^n x}{dt^n} + p_1(t)\frac{d^{n-1}x}{dt^{n-1}} + \cdots + p_{n-1}(t)\frac{dx}{dt} + p_n(t)x = g(t) \tag{2.3.1}$$

with n conditions, such as initial or boundary conditions. For example, initial conditions can be of the form $(x^{(k)} = \frac{d^k x}{dt^k})$

$$x(0) = x_0, \ x^{(1)}(0) = x_1, \ldots, x^{(n-1)}(0) = x_{n-1}.$$

By introducing new variables: $x_1 = x, x_2 = x^{(1)}, \ldots, x_n = x^{(n-1)}$, we can convert the above equations into a system of n first-order equations having a matrix representation as $\mathbf{x}'(t) = \mathbf{A}(t)\mathbf{x}(t) + \mathbf{g}(t)$, where

$$\mathbf{A} = \mathbf{A}(t) = \begin{bmatrix} 0 & 1 & 0 & \cdots & 0 \\ 0 & 0 & 1 & \cdots & 0 \\ \cdots & \cdots & \cdots & \cdots & \cdots \\ 0 & 0 & 0 & \cdots & 1 \\ -p_n(t) & -p_{n-1}(t) & \cdots & \cdots & -p_1(t) \end{bmatrix}$$

and

$$\mathbf{x}(t) = \begin{bmatrix} x_1(t) \\ x_2(t) \\ \cdot \\ \cdot \\ \cdot \\ x_{n-1}(t) \\ x_n(t) \end{bmatrix}, \quad \mathbf{g}(t) = \begin{bmatrix} 0 \\ 0 \\ \cdot \\ \cdot \\ \cdot \\ 0 \\ g(t) \end{bmatrix}, \quad \mathbf{x}(0) = \begin{bmatrix} x_0 \\ x_1 \\ \cdot \\ \cdot \\ \cdot \\ x_{n-2} \\ x_{n-1} \end{bmatrix}$$

Exercise 2.2

(a) The coefficient matrix is $\mathbf{A} = \begin{bmatrix} -1 & -2 \\ 4 & -5 \end{bmatrix}$, whose eigenvalues are the roots of the characteristic equation $\lambda^2 - (-1-5)\lambda + (5+8) = 0$, or $\lambda^2 + 6\lambda + 13 = 0$. Thus, $\lambda = -3 \pm 2i$. As the real parts of the (complex) eigenvalues are negative, we infer that the zero solution is asymptotically stable and it is a spiral. The real and imaginary parts of the eigenvectors[1] are $\begin{bmatrix} 1 \\ 0 \end{bmatrix}$ and $\begin{bmatrix} -1 \\ -2 \end{bmatrix}$, respectively. Thus, we take

$\mathbf{C} = \begin{bmatrix} 1 & -1 \\ 0 & -2 \end{bmatrix}$, and therefore, it follows that

---

[1] If $a + ib$ is a non-real eigenvalue of a *real* matrix $\mathbf{A}$ with the corresponding eigenvector $\mathbf{u} + i\mathbf{v}$, then $\mathbf{A}\mathbf{u} = a\mathbf{u} - b\mathbf{v}$ and $\mathbf{A}\mathbf{v} = b\mathbf{u} + a\mathbf{v}$. Further, $\mathbf{u}, \mathbf{v}$ are linearly independent.

$$\begin{bmatrix} 1 & -1/2 \\ 0 & -1/2 \end{bmatrix} \begin{bmatrix} -1 & -2 \\ 4 & -5 \end{bmatrix} \begin{bmatrix} 1 & -1 \\ 0 & -2 \end{bmatrix} = \begin{bmatrix} -3 & 2 \\ -2 & -3 \end{bmatrix}.$$

The substitution $u = x - y/2$, $v = -y/2$ reduces the given system to $u' = -3u + 2v$, $v' = -2u - 3v$, whose general solution is given by $u(t) = e^{-3t}(u_0 \cos(2t) + v_0 \sin(2t))$ and $v(t) = e^{-3t}(v_0 \cos(2t) - u_0 \sin(2t))$, for arbitrary constants u_0 and v_0. In terms of x, y, we have

$$x(t) = e^{-3t}[(u_0 - v_0)\cos(2t) + (u_0 + v_0)\sin(2t)]$$
$$y(t) = -2e^{-3t}[v_0 \cos(2t) - u_0 \sin(2t)].$$

With these explicit expressions, it is not difficult to draw the phase portrait of this system. A typical orbit is depicted in Figure 2.1(a).

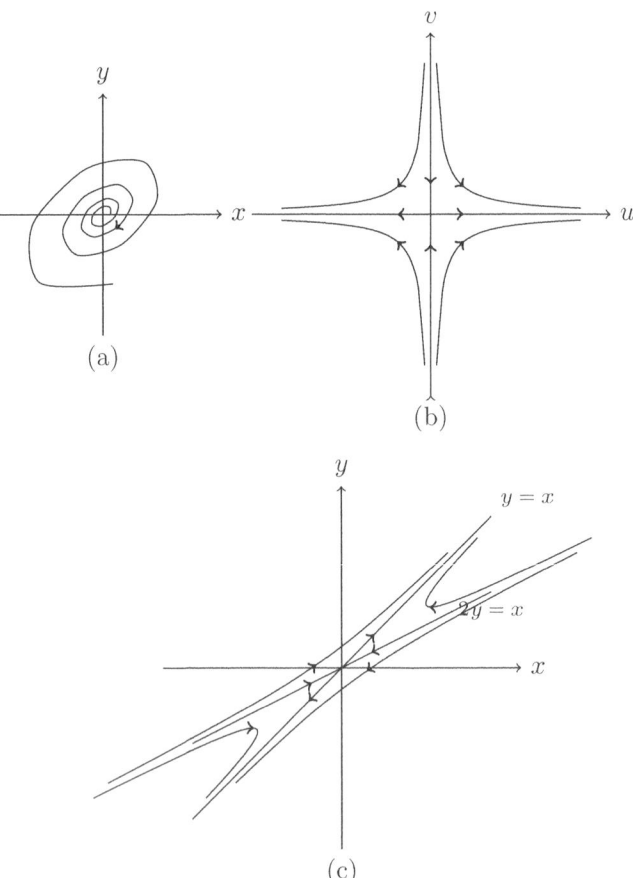

Figure 2.1 | Phase portraits of Exercise 2.2.

(b) The coefficient matrix is $\mathbf{A} = \begin{bmatrix} -3 & 4 \\ -2 & 3 \end{bmatrix}$ whose eigenvalues are the roots of the characteristic equation $\lambda^2 - (-3+3)\lambda + (-9+8) = 0$, or $\lambda = \pm 1$, with the corresponding eigenvectors $\begin{bmatrix} 1 \\ 1 \end{bmatrix}$ and $\begin{bmatrix} 2 \\ 1 \end{bmatrix}$. Thus, we take $\mathbf{C} = \begin{bmatrix} 1 & 2 \\ 1 & 1 \end{bmatrix}$ and the transformation $u = 2y - x$, $v = x - y$ reduces the given system to the diagonal system $u' = u$, $v' = -v$, whose solutions can be readily obtained as $u(t) = c_1 e^t$ and $v(t) = c_2 e^{-t}$ for arbitrary constants c_1, c_2. This in turn gives $x(t) = c_1 e^t + 2c_2 e^{-t}$ and $y(t) = c_1 e^t + c_2 e^{-t}$. Thus, the origin $(0,0)$ or the zero solution is unstable and is a saddle. Since the given system is two dimensional, it is not difficult to obtain the phase portrait. In the $u - v$ plane, both the u-axis and the v-axis are union of orbits. The other orbits are given by the relation $uv = c$, where c is a non-zero constant, which are branches of hyperbolas. The orbits in the $u - v$ plane are depicted in Figure 2.1(b). To obtain the phase portrait in the $x - y$ plane, first observe that the lines $y = x$ and $2y = x$ are union of orbits corresponding to the u-axis and v-axis, respectively. The other orbits are again branches of hyperbolas and are depicted in Figure 2.1(c).

(c) Here the coefficient matrix is $\mathbf{A} = \begin{bmatrix} 4 & -2 \\ 5 & 2 \end{bmatrix}$, whose eigenvalues are $3 \pm 3i$. The analysis is therefore the same as in (a) above, except that the origin is now unstable and a spiral as the real part of the complex eigenvalue is positive. Computing the real and imaginary parts of the complex eigenvector, we take $\mathbf{C} = \begin{bmatrix} 2 & -4 \\ -5 & -5 \end{bmatrix}$ and see that

$$(1/30) \begin{bmatrix} 5 & -4 \\ -5 & -2 \end{bmatrix} \begin{bmatrix} 4 & -2 \\ 5 & 2 \end{bmatrix} \begin{bmatrix} 2 & -4 \\ -5 & -5 \end{bmatrix} = \begin{bmatrix} 3 & 3 \\ -3 & 3 \end{bmatrix}.$$

The substitution $u = x/6 - 2y/15$, $v = -x/6 - y/15$ reduces the given system to $u' = 3u + 3v$, $v' = -3u + 3v$, whose general solution is given by $u(t) = e^{3t}(u_0 \cos(3t) + v_0 \sin(3t))$ and $v(t) = e^{3t}(v_0 \cos(3t) - u_0 \sin(3t))$, for arbitrary constants u_0 and v_0. In terms of x, y, we have

$$x(t) = 2u - 4v = 2e^{3t}[(u_0 - 2v_0)\cos(3t) + (v_0 + 2u_0)\sin(3t)]$$
$$y(t) = -5(u + v) = -5e^{3t}[(u_0 + v_0)\cos(3t) + (v_0 - u_0)\sin(2t)].$$

Exercise 2.3

Indeed, if $\mathbf{A}$ is the identity matrix of order n, then the only eigenvalue 1 has algebraic multiplicity n, and so is its geometric multiplicity. For $n = 2$, we have a standard non-diagonalizable matrix $\mathbf{A} = \begin{bmatrix} 1 & 0 \\ 1 & 1 \end{bmatrix}$ and, the algebraic and geometric multiplicities of the only eigenvalue 1 are, respectively 2 and 1. To find the geometric multiplicity, let $\mathbf{x} = \begin{bmatrix} x_1 & x_2 \end{bmatrix}^T$ be an eigenvector corresponding to the eigenvalue 1. Then, $x_1 + x_2 = x_2$, which implies $x_1 = 0$ and x_2 is arbitrary. Hence, there is only one linearly independent eigenvector, say $\mathbf{e}_2 = \begin{bmatrix} 0 & 1 \end{bmatrix}^T$. If $n = 3$, there are two possibilities, namely

$$\mathbf{A}_1 = \begin{bmatrix} 1 & 0 & 0 \\ 1 & 1 & 0 \\ 0 & 0 & 1 \end{bmatrix} \text{ and } \mathbf{A}_2 = \begin{bmatrix} 1 & 0 & 0 \\ 1 & 1 & 0 \\ 0 & 1 & 1 \end{bmatrix}.$$

The first matrix has two blocks, whereas the second one has a single block. For $\mathbf{A}_1$, we get only one constraint, $x_1 + x_2 = x_2$, for an eigenvector $\begin{bmatrix} x_1 & x_2 & x_3 \end{bmatrix}^T$, leaving x_2, x_3 arbitrary and thus producing exactly two independent eigenvectors. Thus, we can choose $\mathbf{e}_2 = \begin{bmatrix} 0 & 1 & 0 \end{bmatrix}^T$ and $\mathbf{e}_3 = \begin{bmatrix} 0 & 0 & 1 \end{bmatrix}^T$ as the two linearly independent eigenvectors. The geometric multiplicity of the eigenvalue 1 for the matrix $\mathbf{A}_1$ is therefore 2. For the matrix $\mathbf{A}_2$, we get, in the process of determining the eigenvectors, two equations $x_1 + x_2 = x_2$ and $x_2 + x_3 = x_3$, leaving out only x_3 as arbitrary. Thus, the geometric multiplicity of the eigenvalue 1 in this case is 1.

The above ideas can be used to construct the required $n \times n$ matrix $\mathbf{A}_k = [a_{ij}], k = 1, \ldots, n$ with diagonal entries 1 and also in the first sub-diagonal with $n - k$ entries as 1, that is, $a_{ii} = 1$ for $i = 1, \ldots, n$ and $a_{i+1,i} = 1$ for $i = 1, \ldots, n - k + 1$, and $a_{ij} = 0$ for all other i, j. In this case, the geometric multiplicity of the only eigenvalue 1 of the matrix $\mathbf{A}_k$ is k, and its algebraic multiplicity is n.

Exercise 2.4

The matrix $\mathbf{A}$ is $\mathbf{A} = \begin{bmatrix} -1 & 1 \\ 1 & -1 \end{bmatrix}$ whose eigenvalues are $\lambda = 0, -2$ and the corresponding eigenvectors can be chosen as $\begin{bmatrix} 1 & 1 \end{bmatrix}^T$ and $\begin{bmatrix} 1 & -1 \end{bmatrix}^T$. We take $\mathbf{P} = \begin{bmatrix} 1 & 1 \\ 1 & -1 \end{bmatrix}$. Therefore,

$$\mathbf{P}^{-1} = (1/2) \begin{bmatrix} 1 & -1 \\ 1 & 1 \end{bmatrix} \text{ and } \mathbf{P}^{-1}\mathbf{AP} = \mathbf{B} = \begin{bmatrix} 0 & 0 \\ 0 & -2 \end{bmatrix}$$

as expected. Hence, the given system and its equivalent system are $\mathbf{x}' = \mathbf{Ax}$ and $\mathbf{y}' = \mathbf{By}$, where $\mathbf{x} = \mathbf{Py}$, that is, $y_1' = 0$, $y_2' = -2y_2$. The solutions of the latter system are $y_1(t) = c_1$ and $y_2(t) = c_2 e^{-2t}$ for arbitrary constants c_1 and c_2. Consequently, the solutions of the given system are

$$\begin{bmatrix} x_1 \\ x_2 \end{bmatrix} = \mathbf{Py} = \begin{bmatrix} 1 & 1 \\ 1 & -1 \end{bmatrix} \begin{bmatrix} c_1 \\ c_2 e^{-2t} \end{bmatrix} = \begin{bmatrix} c_1 + c_2 e^{-2t} \\ c_1 - c_2 e^{-2t} \end{bmatrix}.$$

Analysis: Since $\mathbf{By} = 0$ implies $y_2 = 0$, we see that all the points on the y_1 axis are equilibrium points and all these points are stable (asymptotically) in the sense that any solution starting from the initial condition (y_{01}, y_{02}) moves towards the equilibrium point $(y_{01}, 0)$ as $t \to \infty$. See Figure 2.2(a). Now coming to the original system, we see that $\mathbf{Ax} = 0$ implies $x_1 = x_2$. Further notice that under the equivalent transformation, the y_2-axis is mapped onto the x_2-axis, whereas the y_1-axis is mapped to the diagonal line $x_1 = x_2$ in the $x_1 - x_2$ plane, and the points on this diagonal are the equilibrium points for the given system. See Figure 2.2(b). Note that the nature of the stability of an equilibrium point does not change under the linear equivalence.

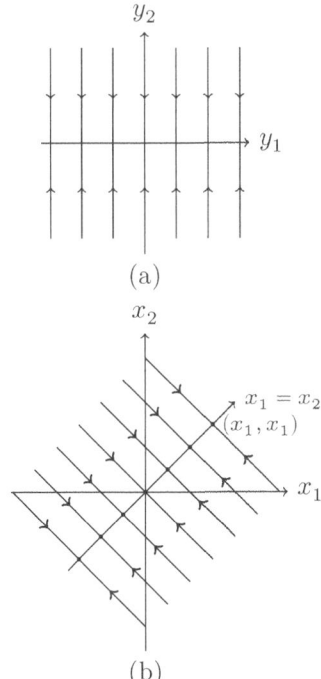

Figure 2.2 | Phase portrait in $y_1 - y_2$ and $x_1 - x_2$ planes in Exercise 2.4.

Exercise 2.5

For this system, the coefficient matrix is $\mathbf{A} = \begin{bmatrix} -1 & 3 \\ 0 & 2 \end{bmatrix}$, whose eigenvalues are -1 and 2. As $\mathbf{A}$ is non-singular, 0 is the only equilibrium point. The analysis is very similar to the one in Exercises 2.2, 2.4 and hence we leave the details to the reader.

Exercise 2.6

Here the matrix is $\mathbf{A} = \begin{bmatrix} 2 & 1 \\ -4 & -2 \end{bmatrix}$. The double eigenvalue is $\lambda = 0$, and we can choose an eigenvector $v_1 = \begin{bmatrix} 1 & -2 \end{bmatrix}^T$. Now consider the equation $\mathbf{A}^2 v = 0$ to find a generalized eigenvector. Since $\mathbf{A}^2 = \mathbf{0}$, the zero matrix, we can choose any vector (choose it independent of v_1) as a generalized vector, say, $v_2 = \begin{bmatrix} 1 & 0 \end{bmatrix}^T$. Take $\mathbf{P} = \begin{bmatrix} 1 & 1 \\ -2 & 0 \end{bmatrix}$, to obtain $\mathbf{P}^{-1} = \begin{bmatrix} 0 & -1/2 \\ 1 & 1/2 \end{bmatrix}$ and $\mathbf{P}^{-1}\mathbf{A}\mathbf{P} = \mathbf{B} = \begin{bmatrix} 0 & 2 \\ 0 & 0 \end{bmatrix}$. The equivalent system is $y_1' = 2y_2$, $y_2' = 0$. Upon solving, we get

$$\begin{bmatrix} y_1 \\ y_2 \end{bmatrix} = \begin{bmatrix} 2c_2 t + c_1 \\ c_2 \end{bmatrix} \quad \text{and} \quad \begin{bmatrix} x_1 \\ x_2 \end{bmatrix} = \mathbf{P}\mathbf{y} = \begin{bmatrix} c_1 + c_2 + 2c_2 t \\ -4c_2 t - 2c_1 \end{bmatrix},$$

for arbitrary constants c_1 and c_2. The readers are advised to find the equilibrium points and do their stability analysis.

Exercise 2.7

The matrix $\mathbf{A}_1$ has one real eigenvalue -3 and two complex eigenvalues $2 \pm i$ and the eigenvectors can be chosen as $\mathbf{u}_1 = \begin{bmatrix} 1 & 0 & 0 \end{bmatrix}^T$ and $\begin{bmatrix} 0 & 2 & 1 \mp i \end{bmatrix}^T$. Put $\mathbf{u}_2 = \begin{bmatrix} 0 & 2 & 1 \end{bmatrix}^T$ and $\mathbf{v}_2 = \begin{bmatrix} 0 & 0 & 1 \end{bmatrix}^T$. Thus, we get

$$\mathbf{P} = \begin{bmatrix} \mathbf{u}_1 & \mathbf{u}_2 & \mathbf{v}_2 \end{bmatrix} = \begin{bmatrix} 1 & 0 & 0 \\ 0 & 2 & 0 \\ 0 & 1 & 1 \end{bmatrix} \quad \text{and} \quad \mathbf{P}^{-1} = \begin{bmatrix} 1 & 0 & 0 \\ 0 & 1/2 & -1/2 \\ 0 & 0 & 1 \end{bmatrix}.$$

Hence, $\mathbf{P}^{-1}\mathbf{A}_1\mathbf{P} = \mathbf{B} = \begin{bmatrix} -3 & 0 & 0 \\ 0 & 2 & -1 \\ 0 & 1 & 2 \end{bmatrix}$. The solution $\mathbf{y}$ of the equivalent system $\mathbf{y}' = \mathbf{B}\mathbf{y}$ is then given by

$$\mathbf{y}(t) = \begin{bmatrix} c_1 e^{-3t} & e^{2t}(c_2 \cos t + c_3 \sin t) & e^{2t}(c_3 \cos t - c_2 \sin t) \end{bmatrix},$$

for arbitrary constants c_i, $i = 1, 2, 3$. Now it is easy to observe that the invariant spaces for the equivalent system are the y_1-axis and the $y_2 - y_3$ plane. An orbit starting on the y_1-axis remains there for all time and tends to 0 as $t \to \infty$. On the other hand, an orbit starting in the $y_2 - y_3$ plane remains there for all time and moves away from 0 as $t \to \infty$. Therefore, the origin, the only equilibrium point, is unstable and of saddle type. However, when an orbit is confined to the $y_2 - y_3$ plane, the origin appears to be an unstable node, as such an orbit spirals around the origin, but moving away from it. A few typical trajectories are depicted in Figure 2.3(a). Thus, the stable manifold E^s is the y_1-axis, and the unstable manifold E^u is the $y_2 - y_3$ plane.

We will now discuss the system with the coefficient matrix $\mathbf{A}_2$, the details of which can be worked out in a similar fashion. The eigenvalues of $\mathbf{A}_2$ are $-2 \pm i$ and 3. For the equivalent system, we now see that the stable manifold E^s is the $y_1 - y_2$ plane, whereas the unstable manifold E^u is the y_3 axis. The orbits are shown schematically in Figure 2.3(b).

What we have discussed is the stable and unstable manifolds of the equivalent system for $\mathbf{y}$. It is remarked that in both cases the stable and unstable manifolds will not change for the original system as the given matrix is already in block diagonal form.

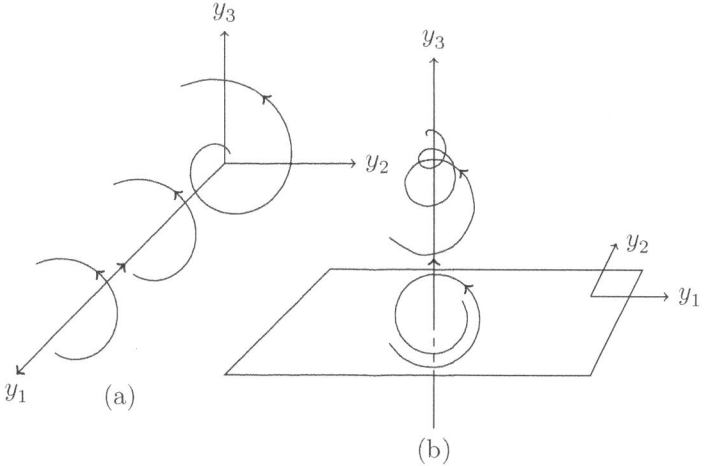

Figure 2.3 | Orbits in Exercise 2.7.

Exercise 2.8

First observe that the system is a pair of decoupled systems, involving two variables each, and hence can be solved for x_1, x_2 and x_3, x_4 separately. All the eigenvalues of the coefficient matrix are complex and are given by $1 \pm i$ and $2 \pm i$, and corresponding eigenvectors can be chosen as $\begin{bmatrix} i & 1 & 0 & 0 \end{bmatrix}^T = \mathbf{u}_1 + i\mathbf{v}_1$ and $\begin{bmatrix} 0 & 0 & 1+i & 1 \end{bmatrix}^T = \mathbf{u}_2 + i\mathbf{v}_2$. Thus, we have

$$\mathbf{P} = \begin{bmatrix} \mathbf{v}_1 & \mathbf{u}_1 & \mathbf{v}_2 & \mathbf{u}_2 \end{bmatrix} = \begin{bmatrix} 1 & 0 & 0 & 0 \\ 0 & 1 & 0 & 0 \\ 0 & 0 & 1 & 1 \\ 0 & 0 & 0 & 1 \end{bmatrix}, \quad \mathbf{P}^{-1} = \begin{bmatrix} 1 & 0 & 0 & 0 \\ 0 & 1 & 0 & 0 \\ 0 & 0 & 1 & -1 \\ 0 & 0 & 0 & 1 \end{bmatrix}$$

and $\mathbf{P}^{-1}\mathbf{AP} = \mathbf{B} = \begin{bmatrix} 1 & -1 & 0 & 0 \\ 1 & 1 & 0 & 0 \\ 0 & 0 & 2 & -1 \\ 0 & 0 & 1 & 2 \end{bmatrix}.$

The solution $\mathbf{y}$ to the equivalent system $\mathbf{y}' = \mathbf{By}$ is given by

$$\mathbf{y}(t) = \begin{bmatrix} e^t \cos t & -e^t \sin t & 0 & 0 \\ e^t \sin t & e^t \cos t & 0 & 0 \\ 0 & 0 & e^{2t} \cos t & -e^{2t} \sin t \\ 0 & 0 & e^{2t} \sin t & e^{2t} \cos t \end{bmatrix} \mathbf{y}_0$$

and $\mathbf{x} = \mathbf{Py}$, where $\mathbf{y}_0 = \mathbf{Px}_0$. Since the system is four dimensional, plotting trajectories will be difficult. However, one can plot the trajectories in $y_1 - y_2$ plane and $y_3 - y_4$ plane separately. These are spirals both of which are diverging (moving away from the origin) as real part of the eigenvalues are non-negative. Thus, the origin, which is the only equilibrium point, is an unstable node.

Exercise 2.9

Obviously the eigenvalues are $1, 2, 2$. An eigenvector corresponding the eigenvalue 1 is $\mathbf{u}_1 = \begin{bmatrix} 1 & 1 & -2 \end{bmatrix}^T$, whereas the geometric multiplicity of the eigenvalue 2 is 1 and a corresponding eigenvector is $\mathbf{u}_2 = \begin{bmatrix} 0 & 0 & 1 \end{bmatrix}^T$. A generalized eigenvector for the eigenvalue 2, which is linearly independent of $\mathbf{u}_1$ and $\mathbf{u}_2$ can be obtained as $\mathbf{u}_3 = \begin{bmatrix} 0 & 1 & 0 \end{bmatrix}^T$, by solving the equation $(\mathbf{A} - 2\mathbf{I})^2 \mathbf{x} = 0$. Thus,

$$\mathbf{P} = \begin{bmatrix} \mathbf{u}_1 & \mathbf{u}_2 & \mathbf{u}_3 \end{bmatrix} = \begin{bmatrix} 1 & 0 & 0 \\ 1 & 0 & 1 \\ -2 & 1 & 0 \end{bmatrix} \quad \text{and} \quad \mathbf{P}^{-1} = \begin{bmatrix} 1 & 0 & 0 \\ 2 & 0 & 1 \\ -1 & 1 & 0 \end{bmatrix}.$$

Hence, $\mathbf{S} = \mathbf{P} \begin{bmatrix} 1 & 0 & 0 \\ 0 & 2 & 0 \\ 0 & 0 & 2 \end{bmatrix} \mathbf{P}^{-1} = \begin{bmatrix} 1 & 0 & 0 \\ -1 & 2 & 0 \\ 2 & 0 & 2 \end{bmatrix}$ and thus

$$\mathbf{N} = \mathbf{A} - \mathbf{S} = \begin{bmatrix} 0 & 0 & 0 \\ 0 & 0 & 0 \\ -1 & 1 & 2 \end{bmatrix} \quad \text{and} \quad \mathbf{N}^2 = \mathbf{0}.$$

The final solution is given by

$$\mathbf{x}(t) = \mathbf{P} \begin{bmatrix} e^t & 0 & 0 \\ 0 & e^{2t} & 0 \\ 0 & 0 & e^{2t} \end{bmatrix} \mathbf{P}^{-1} (\mathbf{I} + t\mathbf{N}) \mathbf{x}_0$$

$$= \begin{bmatrix} e^t & 0 & 0 \\ e^t - e^{2t} & e^{2t} & 0 \\ 2e^t + (2-t)e^{2t} & te^{2t} & e^{2t} \end{bmatrix} \mathbf{x}_0.$$

Exercise 2.10

The given matrix is already in block diagonal form, which has pure imaginary eigenvalues $\pm i$ and a real eigenvalue 2. Hence, the decomposition of $\mathbf{R}^n = E^c \oplus E^u$, where E^c is the centre subspace, which is the $x_1 - x_2$ plane, and the unstable subspace E^u is the x_3-axis. If we change 2 by -2 in the given matrix, E^u would be replaced by the stable subspace E^s. If $\mathbf{x}_0 = (x_{01}, x_{02}, 0)$ is an initial point on the $x_1 - x_2$ plane, then the solution is given $x_1(t) = x_{01} \cos t - x_{02} \sin t$, $x_2(t) = x_{01} \sin t + x_{02} \cos t$ and $x_3(t) = 0$. The solutions are obviously bounded. Indeed, the other solutions with initial conditions outside the $x_1 - x_2$ plane will not be bounded as x_3 is not bounded (see Figure 2.4(a)).

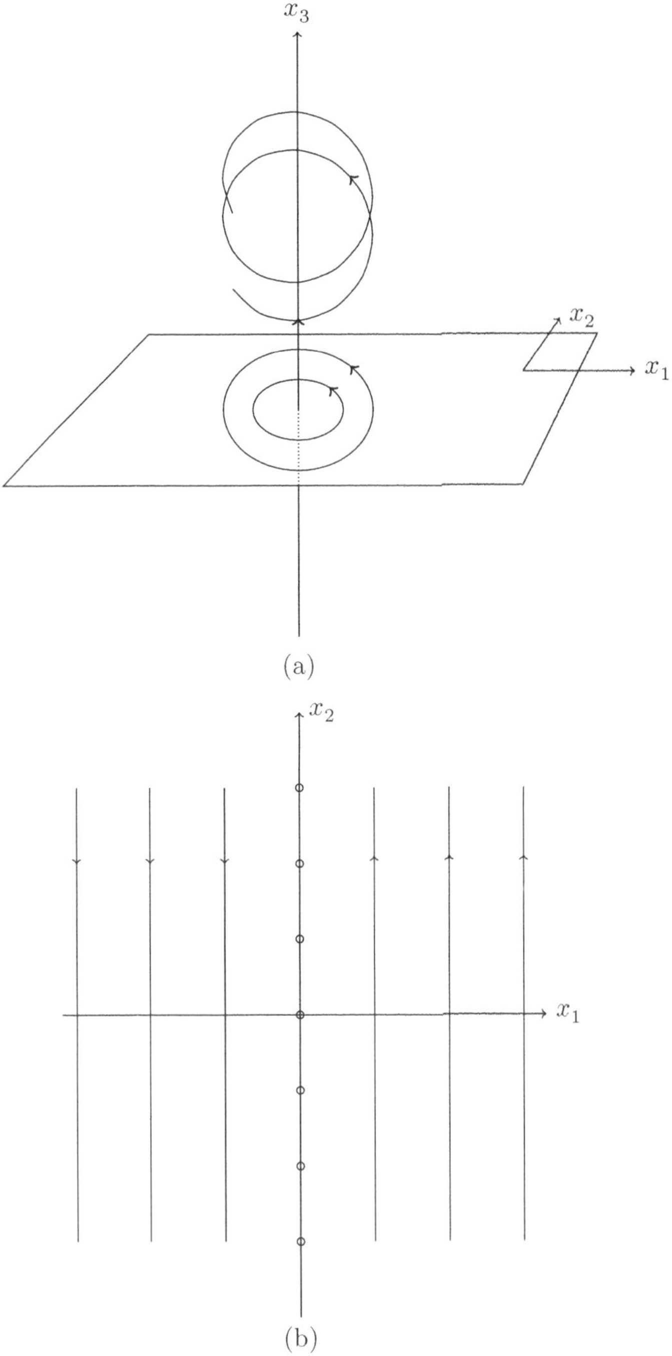

Figure 2.4 | Orbits in Exercise 2.10.

Now consider the 2×2 matrix $\mathbf{A} = \begin{bmatrix} 0 & 0 \\ 1 & 0 \end{bmatrix}$. The solutions to the corresponding ODE system is given by $x_1(t) = x_{01}$, constant and $x_2(t) = x_{01}t + x_{02}$, the straight lines parallel to the x_2 axis (see Figure 2.4(b)). In this case, all the points on the x_2 axis are equilibrium points. Compare the equivalent system in Exercise 2.4 and the Figure 2.2. There is a subtle difference between the two. In the previous problem, the trajectories outside the x_1 axis moves towards one of the equilibrium points. In this case, the other trajectories outside the x_2 axis does not move neither towards nor away to the equilibrium points, it moves parallel to the x_2 axis. Hence, it can be classified the equilibrium points as centres and $E^c = \mathbb{R}^2$ and all solutions are unbounded. See Figure 2.4(b).

Exercise 2.11

The possible values of the geometric multiplicity of λ are $1, 2, 3$ and 4. If it is 1, the deficiency index is 3, and there is only one Jordan block, namely $\begin{bmatrix} \lambda & 1 & 0 & 0 \\ 0 & \lambda & 1 & 0 \\ 0 & 0 & \lambda & 1 \\ 0 & 0 & 0 & \lambda \end{bmatrix}$. If the geometric multiplicity is 2, then the deficiency index is 2, and there are two Jordan blocks. The possible Jordan forms in this case are $\begin{bmatrix} \lambda & 0 & 0 & 0 \\ 0 & \lambda & 1 & 0 \\ 0 & 0 & \lambda & 1 \\ 0 & 0 & 0 & \lambda \end{bmatrix}$ and $\begin{bmatrix} \lambda & 1 & 0 & 0 \\ 0 & \lambda & 0 & 0 \\ 0 & 0 & \lambda & 1 \\ 0 & 0 & 0 & \lambda \end{bmatrix}$. If the geometric multiplicity is 3, then the deficiency index is 1, and there are three Jordan blocks. The only possible Jordan form in this case is $\begin{bmatrix} \lambda & 0 & 0 & 0 \\ 0 & \lambda & 0 & 0 \\ 0 & 0 & \lambda & 1 \\ 0 & 0 & 0 & \lambda \end{bmatrix}$. Finally, if the geometric multiplicity is 4, then the deficiency index is 0, and therefore the matrix is diagonalizable. Thus, the Jordan form in this case is $\begin{bmatrix} \lambda & 0 & 0 & 0 \\ 0 & \lambda & 0 & 0 \\ 0 & 0 & \lambda & 0 \\ 0 & 0 & 0 & \lambda \end{bmatrix}$.

Exercise 2.12

Suppose $\mathbf{B}(t)$ is a matrix-valued differentiable function for $t \in [a, b]$. Then, we have

$$\frac{d}{dt}\left(\mathbf{B}^2(t)\right) = \mathbf{B}(t)\mathbf{B}'(t) + \mathbf{B}'(t)\mathbf{B}(t) = 2\mathbf{B}(t)\mathbf{B}'(t) = 2\mathbf{B}'(t)\mathbf{B}(t),$$

if $\mathbf{B}(t)$ and $\mathbf{B}'(t)$ commute. An induction argument shows that $\frac{d}{dt}\left(\mathbf{B}^k(t)\right) = k\mathbf{B}^{k-1}(t)\mathbf{B}'(t)$ for any positive integer k, if $\mathbf{B}(t)$ and $\mathbf{B}'(t)$ commute. It therefore follows that

$$\frac{d}{dt}\exp\left(\mathbf{B}(t)\right) = \exp\left(\mathbf{B}(t)\right)\mathbf{B}'(t) = \mathbf{B}'(t)\exp\left(\mathbf{B}(t)\right),$$

if $\mathbf{B}(t)$ and $\mathbf{B}'(t)$ commute. Now put $\mathbf{B}(t) = \int_{t_0}^{t} \mathbf{A}(s)\,ds$. Using the hypothesis that $\mathbf{A}(t)$ and $\mathbf{A}(s)$ commute, it follows that $\mathbf{A}(t)$ and $\mathbf{B}(t)$ also commute. But $\mathbf{B}'(t) = \mathbf{A}(t)$, and thus $\mathbf{B}(t)$ and $\mathbf{B}'(t)$ commute. Therefore,

$$\frac{d}{dt}\exp\left(\int_{t_0}^{t} \mathbf{A}(s)\,ds\right) = \mathbf{A}(t)\exp\left(\int_{t_0}^{t} \mathbf{A}(s)\,ds\right).$$

This in turn verifies that the function $\mathbf{x}(t)$ as defined is indeed the solution of the stated IVP.

Exercise 2.13

We have

$$\mathbf{A}(t)\mathbf{A}(s) = \begin{bmatrix} 1 & 1+t \\ 0 & t \end{bmatrix} \begin{bmatrix} 1 & 1+s \\ 0 & s \end{bmatrix} = \begin{bmatrix} 1 & 1+2s+st \\ 0 & st \end{bmatrix}$$

and

$$\mathbf{A}(s)\mathbf{A}(t) = \begin{bmatrix} 1 & 1+s \\ 0 & s \end{bmatrix} \begin{bmatrix} 1 & 1+t \\ 0 & t \end{bmatrix} = \begin{bmatrix} 1 & 1+2t+st \\ 0 & st \end{bmatrix}.$$

Hence, $\mathbf{A}(t)\mathbf{A}(s) \neq \mathbf{A}(s)\mathbf{A}(t)$ if $s \neq t$ and the procedure in Exercise 2.12 is not applicable. Writing $\mathbf{x} = \begin{bmatrix} x \\ y \end{bmatrix}$, the given system can be written as

$$x'(t) = x(t) + (1+t)y(t) \quad \text{and} \quad y'(t) = ty(t).$$

As this system is partially decoupled, we can readily integrate the same to obtain

$$x(t) = e^t \left[x_0 + y_0 \int_0^t (1+s)\exp(s^2/2 - s)\,ds \right] \quad \text{and} \quad y(t) = y_0 \exp(t^2/2),$$

for arbitrary constants x_0 and y_0. Also, it is not difficult to compute the exponential $\exp\left(\int_0^t \mathbf{A}(s)\,ds\right)$. We have

$$\mathbf{B}(t) = \exp\left(\int_0^t \mathbf{A}(s)\,ds\right) = \exp\begin{bmatrix} t & t + t^2/2 \\ 0 & t^2/2 \end{bmatrix} = \begin{bmatrix} e^t & \mu(t)(e^{t^2/2} - e^t) \\ 0 & e^{t^2/2} \end{bmatrix},$$

with $\mu(t) = \dfrac{t+2}{t-2}$ for $t \neq 2$ and taking the limiting value of $\mu(t)(e^{t^2/2} - e^t)$ as $t \to 2$, when $t = 2$. However, we notice that the general solution of the given system is not of the form $\mathbf{B}(t)\begin{bmatrix} x_0 \\ y_0 \end{bmatrix}$.

Exercise 2.14

The given system can be written in matrix form as $\mathbf{B}x'' + \mathbf{A}x = \mathbf{0}$. As the matrix $\mathbf{B}$ is symmetric positive definite and the matrix $\mathbf{A}$ is symmetric, it follows that there is a non-singular matrix $\mathbf{C}$ such that $\mathbf{C}^T\mathbf{B}\mathbf{C} = \mathbf{I}$ and $\mathbf{C}^T\mathbf{A}\mathbf{C} = \operatorname{diag}\begin{bmatrix} \lambda_1 & \cdots & \lambda_n \end{bmatrix}$. This is called *simultaneous diagonalization* of the quadratic forms associated with matrices $\mathbf{A}$ and $\mathbf{B}$. Here is a brief proof of this important fact.

Since $\mathbf{B}$ is symmetric positive definite, there is an orthogonal matrix $\mathbf{V}$ such that $\mathbf{V}^T\mathbf{B}\mathbf{V} = \operatorname{diag}\begin{bmatrix} \mu_1^2 & \cdots & \mu_n^2 \end{bmatrix}$, where μ_i^2 are the eigenvalues of the matrix $\mathbf{B}$. Put $\mathbf{U} = \mathbf{V}\operatorname{diag}\begin{bmatrix} \mu_1^{-1} & \cdots & \mu_n^{-1} \end{bmatrix}$. It follows that $\mathbf{U}^T\mathbf{B}\mathbf{U} = \mathbf{I}$. Next, observe that the matrix $\mathbf{U}^T\mathbf{A}\mathbf{U}$ is also a symmetric matrix as the matrix $\mathbf{A}$ is symmetric. Therefore, there is an orthogonal matrix $\mathbf{D}$ such that $\mathbf{D}^T\mathbf{U}^T\mathbf{A}\mathbf{U}\mathbf{D} = \operatorname{diag}\begin{bmatrix} \lambda_1 & \cdots & \lambda_n \end{bmatrix}$, where λ_i are the eigenvalues of the matrix $\mathbf{U}^T\mathbf{A}\mathbf{U}$, hence real. Finally, put $\mathbf{C} = \mathbf{U}\mathbf{D}$. Since the matrix $\mathbf{D}$ is orthogonal, it follows that $\mathbf{C}^T\mathbf{B}\mathbf{C} = \mathbf{I}$.

It is also easy to find out the connection between the eigenvalues λ_i and the matrices $\mathbf{A}$, $\mathbf{B}$. We have $\mathbf{C}^T = \mathbf{C}^{-1}\mathbf{B}^{-1}$ and therefore, $\mathbf{C}^{-1}\mathbf{B}^{-1}\mathbf{A}\mathbf{C} = \operatorname{diag}\begin{bmatrix} \lambda_1 & \cdots & \lambda_n \end{bmatrix}$ or $\mathbf{A}\mathbf{C} = \mathbf{B}\mathbf{C}\operatorname{diag}\begin{bmatrix} \lambda_1 & \cdots & \lambda_n \end{bmatrix}$. If we denote the columns of the matrix $\mathbf{C}$ by $\mathbf{c}_1, \ldots, \mathbf{c}_n$, it follows that $\mathbf{A}\mathbf{c}_i = \lambda_i \mathbf{B}\mathbf{c}_i$ for $1 \leqslant i \leqslant n$. We call λ_i as the eigenvalues of the matrix $\mathbf{A}$ with respect to the positive definite matrix $\mathbf{B}$. Observe that $(\mathbf{B}\mathbf{c}_i, \mathbf{c}_j) = \delta_{ij}$, so that the vectors $\mathbf{c}_i$ are normalised with respect to the matrix $\mathbf{B}$. Note also that the eigenvalues λ_i are the roots of the characteristic polynomial $\det(\mathbf{A} - \lambda\mathbf{B}) = 0$.

Back to kinematics. With the substitution $\mathbf{x} = \mathbf{Cy}$, the given system of equations decouples into $\mathbf{y}'' + \text{diag}\begin{bmatrix} \lambda_1 & \cdots & \lambda_n \end{bmatrix} \mathbf{y} = \mathbf{0}$ or $y_i'' + \lambda_i y_i = 0$ for $1 \leqslant i \leqslant n$. If $\lambda_i \leqslant 0$ for some i, then the solution y_i grows exponentially or linearly for $t > 0$. Thus, the equilibrium point is not stable. To overcome this situation, we impose an additional restriction on the potential function P, by assuming that P has a strict minimum in a position of equilibrium, that is, $P(\mathbf{x}) > 0$ for $\mathbf{x} \neq \mathbf{0}$. It follows that the matrix $\mathbf{A}$ is also positive definite.[2] With this additional assumption, it now follows that all the eigenvalues λ_i are positive. Therefore, we obtain $y_i(t) = r_i \sin(\sqrt{\lambda_i}\, t + \theta_i)$ or $\mathbf{x} = \sum_{i=1}^{n} r_i \sin(\sqrt{\lambda_i}\, t + \theta_i) \mathbf{c}_i$. Here r_i are the amplitudes and θ_i are the angular shifts.

Exercise 2.15

In both cases, the matrix $\mathbf{A} = \begin{bmatrix} -a & 0 \\ a & -b \end{bmatrix}$ is the same, whereas $\mathbf{B} = \mathbf{B}_1 = \begin{bmatrix} 0 & 1 \end{bmatrix}^T$ for the problem (1) and $\mathbf{B} = \mathbf{B}_2 = \begin{bmatrix} 1 & 0 \end{bmatrix}^T$ for the problem (2). For the system in (1), it is evident that the system is not controllable as $x_1(t) = e^{-at} x_{01}$ fixes the trajectory and it is not influenced by the control appearing in the second equation. Hence, $x_1(t)$ cannot be steered to any arbitrary point, but $x_2(t)$ can be steered. Formally, the lack of controllability can be concluded by computing the rank of the matrix $\begin{bmatrix} \mathbf{A} & \mathbf{AB}_1 \end{bmatrix} = \begin{bmatrix} 0 & 0 \\ 1 & -b \end{bmatrix}$, which is 1.

In the case of the problem (2), both the trajectories are affected by the single control u. But this observation alone is insufficient to conclude the controllability or otherwise. Here, $\begin{bmatrix} \mathbf{A} & \mathbf{AB}_2 \end{bmatrix} = \begin{bmatrix} 1 & -a \\ 0 & -b \end{bmatrix}$ which has rank 2, and hence the system is controllable with one control.

Physical Interpretation: The above-mentioned problems can be obtained via the modelling of controlling the water levels in two tanks. The two tanks, Tank 1 and Tank 2, are filled with water, with levels $x_1(t)$ and $x_2(t)$, respectively. Water flows from Tank 1 into Tank 2 at the rate a (in appropriate units) and water is flowing out of Tank 2 at the rate b; see Figure 2.5. The above-mentioned problems will be respectively formed if we now supply water externally, at the rate u say, to Tank 2 (Figure 2.5(a))) and Tank 1 (Figure 2.5(b)). The signs of the coefficients are well justified. If the external water supply is into Tank 2, then we cannot maintain (control) the water level in Tank 1. But controllability tells us that the water levels in both tanks can be controlled (maintained) by choosing the control appropriately if the external supply is made into Tank 1.

---

[2] Apparently this is a theorem of Dirichlet.

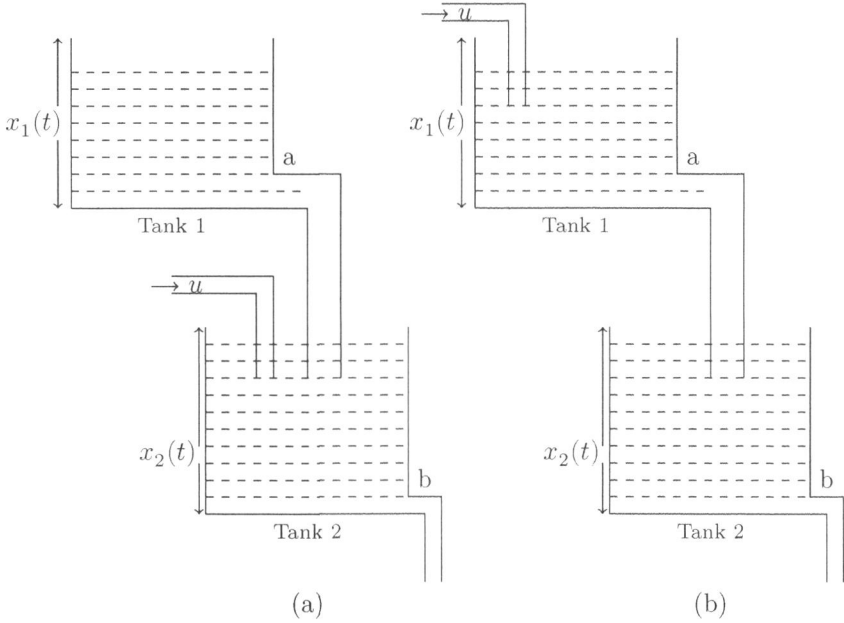

Figure 2.5 | Schematic description in Exercise 2.15.

Exercise 2.16

As in Exercise 2.1, the nth-order equation can be converted to a first system by introducing the variables $x_1, x_2, \cdots x_n$ as defined there. Recall the expression for $\mathbf{A}$ from Exercise 2.1 and, of course, $\mathbf{B} = \begin{bmatrix} 0 & \cdots & 0 & 1 \end{bmatrix}^{\mathrm{T}}$. Again, observe that all the variables $x_1, x_2, \ldots, x_n$ are influenced by the single control $u = g$ iteratively via the connection $x^{(k-1)} = x_k$ for $k = n, \ldots, 2$ in the reverse order. Hence, there is a possibility of getting controllability. To arrive at this conclusion, we may check the rank condition. We will do it for $n = 2$ and $n = 3$ and the general case can be worked out by the reader. For the case $n = 2$, the matrix $\begin{bmatrix} \mathbf{A} & \mathbf{AB} \end{bmatrix} = \begin{bmatrix} 0 & 1 \\ 1 & -p_1 \end{bmatrix}$, which has full rank 2. When $n = 3$, the matrix $\begin{bmatrix} \mathbf{A} & \mathbf{AB} & \mathbf{A}^2\mathbf{B} \end{bmatrix} = \begin{bmatrix} 0 & 0 & 1 \\ 0 & 1 & -p_1 \\ 1 & -p_1 & -p_2 + p_1^2 \end{bmatrix}$, which has full rank 3.

Exercise 2.17

Trivially (i) implies both (ii) and (iii), since the latter two statements are special cases of the first one. The proof of other equivalences is easy using the representation (2.2.3). To see this, assume (ii) holds true and let $\mathbf{x}_0, \mathbf{x}_1 \in \mathbb{R}^n$ be any two arbitrary points. Then equation (2.2.3) suggests to consider the initial point $\tilde{\mathbf{x}}_0 = \mathbf{x}_0 - \boldsymbol{\Phi}^{-1}(T, t_0)\mathbf{x}_1$ and let $\mathbf{u}$ be the control that steers the system from $\tilde{\mathbf{x}}_0$ to $\mathbf{0}$, by the assumption (ii). Then using equation (2.2.3), we get

$$0 = \boldsymbol{\Phi}(T, t_0)\tilde{\mathbf{x}}_0 + \int_{t_0}^{T} \boldsymbol{\Phi}(T, s)\mathbf{B}(s)\mathbf{u}(s) \, ds,$$

which is the same as equation (2.2.3). Hence, (i) holds. If we assume (iii) holds and if $\mathbf{x}_0, \mathbf{x}_1 \in \mathbb{R}^n$ are arbitrary points, we take $\mathbf{u}$ as the control that steers the system from $\mathbf{0}$ to $\tilde{\mathbf{x}}_1 = \mathbf{x}_1 - \boldsymbol{\Phi}^{-1}(T, t_0)\mathbf{x}_0$. Then using equation (2.2.3) again, we get

$$\mathbf{x}_1 - \boldsymbol{\Phi}(T, t_0)\mathbf{x}_0 = \mathbf{0} + \int_{t_0}^{T} \boldsymbol{\Phi}(T, s)\mathbf{B}(s)\mathbf{u}(s) \, ds = \int_{t_0}^{T} \boldsymbol{\Phi}(T, s)\mathbf{B}(s)\mathbf{u}(s) \, ds,$$

which is nothing but equation (2.2.3), showing that (i) holds.

Exercise 2.18

The duality and inner product have to be understood appropriately since we use the same notation $\langle \cdot, \cdot \rangle$ for the inner product and the duality. Since $C : X = L^2(0, T; \mathbb{R}^m) \to \mathbb{R}^n$ and $C^* : \mathbb{R}^n \to X$, we have for $\mathbf{u} \in X, \mathbf{y} \in \mathbb{R}^n$,

$$\langle \mathbf{u}, C^*\mathbf{y} \rangle_X = \langle C\mathbf{u}, \mathbf{y} \rangle_{\mathbb{R}^n} = \langle \int_{t_0}^{T} \boldsymbol{\Phi}(T, s)\mathbf{B}(s)\mathbf{u}(s) \, ds, \mathbf{y} \rangle_{\mathbb{R}^n}$$

$$= \int_{t_0}^{T} \langle \boldsymbol{\Phi}(T, s)\mathbf{B}(s)\mathbf{u}(s), \mathbf{y} \rangle_{\mathbb{R}^n} \, ds$$

$$= \int_{t_0}^{T} \langle \mathbf{u}(s), \mathbf{B}^*(s)\boldsymbol{\Phi}^*(T, s)\mathbf{y} \rangle_{\mathbb{R}^m} \, ds$$

$$= \langle \mathbf{u}(\cdot), \mathbf{B}^*(\cdot)\boldsymbol{\Phi}^*(T, \cdot)\mathbf{y} \rangle_X.$$

Hence, $(C^*\mathbf{y})(t) = \mathbf{B}^*(t)\boldsymbol{\Phi}^*(T, t)\mathbf{y}$, which in turn gives the Gramian matrix as

$$CC^* = \int_{t_0}^{T} \boldsymbol{\Phi}(T, s)\mathbf{B}(s)\mathbf{B}^*(s)\boldsymbol{\Phi}^*(T, s) \, ds.$$

Next, we compute L^*. The operator $L:\mathbb{R}^n \to X$ is defined as $(L\mathbf{y}_0)(t) = \mathbf{H}(t)\mathbf{\Phi}(t, t_0)\mathbf{y}_0$ and $L^*: X \to \mathbb{R}^n$. Now, for any $\mathbf{v} \in X$, $\mathbf{y} \in \mathbb{R}^n$, we have

$$\langle \mathbf{y}, L^*\mathbf{v} \rangle_{\mathbb{R}^n} = \langle L\mathbf{y}, \mathbf{v} \rangle_X = \int_{t_0}^{T} \langle \mathbf{H}(s)\mathbf{\Phi}(s, t_0)\mathbf{y}, \mathbf{v}(s) \rangle_{\mathbb{R}^m} \, ds$$

$$= \int_{t_0}^{T} \langle \mathbf{y}, \mathbf{\Phi}^*(s, t_0)\mathbf{H}^*(s)\mathbf{v}(s) \rangle_{\mathbb{R}^n} \, ds$$

$$= \langle \mathbf{y}, \int_{t_0}^{T} \mathbf{\Phi}^*(s, t_0)\mathbf{H}^*(s)\mathbf{v}(s) \, ds \rangle_{\mathbb{R}^n}$$

which implies $L^*\mathbf{v} = \int_{t_0}^{T} \mathbf{\Phi}^*(t, t_0)\mathbf{H}^*(t)\mathbf{v}(t) \, dt$. The observability matrix is given by

$$\mathbf{M} = \mathbf{M}(t_0, T) = L^*L = \int_{t_0}^{T} \mathbf{\Phi}^*(t, t_0)\mathbf{H}^*(t)\mathbf{H}(t)\mathbf{\Phi}(t, t_0) \, dt.$$

Exercise 2.19

We establish the result in the case of controllability. The observability case can be handled analogously. If the linear system (2.2.2) is controllable, then the operator C is onto. We now prove that $\mathbb{R}^n = \text{im}(C) = \text{im}[\mathbf{A} :: \mathbf{AB}]$, where the notation means the block matrix $[\mathbf{A} :: \mathbf{AB}] = \begin{bmatrix} \mathbf{A} & \mathbf{AB} & \cdots & \mathbf{A}^{n-1}\mathbf{B} \end{bmatrix}$. Thus, $\text{rank}([\mathbf{A} :: \mathbf{AB}]) = n$, which is the maximum possible rank. For the autonomous system, we have $\mathbf{\Phi}(t, t_0) = e^{(t-t_0)\mathbf{A}}$. For any $\mathbf{x} \in \mathbb{R}^n$, by controllability from the point $\mathbf{0}$ to $\mathbf{x}$, there is a control $\mathbf{u} \in X$ such that $\int_{t_0}^{T} e^{(T-s)\mathbf{A}} \mathbf{B}\mathbf{u}(s) \, ds = \mathbf{x}$. Expanding $e^{(T-s)\mathbf{A}}$ and using the Cayley–Hamilton theorem, we arrive at the result

$$\int_{t_0}^{T} (P_0 + P_1\mathbf{A} + \cdots + P_{n-1}\mathbf{A}^{n-1})\mathbf{B}\mathbf{u}(s) \, ds = \mathbf{x},$$

where the coefficients P_i are functions of s and T. This in turn shows that $\mathbf{x} \in \text{im}[\mathbf{A} :: \mathbf{AB}]$. Conversely, if the system is not controllable, then $\text{rank}(\mathbf{W}(t_0, T)) < n$, assuming the Kalman rank condition. Hence, there exists a non-zero $\mathbf{v} \in \mathbb{R}^n$ such that $\mathbf{W}(t_0, T)\mathbf{v} = \mathbf{0}$, which implies $\mathbf{v}^T \mathbf{W}(t_0, T)\mathbf{v} = \mathbf{0}$. Thus, we get

$$\int_{t_0}^{T} \|\mathbf{B}^*\mathbf{\Phi}^*(T,s)\mathbf{v}\|^2 = \int_{t_0}^{T} \mathbf{v}^T \mathbf{\Phi}(T,s)\mathbf{B}\mathbf{B}^*\mathbf{\Phi}^*(T,s)\mathbf{v} = 0.$$

Hence, we conclude that $\mathbf{B}^*\mathbf{\Phi}^*(T,s)\mathbf{v} = 0$ or, equivalently, $\mathbf{v}^T e^{(T-s)\mathbf{A}}\mathbf{B} = \mathbf{v}^T \mathbf{\Phi}(T,s)\mathbf{B} = 0$ for all $s \in [t_0, T]$. Taking $s = T$, we get $\mathbf{v}^T \mathbf{B} = 0$. Next, differentiating the expression $\mathbf{v}^T e^{(T-s)\mathbf{A}}\mathbf{B}\mathbf{v} = 0$, sufficient number of times, with respect to s, and taking $s = T$, we obtain recursively that $\mathbf{v}^T \mathbf{A}\mathbf{B} = \cdots = \mathbf{v}^T \mathbf{A}^{n-1}\mathbf{B} = 0$. Hence, $\mathbf{v}$ is orthogonal to $\mathrm{im}[\mathbf{A} :: \mathbf{A}\mathbf{B}]$ implying that $\mathrm{rank}[\mathbf{A} :: \mathbf{A}\mathbf{B}] < n$.

Exercise 2.20

This indeed is a special case of the general nth-order equation, and hence we know the controllability. The computations are not that trivial even in simple cases. Put $x_1 = x$, $x_2 = x_1' = x'$. Then $x_2' = -x_1 + u$. Thus, $\mathbf{A} = \begin{bmatrix} 0 & 1 \\ -1 & 0 \end{bmatrix}$ and $\mathbf{B} = \begin{bmatrix} 0 & 1 \end{bmatrix}^T$. It is easy to see the pattern of $\mathbf{A}^k$ as $\mathbf{A}^2 = -\mathbf{I}$. Thus, $\mathbf{A}^{2k+1} = (-1)^k \mathbf{A}$ and $\mathbf{A}^{2k} = (-1)^k \mathbf{I}$ for $k = 0, 1, 2, \ldots$. Therefore, $e^{t\mathbf{A}} = \begin{bmatrix} \cos t & \sin t \\ -\sin t & \cos t \end{bmatrix}$ and we take $t_0 = 0$ in further computations. We now compute $\mathbf{\Phi}(T,t)\mathbf{B} = e^{(T-t)\mathbf{A}}\mathbf{B} = \begin{bmatrix} \sin(T-t) & \cos(T-t) \end{bmatrix}^T$ and $\mathbf{B}^*\mathbf{\Phi}^*(T,t) = \begin{bmatrix} \sin(T-t) & \cos(T-t) \end{bmatrix}$. Thus, the Gramian matrix can be computed as

$$\mathbf{W}(0,T) = \int_0^T \mathbf{\Phi}(T,t)\mathbf{B}\mathbf{B}^*\mathbf{\Phi}^*(T,t)\, dt = \begin{bmatrix} \alpha & \beta \\ \beta & \gamma \end{bmatrix},$$

where $\alpha = (1/2)(T + \sin(2T)/2)$, $\beta = -(1/4)(1 - \cos(2T))$, $\gamma = (1/2)(T - \sin(2T)/2)$. Put $\delta = \alpha\gamma - \beta^2 = (1/4)(T^2 - 1) + (1/8)(1 + \cos(2T)) > 0$, if $T > 1$. Furthermore, $\mathbf{W}^{-1}(0,T) = (1/\delta)\begin{bmatrix} \gamma & -\beta \\ -\beta & \alpha \end{bmatrix}$. A control for a terminal point can be computed using the formula

$$u(t) = \mathbf{B}^*(t)\mathbf{\Phi}^*(T,t)\mathbf{W}^{-1}(\mathbf{x}_1 - \mathbf{\Phi}(T,0)\mathbf{x}_0).$$

Remark: The time T appearing in control problems is known as controllability time. If a certain T works for all the initial and final values, we say that the system is uniformly exactly controllable. What we have seen so far is that the linear systems are uniformly exactly controllable under the rank condition or the invertibility of controllability Gramian matrix. It is quite natural to expect that we cannot steer a system from one position to another instantaneously or even in a very small time. Thus, it may take some time, depending on the dynamics of the system, to move from one position to another. This is

Linear Systems | 81

what is indicated by the requirement $T > 1$ in this example. This phenomenon occurs in PDEs as well. For example, for the controllability of the wave equation, due to the finite speed of propagation, the controllability happens at a large time, which is essentially related to the diameter of the domain under consideration. Interestingly, there are examples of PDE systems where the controllability happens at arbitrarily small time. Note that the first-order PDE (of course with real coefficients) are hyperbolic, more like the wave equation.

Exercise 2.21

(a) If $\hat{\mathbf{y}}_T$ is a minimizer of J, then $J(\hat{\mathbf{y}}_T + h\mathbf{y}_T) - J(\hat{\mathbf{y}}_T) \geq 0$ for any $h \in \mathbb{R}$ and $\mathbf{y}_T \in \mathbb{R}^n$. Choosing $h > 0$ and dividing by h throughout, we see that

$$0 \leq \frac{1}{h}\left[\int_{t_0}^T h\langle \mathbf{B}^*(t)\hat{\mathbf{y}}(t), \mathbf{B}^*(t)\mathbf{y}(t)\rangle\, dt + h^2|\mathbf{B}^*\mathbf{y}|^2 + h\langle \mathbf{x}_0, \mathbf{y}(t_0)\rangle\right],$$

and letting $h \to 0$, we get

$$\int_{t_0}^T \langle \mathbf{B}^*(t)\hat{\mathbf{y}}(t), \mathbf{B}^*(t)\mathbf{y}(t)\rangle + \langle \mathbf{x}_0, \mathbf{y}(t_0)\rangle \geq 0.$$

Similarly, by taking $h < 0$ and letting again $h \to 0$, we get the reverse inequality.

(b) The equivalence is trivial by considering the map $f : \mathbb{R}^n \to \mathbb{R}^n$ defined by $f(\mathbf{y}_T) = \mathbf{y}(t_0)$. This is a bounded linear operator with a bounded inverse.

(c) Indeed, observability inequality implies **u.c.p**. To get the converse, introduce a new norm in $\mathbb{R}^n$, namely $|\mathbf{y}_T|^2_{\text{new}} := \int_{t_0}^T \|\mathbf{B}^*\mathbf{y}_T\|^2$. We conclude from **u.c.p.** that $|\mathbf{y}_T|_{\text{new}}$ is indeed a norm $\mathbb{R}^n$. Hence, it is equivalent to the standard norm as $\mathbb{R}^n$ is finite dimensional. Hence, we get the inequality in equation (2.3.2), which is equivalent to the observability inequality because of (b).

Remark: However, in the context of PDEs, observability inequality is a stronger concept than **u.c.p.**, as the analysis is done in an infinite-dimensional space. Hence, for PDE, we need to prove observability inequalities to achieve controllability. Hence, the analysis is more delicate.

Exercise 2.22

If equation (2.3.2) holds, then J is coercive in the sense that $J(\mathbf{y}_T) \to \infty$ as $|\mathbf{y}_T| \to \infty$. Then, there is a ball $B_0(R)$ in $\mathbb{R}^n$ of radius R such that

$$\inf_{\mathbf{y}_T \in \mathbb{R}^n} J(\mathbf{y}_T) = \inf_{\mathbf{y}_T \in \overline{B_0(R)}} J(\mathbf{y}_T).$$

The latter minimization problem is on a compact set, and hence a minimum exists.

Exercise 2.23

Consider $\mathbf{A}(t) = \begin{bmatrix} a\cos^2 t - 1 & 1 - a\sin t \cos t \\ -1 - a\sin t \cos t & a\sin^2 t - 1 \end{bmatrix}$, where a, b are real constants. It is easy to see that the eigenvalues are given by $(1/2)[(a-2) \pm \sqrt{a^2-4}]$, for all t. Therefore, the real parts of the eigenvalues are negative if $|a| < 2$. As in the case of an autonomous linear system (though the system in question is non-autonomous), we seek a solution in the form:

$$x(t) = e^{\lambda t}[c_1 \cos t + c_2 \sin t], \quad y(t) = e^{\lambda t}[c_2 \cos t - c_1 \sin t],$$

where the constant λ and the coefficients c_1, c_2 are to be determined. A straight-forward computation shows that the functions x and y indeed satisfy the system in question, provided that $\lambda c_1 = (a-1)c_1$ and $\lambda c_2 = -c_2$. For a non-trivial solution, c_1, c_2 are not both zero. This condition gives two possible values of λ, namely $\lambda = a - 1$ and $\lambda = -1$. Thus, we obtain the following pair of solutions: $x(t) = e^{(a-1)t} \cos t$, $y(t) = -e^{(a-1)t} \sin t$ and $x(t) = e^{-t} \sin t$, $y(t) = e^{-t} \cos t$. It is easy to see that these solutions are linearly independent. Therefore, the solution of the system satisfying $x(0) = x_0$ and $y(0) = y_0$ is given by

$$\begin{bmatrix} x(t) \\ y(t) \end{bmatrix} = \begin{bmatrix} e^{(a-1)t} \cos t & e^{-t} \sin t \\ -e^{(a-1)t} \sin t & e^{-t} \cos t \end{bmatrix} \begin{bmatrix} x_0 \\ y_0 \end{bmatrix}.$$

On the other hand, both $|x(t)|$ and $|y(t)|$ are unbounded for $t > 0$, if $a > 1$, unless $x_0 = y_0 = 0$. Therefore, we conclude that the zero solution is unstable if $1 < a < 2$, though the real parts of the eigenvalues of the matrix in question are negative.

For another example, consider the two-dimensional system $\begin{bmatrix} x' \\ y' \end{bmatrix} = \mathbf{A} \begin{bmatrix} x \\ y \end{bmatrix}$, with

$$\mathbf{A} = \begin{bmatrix} a + b\sin(2\alpha t) & b\cos(2\alpha t) \\ b\cos(2\alpha t) & a - b\sin(2\alpha t) \end{bmatrix},$$

where a, b, α are fixed real numbers. The eigenvalues of $\mathbf{A}$ can be easily computed and found to be $a \pm b$. Note that the eigenvalues do not depend on t and α and, we choose a, b so that at least one of the eigenvalues is positive. As in the previous example, we seek a solution in the form

$$x(t) = e^{\lambda t}[c_1 \cos(\alpha t) + c_2 \sin(\alpha t)], \quad y(t) = e^{\lambda t}[c_2 \cos(\alpha t) - c_1 \sin(\alpha t)],$$

where the constant λ and the coefficients c_1, c_2 are to be determined. A straight-forward computation shows that the functions x and y indeed satisfy the system in question, provided that

$$(\lambda - a)c_1 = (b - \alpha)c_2, \quad (\lambda - a)c_2 = (b - \alpha)c_1.$$

For a non-trivial solution, c_1, c_2 are not both zero. This imposes the following condition on λ: $(\lambda - a)^2 = \alpha^2 - b^2$. As we are looking for real λ, we impose the condition that $|b| < |\alpha|$. Thus, we have $\lambda = a \pm \sqrt{\alpha^2 - b^2}$. By further choosing a, b and α suitably, we can make the values of λ distinct and negative. The coefficients c_1, c_2 can now be calculated using the above-mentioned relations. In this manner, we do obtain two linearly independent solutions of the system, both tending to zero as $t \to \infty$. Thus, the zero solution is asymptotically stable, though an eigenvalue of the coefficient matrix is positive.

3
Series Solutions: Frobenius Theory

3.1 Introduction

In Chapter 1 on first- and second-order equations, we have seen that the solutions of linear first-order equations can be obtained in explicit form by converting the problem essentially to an integral calculus problem. We have also seen that there is no procedure in general to obtain the solutions of linear second-order equations with variable coefficients in explicit form. Nevertheless, we could obtain valuable information about the solutions by exploiting the linearity, superposition principle and so on. In this chapter, we consider a class of linear second-order equations whose solutions may be written down in explicit form. Since these solutions will be in the form of an infinite (power) series, eliciting the qualitative behaviour of solutions near some specified point or at infinity will be an important aspect. The results of this chapter are collectively called *Frobenius theory*. Some important equations, such as Bessel's equation, Hermite equation, Chebyshev equation, Laguerre equation, Airy equation and so on, are included in the class of equations considered here. Owing to the importance of these equations, which appear in applications frequently, the major properties of their solutions have been tabulated in mathematical handbooks. The interested reader may refer to Ref. [1]. We restrict the discussion to the real domain, though it is more advantageous to work in a complex domain. For example, the function $(1+x^2)^{-1}$ as a function of the real variable x is smooth and has no singularity. However, when x is considered in the complex domain, this function has singularities at the points $x = \pm i$. The discussion in the complex domain also requires tools from complex analysis. For these interesting and important results for the equations in the complex domain, the interested reader is referred to Refs. [7], [15], [29], [54], among others.

3.2 Real Analytic Functions

The class of equations we consider will have *analytic* coefficients. Roughly speaking, analyticity means convergent power series. We are familiar with power series in the context of Taylor's series and Maclaurin's series in calculus.

Definition 3.1 (Analyticity). A function $f : (a, b) \to \mathbb{R}$, where (a, b) is an open interval in $\mathbb{R}$, is said to be *real analytic* or simply *analytic* at $t_0 \in (a, b)$ if there exists $\delta > 0$ such that $(t_0 - \delta, t_0 + \delta) \subset (a, b)$ and

$$f(t) = \sum_{n=0}^{\infty} a_n (t - t_0)^n,$$

for all $t \in (t_0 - \delta, t_0 + \delta)$, where a_n's are real numbers, that is, $f(t)$ is represented as a convergent power series in $t - t_0$ in a neighbourhood of t_0. If f is analytic at every point in the interval (a, b), we say that f is analytic in (a, b).

We now recall certain facts about convergent power series, which will be needed in what follows. For details, see Refs. [2], [50].

Consider a real power series $\sum_{n=0}^{\infty} a_n t^n$ and put $R^{-1} = \limsup_{n \to \infty} \sqrt[n]{|a_n|}$. Then, the given power series converges for all t satisfying $|t| < R$ and diverges for $|t| > R$; the case of $|t| = R$ is, in general, inconclusive. The number R is called the *radius of convergence* of the power series. Note that, R can also take the value 0 or ∞. Put $f(t) = \sum_{n=0}^{\infty} a_n t^n$ for $t \in (-R, R)$. The following statements hold:

(i) The series converges uniformly in any compact subset of $(-R, R)$.

(ii) The function f is infinitely differentiable in $(-R, R)$ and

$$f^{(k)}(t) = \sum_{n=0}^{\infty} (n+1)(n+2) \cdots (n+k) a_{n+k} t^n,$$

for $k = 1, 2, \cdots$. The above series converges for all $t \in (-R, R)$.

(iii) In particular, $f^{(k)}(0) = k! a_k$, for $k = 0, 1, 2, \ldots$.

Remark 3.2. If $a_n \neq 0$ after a certain stage and $\lim_{n \to \infty} \frac{|a_{n+1}|}{|a_n|} = \ell$, then it is well known that $\lim_{n \to \infty} \sqrt[n]{|a_n|}$ also equals ℓ. Thus, we have an alternative way of calculating the radius of convergence, when applicable.

We may consider a power series around any given t_0 by replacing t by $t - t_0$ in the above discussion. We now give several examples of analytic functions, which are familiar to us from calculus. The functions $\sin t, \cos t$ and e^t are analytic in $\mathbb{R}$. The function $\log t$ is analytic in $(0, \infty)$, and $t^{1/3}$ is analytic at any $t_0 > 0$ as follows from the binomial series. The function $(t-a)^{-1}$ is analytic everywhere except at the point a. A point where a function is not analytic is termed as a *singular point* of the function.

Denote by $\mathcal{A}(a,b)$, the set of all analytic functions in (a,b). This is a real vector space. It is also closed under multiplication and thus becomes an *algebra*. The composition of two analytic functions, when defined, is also analytic. If f is analytic at t_0 and $f^{(k)}(t_0) = 0$, for $k = 0, 1, 2, ...$, it follows that $f \equiv 0$ in a neighbourhood of t_0. This property distinguishes an analytic function from a mere infinitely differentiable function. For example, if f is analytic in (a,b), then f cannot be compactly supported in (a,b), that is, f cannot vanish outside any $[c,d] \subset (a,b)$.

Example 3.3. Consider the function $f : \mathbb{R} \to \mathbb{R}$ defined by

$$f(t) = \begin{cases} \exp(-1/t) & \text{if } t > 0, \\ 0 & \text{if } t \leq 0. \end{cases}$$

It is not difficult to check that f is in $C^\infty(\mathbb{R})$ (verification needed only at $t = 0$), but f is not analytic at $t = 0$.

The natural question that arises is: which C^∞ functions are analytic? We state the following result without proof.

Theorem 3.4. A function f defined in a neighbourhood of t_0 is analytic at t_0 if and only if

(i) f is C^∞ in a neighbourhood of t_0, and

(ii) there exist positive δ and M such that for any $t \in (t_0 - \delta, t_0 + \delta)$, the inequality

$$|f^{(k)}(t)| \leq M \frac{k!}{\delta^k}, \qquad (3.2.1)$$

holds for $k = 0, 1, 2, ...$

Remark 3.5. If we replace $k!$ in equation (3.2.1) by a weaker condition, $(k!)^s$, where $s \geq 1$, we obtain a class of C^∞ functions called the *Gevrey class* of *index s*. For $s = 1$, we recover the functions that are analytic. When $s > 1$, it is interesting to note that the Gevrey class contains functions with compact support. The definition of the Gevrey class of functions may easily be extended to open subsets of $\mathbb{R}^n$. These functions play an important role in obtaining the regularity of weak solutions to linear parabolic equations and weakly hyperbolic systems in the theory of partial differential equations.

3.3 Equations with Analytic Coefficients

We begin with a familiar example. Consider the following equation

$$y'' + y = 0. \tag{3.3.1}$$

We seek an analytic solution y of equation (3.3.1) around $t = 0$ in the form

$$y(t) = \sum_{n=0}^{\infty} a_n t^n. \tag{3.3.2}$$

Assuming the convergence of the above series in an interval $(-R, R)$, we obtain by term-by-term differentiation that

$$y'(t) = \sum_{n=0}^{\infty} (n+1) a_{n+1} t^n, \tag{3.3.3}$$

and

$$y''(t) = \sum_{n=0}^{\infty} (n+1)(n+2) a_{n+2} t^n, \tag{3.3.4}$$

Substituting the expressions in equations (3.3.4) and (3.3.2) into (3.3.1), we obtain

$$\sum_{n=0}^{\infty} [(n+1)(n+2) a_{n+2} + a_n] t^n = 0. \tag{3.3.5}$$

Therefore, we have

$$(n+1)(n+2) a_{n+2} + a_n = 0, \; n = 0, 1, 2, \ldots.$$

We thus recursively obtain the following:

$$a_2 = -\frac{a_0}{2}, \; a_3 = -\frac{a_1}{3!}, \; a_4 = -\frac{a_2}{3 \cdot 4} = \frac{a_0}{4!}, \; a_5 = -\frac{a_3}{4 \cdot 5} = \frac{a_1}{5!}, \ldots$$

Substituting these expressions in equation (3.3.2), the expression for y becomes

$$y(t) = a_0 \sum_{n=0}^{\infty} (-1)^n \frac{t^{2n}}{(2n)!} + a_1 \sum_{n=0}^{\infty} (-1)^n \frac{t^{2n+1}}{(2n+1)!}. \tag{3.3.6}$$

The two power series in equation (3.3.6) are very familiar and represent $\cos t$ and $\sin t$, respectively. Thus,

$$y(t) = a_0 \cos t + a_1 \sin t,$$

where a_0 and a_1 are arbitrary real constants. Thus, the series (3.3.2) for y converges for all $t \in \mathbb{R}$. This is expected as the coefficients in equation (3.3.1) are analytic in $\mathbb{R}$.

Of course, we would have obtained the above solution without going into the exercise of power series, as (3.3.1) is an equation with constant coefficients. Nevertheless, this exercise contains all the ingredients of a general procedure to obtain series solutions to linear equations with analytic coefficients. However, it is not always possible to recognize a power series in general in terms of elementary functions, like an exponential function, trigonometric functions, logarithmic functions and so on.

We consider one more example before stating the general result. The second-order equation

$$y'' - 2ty' + 2py = 0, \tag{3.3.7}$$

where p is a real constant, is termed as the *Hermite's equation*. If we again assume the solution in the form (3.3.2) and substitute the expressions in equations (3.3.2), (3.3.3) and (3.3.4) into (3.3.7), then we obtain

$$\sum_{n=0}^{\infty} [(n+1)(n+2)a_{n+2} - 2na_n + 2pa_n]t^n = 0. \tag{3.3.8}$$

Equating each coefficient to zero in the above series, we get

$$(n+1)(n+2)a_{n+2} = -2(p-n)a_n, \; n = 0, 1, 2, \ldots,$$

similar to the expression obtained in the previous example. Therefore, the solution y may be written as

$$y(t) = a_0 y_1(t) + a_1 y_2(t), \tag{3.3.9}$$

where y_1 and y_2 are given by the following series

$$y_1(t) = 1 - \frac{2p}{2!}t^2 + \frac{2^2 p(p-2)}{4!}t^4 - \frac{2^3 p(p-2)(p-4)}{6!}t^6 + \cdots \tag{3.3.10}$$

and

$$y_2(t) = t - \frac{2(p-1)}{3!}t^3 + \frac{2^2(p-1)(p-3)}{5!}t^5$$
$$- \frac{2^3(p-1)(p-3)(p-5)}{7!}t^7 + \cdots \tag{3.3.11}$$

By the ratio test, it is straight forward to verify that both the above series converge for all $t \in \mathbb{R}$. It is also not difficult to see that they are linearly independent, and hence they span the solution space of the Hermite's equation. We now make the following observations.

First, note that unless p is a non-negative integer, the infinite series for y_1 and y_2 do not terminate. If p is a non-negative even integer, the series for y_1 terminates so that y_1 becomes a polynomial of degree p. Similarly, if p is a non-negative odd integer, y_2 becomes a polynomial of degree p. Any other polynomial solution of the Hermite's equation is a multiple of one of the above polynomials. It is not difficult to compute these polynomials for small p. For example, when $p = 0, 1, 2, 3$, the respective polynomials are given by 1, t, $1 - 2t^2$, $t - \frac{2}{3}t^3$.

Since any constant multiple of these polynomials is also a solution of the Hermite's equation with p a non-negative integer, it is customary to take the coefficient of t^n, the leading term, as 2^n. The resulting polynomials are then termed as *Hermite polynomials* and are denoted by $H_n(t)$. Thus, $H_0(t) = 1$, $H_1(t) = 2t$ and $H_3(t) = 8t^3 - 12t$. Hermite polynomials appear frequently in several applications, especially in quantum mechanics.

The following interesting formula for H_n may be deduced from the expressions (3.3.10) and (3.3.11):

$$H_n(t) = (-1)^n e^{t^2} \frac{d^n}{dt^n}\left(e^{-t^2}\right).$$

Theorem 3.6. Consider the second-order linear equation

$$y'' + P(t)y' + Q(t)y = 0, \tag{3.3.12}$$

where P and Q are analytic functions at $t_0 \in \mathbb{R}$ and are expressed in convergent power series in $t - t_0$, $t \in (t_0 - R, t_0 + R)$ for some $R > 0$. Then, given any arbitrary real numbers a_0 and a_1, there exists a unique analytic solution y of equation (3.3.12) satisfying the initial conditions $y(t_0) = a_0$ and $y'(t_0) = a_1$. Further, the solution y is also be expressed as a convergent power series in $t - t_0$ for $t \in (t_0 - R, t_0 + R)$.

The point t_0 is referred to as an *ordinary point*. We may take $t_0 = 0$, by changing the variable, if necessary. If P and Q have the power series representations

$$P(t) = \sum_{n=0}^{\infty} p_n t^n \text{ and } Q(t) = \sum_{n=0}^{\infty} q_n t^n, \tag{3.3.13}$$

for $t \in (-R, R)$, respectively, then the solution y of equation (3.3.12) has the following power series representation

$$y(t) = \sum_{n=0}^{\infty} a_n t^n, \tag{3.3.14}$$

where the coefficients a_n's satisfy the following recursion relations:

$$(n+1)(n+2)a_{n+2} = -\sum_{k=0}^{n}[(k+1)p_{n-k}a_{k+1} + q_{n-k}a_k]. \tag{3.3.15}$$

Thus, a_n's are determined in terms of a_0, a_1, p_n's and q_n's.

3.4 Regular Singular Points

Many second-order linear equations that make frequent appearance in applications do not have analytic coefficients. However, the singularities present are isolated and *regular*, which will be defined below. We have already had an idea in Chapter 1 how a singularity in the coefficients may influence the solution in question. We consider the following simple example to see what we can expect when singularities are present in the coefficients.

Example 3.7. Consider the second-order equation $y'' + \dfrac{k}{t^2}y = 0$, for $t > 0$. Here, k is a real constant. Notice that $t = 0$ is a singular point of the coefficient of y. It is not hard to write down the general solution of this equation. We have

$$y(t) = t^{1/2}(c_1 \sin(\mu \log t) + c_2 \cos(\mu \log t)), \ t > 0, \ \text{if } k > 1/4,$$
$$y(t) = t^{1/2}(c_1 \log t + c_2), \ t > 0, \ \text{if } k = 1/4,$$
$$y(t) = t^{1/2}(c_1 t^{\mu} + c_2 t^{-\mu}), \ t > 0, \ \text{if } k < 1/4.$$

Here, $\mu = \sqrt{k - 1/4}$ if $k > 1/4$ and $\mu = \sqrt{1/4 - k}$ if $k < 1/4$; c_1 and c_2 are arbitrary constants. Observe that any solution, except one, is defined only for $t > 0$ and possesses a singularity at $t = 0$. One solution, which is a multiple of $t^{1/2}$ when $k = 1/4$ is defined for $t \geq 0$. Even in this case, we cannot arbitrarily prescribe the initial data at $t = 0$.

The above example is an Euler equation, which was discussed in an exercise in Chapter 1. The general situation is going to be somewhat similar. Here we list some important equations that fall in this category.

(i) *Bessel's Equation of Order Zero*: $ty'' + y' + ty = 0$.
Observe that $t = 0$ is a singularity.

(ii) *Bessel's Equation of Order p*: $t^2 y'' + ty' + (t^2 - p^2)y = 0$,
where p is a non-negative real number. Again $t = 0$ is a singularity.

(iii) *Legendre's Equation*: $(1 - t^2)y'' - 2ty' + p(p+1)y = 0$,
where p is a real constant. Here $t = \pm 1$ are the singular points.

(iv) *Chebyshev's Equation*: $(1-t^2)y'' - ty' + p^2 y = 0$,
where p is a real constant. Again, $t = \pm 1$ are the singular points.

(v) *Gauss's Hypergeometric Equation*: $t(1-t)y'' + [c - (a+b+1)t]y' - aby = 0$,
where a, b and c are real constants. Here $t = 0, 1$ are the singular points.

(vi) *Airy Equation*: $y'' = ty$.

3.4.1 Equations with Regular Singular Points

We now consider the general second-order equation

$$y'' + P(t)y' + Q(t)y = 0. \qquad (3.4.1)$$

We assume that the functions P and Q in equation (3.4.1) have a singular point t_0 in $\mathbb{R}$. The singular point t_0 is called a *regular singular point* of equation (3.4.1) if the functions $(t-t_0)P(t)$ and $(t-t_0)^2 Q(t)$ are analytic at $t = t_0$. If t_0 is neither an ordinary point nor a regular singular point, then it is called an *irregular singular point*. The regular singular point and irregular singular point are also called *singular points of first* and *second kind*, respectively. We are going to obtain a convergent series solution y of equation (3.4.1) near a regular singular point. However, we will not have freedom to arbitrarily fix the initial conditions at a regular singular point. For the ease of writing, we assume that $t = 0$ is a regular singular point of equation (3.4.1); the general case follows from a change of variable.

From the hypothesis, it follows that P and Q are of the form

$$P(t) = \frac{p_0}{t} + p_1 + p_2 t + \cdots \qquad (3.4.2)$$

and

$$Q(t) = \frac{q_0}{t^2} + \frac{q_1}{t} + q_2 + q_3 t + \cdots \qquad (3.4.3)$$

We assume a solution y of equation (3.4.1) in the form of a 'quasi power series'

$$y(t) = t^m (a_0 + a_1 t + a_2 t^2 + \cdots) \qquad (3.4.4)$$

where $a_0 \neq 0$. The determination of the exponent m is part of the problem. The equation satisfied by the index m, the *indicial equation*, will be a quadratic equation reflecting the order of equation (3.4.1). The nature of the indicial equation is going to produce solutions with different behaviours at $t = 0$.

Assuming that the series for y is convergent in an interval $(0, T)$, we obtain the following (see the previous section):

$$y' = \sum_{n=0}^{\infty} a_n(m+n)t^{m+n-1},$$

$$y'' = \sum_{n=0}^{\infty} a_n(m+n)(m+n-1)t^{m+n-2}$$

$$= t^{m-2} \sum_{n=0}^{\infty} a_n(m+n)(m+n-1)t^n.$$

For the terms $P(t)y'$ and $Q(t)y$, using equations (3.4.2) and (3.4.3), we get

$$P(t)y' = \frac{1}{t}\left(\sum_{n=0}^{\infty} p_n t^n\right)\left[\sum_{n=0}^{\infty} a_n(m+n)t^{m+n-1}\right]$$

$$= t^{m-2} \sum_{n=0}^{\infty}\left[\sum_{k=0}^{n} p_{n-k}a_k(m+k)\right] t^n$$

$$= t^{m-2} \sum_{n=0}^{\infty}\left[\sum_{k=0}^{n-1} p_{n-k}a_k(m+k) + p_0 a_n(m+n)\right] t^n$$

and

$$Q(t)y = \frac{1}{t^2}\left(\sum_{n=0}^{\infty} q_n t^n\right)\left[\sum_{n=0}^{\infty} a_n t^{m+n}\right]$$

$$= t^{m-2} \sum_{n=0}^{\infty}\left[\sum_{k=0}^{n} q_{n-k}a_k\right] t^n$$

$$= t^{m-2} \sum_{n=0}^{\infty}\left[\sum_{k=0}^{n-1} q_{n-k}a_k + q_0 a_n\right] t^n.$$

After the substitution of these expressions for y'', $P(t)y'$ and $Q(t)y$ in equation (3.4.1) and cancelling the common factor t^{m-2} throughout, we obtain

$$\sum_{n=0}^{\infty}\left[a_n\{(m+n)(m+n-1) + (m+n)p_0 + q_0\}\right.$$
$$\left. + \sum_{k=0}^{n-1} a_k\{(m+k)p_{n-k} + q_{n-k}\}\right]t^n = 0. \qquad (3.4.5)$$

By equating the coefficients of t^n in equation (3.4.5) to zero and putting $f(m) = m(m-1) + mp_0 + q_0$ we successively obtain

$$a_0 f(m) = 0,$$
$$a_1 f(m+1) + a_0(mp_1 + q_1) = 0,$$
$$a_2 f(m+2) + a_0(mp_2 + q_2) + a_1[(m+1)p_1 + q_1] = 0,$$
$$\cdots \cdots \cdots \cdots \cdots \cdots$$
$$a_n f(m+n) + a_0(mp_n + q_n) + \cdots$$
$$+ a_{n-1}[(m+n-1)p_1 + q_1] = 0, \qquad (3.4.6)$$
$$\cdots \cdots \cdots \cdots \cdots \cdots$$

Since $a_0 \neq 0$, it follows that $f(m) = 0$, that is,

$$m(m-1) + mp_0 + q_0 = 0. \qquad (3.4.7)$$

This is called the *indicial equation*, which determines the possible values of the exponent m in the assumed expression for the solution y. Let m_1 and m_2 be the roots of equation (3.4.7). If we choose $m = m_1$, then, from the above expressions, we see that a_n is determined in terms of $a_0, a_1, \cdots, a_{n-1}$, successively for $n = 1, 2, \cdots$, provided that $f(m+n) \neq 0$. The process breaks off if $f(m+n) = 0$. Thus, if $m_1 = m_2 + n$, for some positive integer n, the choice $m = m_1$ gives a formal solution, but in general the choice $m = m_2$ does not, since $f(m_2 + n) = f(m_1) = 0$. If $m_1 = m_2$, then also we obtain only one formal solution. In all the other cases, when the roots of the indicial equation are real, we obtain two linearly independent formal solutions.

The roots of the indicial equation may also be complex, and therefore, the above procedure leads to a formal series with complex coefficients. Since we are only interested in real solutions, we need to consider real and imaginary parts of these formal solutions, which in general is quite complicated and requires tools from complex analysis. We will not pursue these topics here, and the interested reader may refer to Refs. [7], [15], [29], [54] for a discussion on differential equations in the complex domain.

Assume that $t = 0$ is a regular singular point of equation (3.4.1) and that the power series for $tP(t)$ and $t^2 Q(t)$ given, respectively, by equations (3.4.2) and (3.4.3) converge for $t \in (-R, R)$ for some $R > 0$. Assume that the roots m_1 and m_2 of the indicial equation (3.4.7) are real with $m_2 \leqslant m_1$. Then, the equation (3.4.1) has at least one solution given by

$$y_1 = t^{m_1} \sum_{n=0}^{\infty} a_n t^n \quad (a_0 \neq 0) \qquad (3.4.8)$$

on the interval $0 < t < R$, where a_n's are determined in terms of a_0 by the recursion formula (3.4.6) with m replaced by m_1. Also, the series $\sum_{n=0}^{\infty} a_n t^n$ converges on the interval $(-R, R)$. Furthermore, if $m_1 - m_2$ is not a non-negative integer, then equation (3.4.1) has a second linearly independent solution

$$y_2 = t^{m_2} \sum_{n=0}^{\infty} a_n t^n \quad (a_0 \neq 0) \tag{3.4.9}$$

on the same interval, where now a_n's are determined using the recursion relation (3.4.6) in terms of a_0 and m replaced by m_2. Again, the series $\sum_{n=0}^{\infty} a_n t^n$ converges on the interval $(-R, R)$.

The series in equations (3.4.8) and (3.4.9) are called *Frobenius series*. In a specific problem, it is much preferable to start with a series of the form (3.4.4) and derive the indicial equation and recursion relations.

The above procedure leaves unanswered the cases of $m_1 = m_2 + k$, where k is a non-negative integer. In the exercises, we get to see different possibilities in these situations.

Suppose $m_1 = m_2$ and y_1 be a solution given by the Frobenius series. We may now proceed to find a second independent solution by the procedure described in Chapter 1. Let $y_2 = y_1 v$ be another solution, where v is a non-constant function. Then,

$$v' = \frac{1}{y_1^2} \exp\left(-\int^t P(t)\, dt\right)$$

$$= \frac{1}{t^{2m_1}(a_0 + a_1 t + \cdots)^2} \exp\left(-\int^t \left[\frac{p_0}{t} + p_1 + \cdots\right] dt\right)$$

$$= \frac{1}{t^{2m_1}(a_0 + a_1 t + \cdots)^2} \exp\left(-p_0 \log t - p_1 t - \cdots\right)$$

$$= \frac{1}{t(a_0 + a_1 t + \cdots)^2} \exp\left(-p_1 t - \cdots\right)$$

$$= \frac{1}{t} g(t), \text{ say,}$$

where we have used the fact that $2m_1 + p_0 = 1$ when $m_1 = m_2$ and g is an analytic function at $t = 0$ with $g(0) = \frac{1}{a_0^2}$ and $g(t) = b_0 + b_1 t + \cdots$. Therefore, we have $v(t) = b_0 \log t + b_1 t + \cdots$. Of course, it may not be easy to determine the coefficients in the power series expansion of g. When $m_1 - m_2$ is a positive integer k, then the expression for v' is $\frac{1}{t^k} g(t)$. In this case, the function v may or may not contain a logarithmic term.

Remark 3.8. A word of caution. The indicial equation is a quadratic equation, and as such, its roots may not be integers and may even be complex. Thus, we need to properly indicate what is meant by t^m in the expressions above considered for solutions, especially when t assumes negative values. We replace t^m by $|t|^m$ for real m, and when $m = a + ib$ is a complex number,[1] we put

$$|t|^m = \exp(a \log |t|)[\cos(b|t|) + \mathrm{i}\sin(b|t|)].$$

With this remark, all the above procedures described for the method of Frobenius goes through. When the roots of the indicial equation are complex, we consider the real and imaginary parts of the quasi-power series for the two linearly independent solutions of the given equation. We again emphasize that when the roots of the indicial equation are equal, then one solution is given by the quasi-power series and another (linearly independent) solution is obtained by the method of variation of constants; the latter solution contains a logarithmic term. On the other hand, if the roots of the indicial equation differ by a positive integer, then one solution is given by the quasi-power series and another (linearly independent) solution is either given by a quasi-power series or obtained by the method of variation of constants; the latter solution may or may not contain a logarithmic term.

We now illustrate the above procedure, in a somewhat different way, by considering the Bessel's equation of order zero:

$$ty'' + y' + ty = 0. \tag{3.4.10}$$

It is easy to see that $t = 0$ is a regular singular point of the above equation. Let us consider the Frobenius series

$$y = t^m \sum_{n=0}^{\infty} a_n t^n \tag{3.4.11}$$

for a solution of equation (3.4.10), with $a_0 \neq 0$. We may take $a_0 = 1$. We obtain, after collecting the like terms,

$$\begin{aligned} ty'' + y' + ty &= m^2 t^{m-1} + (m+1)^2 a_1 t^m + \{(m+2)^2 a_2 + 1\} t^{m+1} \\ &\quad + \{(m+3)^2 a_3 + a_1\} t^{m+2} + \cdots \end{aligned} \tag{3.4.12}$$

---

[1] In the study of one variable complex analysis, recall how the power z^w is defined for z and w complex.

Now, let $a_1, a_2, \ldots$ be chosen to satisfy the following relations

$$(m+1)^2 a_1 = 0,$$
$$(m+2)^2 a_2 + 1 = 0,$$
$$(m+3)^2 a_3 + a_1 = 0,$$
$$\ldots \qquad \ldots$$

Then, unless m is a negative integer, we have $a_k = 0$ for *odd* integers k and

$$a_2 = -\frac{1}{(m+2)^2},$$
$$a_4 = -\frac{a_2}{(m+4)^2} = \frac{1}{(m+2)^2(m+4)^2},$$
$$\ldots\ldots\ldots$$

Substituting the above values in equations (3.4.11) and (3.4.12), we infer that if

$$y = t^m \left\{ 1 - \frac{t^2}{(m+2)^2} + \frac{t^4}{(m+2)^2(m+4)^2} - \cdots \right\} \tag{3.4.13}$$

and if m is not a negative integer, then

$$ty'' + y' + ty = m^2 t^{m-1}. \tag{3.4.14}$$

Choosing $m = 0$ in equations (3.4.13) and (3.4.14), we see that

$$y = 1 - \frac{t^2}{2^2} + \frac{t^4}{2^2 \cdot 4^2} - \cdots \tag{3.4.15}$$

is a solution of Bessel's equation

$$ty'' + y' + ty = 0. \tag{3.4.16}$$

The series in equation (3.4.15) is denoted by $J_0(t)$ and is called *Bessel's function of zero order of the first kind*. It is easy to see that $J_0(t)$ is an even function of t and converges for all $t \in \mathbb{R}$ with $J_0(0) = 1$. We can also see that the indicial equation for Bessel's equation is given by $m^2 = 0$; thus, its roots are equal and equal to zero. We now proceed to find another independent solution to Bessel's equation. The general procedure tells us that the second solution involves a logarithm term. Using equation (3.4.16) we are going to derive an expression for the same. Differentiating both sides of equation (3.4.16) with respect to m and then letting $m = 0$, we obtain

$$tY_0'' + Y_0' + tY_0 = 0,$$

where $Y_0 = \dfrac{\partial y}{\partial m}$ evaluated at $m = 0$. Now, from equation (3.4.13),

$$\begin{aligned}\frac{\partial y}{\partial m} &= t^m \log t \left\{1 - \frac{t^2}{(m+2)^2} + \frac{t^4}{(m+2)^2(m+4)^2} - \cdots \right\} \\ &+ t^m \left\{\frac{2t^2}{(m+2)^2}\frac{1}{m+2} - \frac{2t^4}{(m+2)^2(m+4)^2}\left(\frac{1}{m+2}+\frac{1}{m+4}\right)\right. \\ &+ \left.\frac{2t^6}{(m+2)^2(m+4)^2(m+6)^2}\left(\frac{1}{m+2}+\frac{1}{m+4}+\frac{1}{m+6}\right) - \cdots \right\}.\end{aligned}$$

Hence, substituting $m = 0$, we obtain

$$Y_0(t) = J_0(t) \log t$$
$$- \frac{t^2}{2^2} + \frac{t^2}{2^2 \cdot 4^2}\left(1+\frac{1}{2}\right) + \frac{t^6}{2^2 \cdot 4^2 \cdot 6^2}\left(1+\frac{1}{2}+\frac{1}{6}\right) - \cdots, \qquad (3.4.17)$$

which is called the *Bessel's function of the second kind of order zero*. Using

$$1 + \frac{1}{2} + \cdots + \frac{1}{n} = \log n + \gamma + \varepsilon_n,$$

where γ is the Euler's constant and $\varepsilon_n \to 0$ as $n \to \infty$, it is straight forward to check that the power series in equation (3.4.17) (excluding the term $J_0(t) \log t$) converges for all values of t. It follows that the general solution of Bessel's equation is given by

$$y = AJ_0 + BY_0,$$

for arbitrary constants A and B.

3.4.2 Singularity at Infinity

In quite many situations, the point at infinity also plays an important role, and its analysis as an ordinary point or a singular point calls for attention. This analysis is particularly useful in understanding the behaviour of solutions for large t. This case may be easily handled as follows. Consider a second-order linear equation

$$y'' + P(t)y' + Q(t)y = 0 \qquad (3.4.18)$$

and make a change of the independent variable t to $\tau = 1/t$. The above equation transforms into[2]

$$\tau^4 \frac{d^2y}{d\tau^2} + [2\tau^3 - \tau^2 P(1/\tau)]\frac{dy}{d\tau} + Q(1/\tau)y = 0. \qquad (3.4.19)$$

---

[2] We continue to use y as a dependent variable and avoid introducing a new variable $z(\tau) = y(t)$.

We say the point at infinity, or $t = \infty$ is an ordinary point or a singular point for equation (3.4.18) if the point $\tau = 0$ is an ordinary point or a singular point or equation (3.4.19), respectively. Thus, if we want to apply the method of Frobenius to the point at infinity, we seek a solution in the form of a power series in $1/t$, rather than t. For example, the constant coefficient equation $y'' + py' + qy = 0$ transforms into $\tau^4 \dfrac{d^2 y}{d\tau^2} + [2\tau^3 - p\tau^2] \dfrac{dy}{d\tau} + qy = 0$. Thus, $t = \infty$ is an irregular singular point, unless $p = q = 0$. The Airy equation $y'' = ty$ transforms into the equation

$$\tau^4 \frac{d^2 y}{d\tau^2} - 2\tau^3 \frac{dy}{d\tau} - \frac{1}{\tau} y = 0.$$

in the variable $\tau = 1/t$. Thus, $t = \infty$ is an irregular singular point for the Airy equation.

Near an irregular singular point, the method of Frobenius does not give, in general, a solution in a power series or quasi-power series. We discuss two examples that give two very different possibilities. The analysis near an irregular singular point is more difficult, and there is no general procedure to obtain a solution. The interested reader may look into Refs. [7], [15], [29], [54] for more information.

Example 3.9. Consider the equation $t^3 y'' + t^2 y' + y = 0$.

Observe that $t = 0$ is an irregular singular point. Let us attempt for a solution of this equation by the method of Frobenius. Assume that

$$y(t) = t^m [a_0 + a_1 t + a_2 t^2 + \cdots].$$

We have

$$y'(t) = m t^{m-1} [a_0 + a_1 t + a_2 t^2 + \cdots] + t^m [a_1 + 2 a_2 t + 3 a_3 t^2 + \cdots]$$

and

$$\begin{aligned} y''(t) &= m(m-1) t^{m-2} [a_0 + a_1 t + a_2 t^2 + \cdots] \\ &\quad + 2 m t^{m-1} [a_1 + 2 a_2 t + 3 a_3 t^2 + \cdots] + t^m [2 a_2 + 6 a_3 t + \cdots]. \end{aligned}$$

Substituting these expressions into the given equation, we obtain recursively $a_0 = 0$, $m(m-1) a_0 + m a_0 + a_1 = 0$, ... Therefore, $a_k = 0$ for all $k = 0, 1, 2, \ldots$ and we see that the Frobenius method gives only the trivial solution. But, for any $t_0 > 0$, the general theory gives a unique solution of the equation with prescribed initial conditions $y(t_0)$ and $y'(t_0)$.

Example 3.10. Consider the equation $t^2 y'' + (1 + 3t) y' + y = 0$. It is straight forward to show that $t = 0$ is an irregular singular point. Let us attempt for a solution of this equation by the method of Frobenius. Assume the quasi-power series $y(t) = t^m \sum_{n=0}^{\infty} a_n t^n$ for a possible solution. Substituting these expressions into the given equation, we obtain

$$t^2 \sum_{n=0}^{\infty} (n+m)(n+m-1) a_n t^{n+m-2} +$$

$$+ (1+3t) t^m \sum_{n=0}^{\infty} (n+m) a_n t^{n+m-1} + \sum_{n=0}^{\infty} a_n t^{n+m} = 0.$$

We now equate the coefficients of various powers of t to zero. For the coefficient of t^{m-1}, we get $m a_0 = 0$. For the coefficient of t^{n+m}, we obtain recursively

$$(n+m+1) a_{n+1} + (n+m+1)^2 a_n = 0, \text{ for } n = 0, 1, 2, \ldots.$$

If we choose $a_0 = 0$, then it follows that $a_n = 0$ for all $n \geq 0$. So we get the trivial solution. On the other hand, if we choose $m = 0$ and $a_0 \neq 0$, then $a_{n+1} = -(n+1) a_n$. Therefore, $a_n = (-1)^n n! a_0$ for $n = 0, 1, 2, \ldots$ and the power series we obtain diverges at every $t \neq 0$! Thus, the method of Frobenius does not give us a solution. As in the previous example, for any $t_0 > 0$, the general theory gives a unique solution of the equation with the prescribed initial conditions $y(t_0)$ and $y'(t_0)$.

Perhaps it was Poincarè who first pointed out that even these divergent power series have a meaning and called the series an *asymptotic series*. The subject of asymptotic series is quite interesting and has many applications not only in differential equations but also in integrals having highly oscillating terms in the integrand. The interested reader can look into the references cited above and further references therein.

3.5 Exercises

Exercise 3.1

Show that the function f in Example 3.3 is a $C^{\infty}(\mathbb{R})$ function with $f^{(n)}(0) = 0$ for $n = 1, 2, \ldots$.

Exercise 3.2

Let $y_1(t) = t^{r_1} \sum_{n=0}^{\infty} a_n t^n$ and $y_2(t) = t^{r_2} \sum_{n=0}^{\infty} b_n t^n$ for $t > 0$. Assume that r_1 and r_2 are distinct real numbers, and $a_0 = b_0 = 1$. Show that y_1 and y_2 are linearly independent.

Exercise 3.3

Consider the second-order Euler equation $t^2 y'' + aty' + by = 0$, where a, b are real numbers. Find two linearly independent solutions of the equation in each of the following cases by applying the method of Frobenius:

(i) $a = 1/2$, $b = -1/2$.   (ii) $a = -5$, $b = 9$.

Exercise 3.4

Consider the Legendre's equation $(1 - t^2)y'' - 2ty' + p(p+1)y = 0$ and discuss its solutions in the neighbourhoods of $t = 1$ and $t = -1$.

Exercise 3.5

For each of the following equations, write down the indicial equation and find its roots. Write the form of two linearly independent solutions without computing the coefficients, and discuss the limiting behaviour of the solutions as $t \to 0$.

(a) $y'' + \dfrac{5}{2t}y' + \dfrac{1}{2t}y = 0$.

(b) $t^2 y'' + 4ty' + (2 - t)y = 0$.

(c) $t^2 y'' + (1 - 6t)y = 0$.

Exercise 3.6

In each of the following equations, locate all the singular points by describing whether they are regular or irregular.

(a) $(t - 2)(t + 3)^2 y'' + 3t^2 y' - 2(t + 3)y = 0$.

(b) $t^2 y'' + (\sin t)y' + (\cos t)y = 0$.

(c) $(e^t - 1)^2 y'' + 2(\sin t)y' + 3y = 0$.

(d) $y'' + 3y' + t^{1/2} y = 0$, $t \geqslant 0$.

Exercise 3.7

Consider the Bessel's equation of order $1/2$: $t^2 y'' + ty' + \left(t^2 - \dfrac{1}{4}\right) y = 0$.

(a) Verify that the functions $y_1(t) = \sqrt{\dfrac{2}{\pi t}} \sin t$ and $y_2(t) = \sqrt{\dfrac{2}{\pi t}} \cos t$ are linearly independent solutions for $t > 0$.

(b) Show that the indicial equation has roots $m_1 = 1/2$ and $m_2 = -1/2$, whose difference is the positive integer 1. Show, nevertheless, that both the solutions y_1, y_2 can be derived from the method of Frobenius without introducing a logarithm term.

Exercise 3.8

For the following equations, determine all the ordinary, regular singular or irregular singular points, if any, including the point at infinity.

(a) $ty'' + 3y' + 2ty = 0$.
(b) $t^5 y'' + 2t^4 y' + y = 0$.
(c) (Bessel's equation of order p) $t^2 y'' + ty' + (t^2 - p^2)y = 0$.

3.6 Solutions

Exercise 3.1

We have $f(t) = \exp(-1/t)$ for $t > 0$ and 0 elsewhere. We need to check the smoothness of this function only at $t = 0$. For $t > 0$, we have[3]

$$\frac{d^n}{dt^n}(\exp(-1/t)) = \exp(-1/t) P(1/t),$$

where P is a polynomial. Therefore, it suffices to show that $t^{-k} \exp(-1/t)$ tends to 0 as $t \to 0+$, for every positive integer k. Now

$$\frac{\exp(-1/t)}{t^k} = \frac{t^{-k}}{\exp(1/t)} \leqslant (k+1)! \frac{t^{-k}}{(1/t)^{k+1}} \to 0$$

as $t \to 0+$, using the fact that $e^a > a^k/k!$ for $a > 0$ and $k = 1, 2, \ldots$ This proves the smoothness of the function f to all orders. Hence, $f^{(n)}(0) = 0$ for $n = 0, 1, 2, \ldots$

Exercise 3.2

First assume that $r_1 > r_2 > 0$. Suppose y_1 and y_2 are linearly dependent. Then there are constants α_1 and α_2, not both zero, such that

$$\alpha_1 y_1(t) + \alpha_2 y_2(t) = 0 \quad \text{for } t > 0.$$

---

[3] *Faà di Bruno's formula ([1])*

$$\frac{d^n}{dt^n} f(g(t)) = \sum_{k=0}^{n} f^{(k)}(g(t)) \sum (n; \alpha_1, \ldots, \alpha_n)' [g'(t)]^{\alpha_1} \cdots [g^{(n)}(t)]^{\alpha_n}$$

where the inner summation is taken over all non-negative integers $\alpha_1, \ldots, \alpha_n$ satisfying the conditions

$$\alpha_1 + \cdots + \alpha_n = k \quad \text{and} \quad \alpha_1 + 2\alpha_2 + \cdots + n\alpha_n = n$$

and $(n; \alpha_1, \ldots, \alpha_n)' = n! \left(\prod_{k=1}^{n} (k!)^{\alpha_k} \alpha_k! \right)^{-1}.$

Since $a_0 = b_0 = 1$, it follows that both α_1 and α_2 are non-zero. So

$$\alpha_1 t^{-r_2} y_1(t) + \alpha_2 t^{-r_2} y_2(t) = 0 \quad \text{for } t > 0.$$

As $t \to 0+$, the first term on the left-hand side tends to zero, and the second term tends to $\alpha_2 b_0 \neq 0$. This contradiction proves the assertion. ∎

Exercise 3.3

First observe that the Euler equation can be transformed into a linear equation with *constant* coefficients by changing the independent variable to $\tau = \log t$, $t > 0$. Here, we apply the method of Frobenius to find the solutions. Note that $t = 0$ is the only singularity of the equation, and it is a regular singular point. Rewrite the equation as $y'' + P(t)y' + Q(t)y = 0$, with $P(t) = a/t$ and $Q(t) = b/t^2$. Therefore, $tP(t) = a$ and $t^2 Q(t) = b$. The indicial equation is $m(m-1) + am + b = 0$.

If $a = 1/2$ and $b = -1/2$, the roots of the indicial equation are $m_1 = -1/2$ and $m_2 = 1$. These are distinct and do not differ by a positive integer. Hence, the Frobenius method gives two linearly independent solutions. First consider the root $-1/2$ and look for a solution in the form

$$y(t) = t^{-1/2} \sum_{n=0}^{\infty} a_n t^n.$$

By computing the first and second derivatives of y and substituting them into the equation, we get (remember $a = 1/2$ and $b = -1/2$)

$$\sum_{n=0}^{\infty} [(n - 1/2)(n - 3/2) + (1/2)(n - 1/2) - (1/2)] a_n t^{n-1/2} = 0.$$

The coefficient of the power $t^{-1/2}$ is automatically zero, giving the coefficient a_0 arbitrary. Equating the coefficients of the powers $t^{n-1/2}$, $n \geq 1$ to zero give, that $a_n = 0$ for $n \geq 1$. Thus, the solution in this case is a multiple of $t^{-1/2}$.

For the root 1, we now look for a solution of the form

$$y(t) = t \sum_{n=0}^{\infty} a_n t^n.$$

Similar computations now give

$$\sum_{n=0}^{\infty} [(n+1)(n) + (1/2)(n+1) - (1/2)] a_n t^{n+1} = 0.$$

The coefficient of the power t is automatically zero, giving the coefficient a_0 arbitrary. Equating the coefficients of the powers t^{n+1}, $n \geq 1$ to zero gives that $a_n = 0$ for $n \geq 1$. Thus, the solution in this case is a multiple of t. Therefore, the general solution of the equation is given by $y(t) = c_1 t^{-1/2} + c_2 t$ when $a = 1/2$ and $b = -1/2$, where the constants c_1 and c_2 are arbitrary.

Next, consider the cases $a = -5$ and $b = 9$. The indicial equation now reduces to $(m-3)^2 = 0$. Thus, the roots are equal and equal to 3. The method of Frobenius certainly gives one solution of the form

$$y(t) = t^3 \sum_{n=0}^{\infty} a_n t^n.$$

Similar computations as in the previous case now give

$$\sum_{n=0}^{\infty} [(n+3)(n+2) - 5(n+3) + 9] a_n t^{n+3} = 0.$$

The coefficient of the power t^3 is automatically zero, giving the coefficient a_0 arbitrary. Equating the coefficients of the powers t^{n+3}, $n \geq 1$ to zero gives that $a_n = 0$ for $n \geq 1$. Thus, the solution in this case is a multiple of t^3. The other linearly independent solution can be found using the method of variation of parameters, which involves a logarithmic term. Thus, the general solution in this case is given by $y(t) = t^3(c_1 + c_2 \log t)$ for arbitrary the constants c_1 and c_2.

Exercise 3.4

Observe that both the points $t = \pm 1$ are regular singular points, and so the method of Frobenius is applicable. First, consider the case of $t = 1$. It is convenient to change the variable t to $\xi = t - 1$. The Legendre's equation becomes

$$\xi(\xi + 2) \frac{d^2 y}{d\xi^2} + 2(\xi + 1) \frac{dy}{d\xi} - p(p+1) y = 0. \tag{3.5.1}$$

Here $P(\xi) = \dfrac{2(\xi+1)}{\xi(\xi+2)}$ and $Q(\xi) = -\dfrac{p(p+1)}{\xi(\xi+2)}$. For this transformed equation, $\xi = 0$ is a regular singular point. Next

$$\xi P(\xi) = \frac{2(\xi+1)}{\xi+2} = (\xi+1) \sum_{n=0}^{\infty} (-1)^n \frac{\xi^n}{2^n},$$

the series being convergent for $|\xi| < 2$. The constant term in the series is $p_0 = -1$. Similarly

$$\xi^2 Q(\xi) = -\xi \frac{p(p+1)}{\xi+2} = -p(p+1) \xi \sum_{n=0}^{\infty} (-1)^n \frac{\xi^n}{2^n},$$

where the constant term is $q_0 = 0$. Thus, the indicial equation is $m(m-1) + p_0 m + q_0 = 0$ or $m^2 = 0$, whose roots are $m = 0, 0$. The roots are equal and equal to zero. Thus, there is only one power series, and the other linearly independent solution contains a logarithmic term. Let us derive the power series solution and consider the series $y(\xi) = \sum_{n=0}^{\infty} a_n \xi^n$ for a possible solution. Computing $\frac{dy}{d\xi}, \frac{d^2y}{d\xi^2}$ and substituting into equation (3.5.1), we get

$$\sum_{n=0}^{\infty} [n(n-1) + 2n - p(p+1)] a_n \xi^n + 2 \sum_{n=0}^{\infty} [n(n-1) + n] a_n \xi^{n-1} = 0.$$

Simplifying we obtain

$$\sum_{n=0}^{\infty} \left([n(n+1) - p(p+1)] a_n - 2(n+1)^2 a_{n+1} \right) \xi^n = 0.$$

Equating the coefficient of ξ^n to zero for $n = 0, 1, 2, \ldots$, we obtain the recursion relations

$$a_{n+1} = -\frac{n(n+1) - p(p+1)}{2(n+1)^2} a_n = -\frac{(n-p)(n+p+1)}{2(n+1)^2} a_n, \quad (3.5.2)$$

for $n = 0, 1, 2, \ldots$. Here a_0 is arbitrary, and all the coefficients are determined in terms of a_0 using equation (3.5.2). The ratio test shows that the series in question converges for $|\xi| < 2$, in tune with the general theory. We also observe that when p is a positive integer k, equation then (3.5.2) implies that $a_j = 0$ for all $j > k$, and thus the power series becomes a polynomial.

Next, consider the case of $t = -1$ and introduce the change of variable $\eta = -(t+1)$. The Legendre's equation becomes

$$\eta(\eta + 2) \frac{d^2y}{d\eta^2} + 2(\eta - 1) \frac{dy}{d\eta} - p(p+1) y = 0,$$

which is exactly the same as equation (3.5.1)! We therefore conclude that the Legendre's equation has power series solutions near $t = \pm 1$, which are given by

$$y(t) = \sum_{n=0}^{\infty} a_n (t-1)^n \quad \text{and}$$

$$y(t) = \sum_{n=0}^{\infty} (-1)^n a_n (t+1)^n,$$

where the coefficients a_n satisfy the recursion relations (3.5.2) with a_0 being arbitrary. Both the series have a radius of convergence 2. Also, when p is a positive integer, both the series become polynomials.

In passing, we remark that near the ordinary point $t = 0$ (and by translation near any point $t = t_0 \neq \pm 1$), the Legendre's equation has power series solutions given by

$$y(t) = \sum_{n=0}^{\infty} (-1)^n b_n t^n,$$

where the coefficients b_k satisfy the recursion relations

$$b_{n+2} = \frac{(n-p)(n+p+1)}{(n+2)(n+1)} b_n, \quad n = 0, 1, 2, ...$$

with b_0 and b_1 being arbitrary. This series has a radius of convergence 1. Thus, Legendre's equation has two linearly independent solutions in the form of a power series near $t = 0$. Again, when p is a positive integer, there is a solution in the form of a polynomial. In fact, among these polynomials, there are the ones that have either only *even* powers of t or only *odd* powers of t. These polynomials with a normalizing constant are called the *Legendre polynomials*.

Exercise 3.5

(a) Here $P(t) = 5/(2t)$ and $Q(t) = 1/(2t)$. Hence, $t = 0$ is a regular singular point and $tP(t) = 5/2$ and $t^2 Q(t) = t/2$. Therefore, the indicial equation is given by $m(m-1) + (5/2)m = 0$; its roots are given by $m_1 = 0$ and $m_2 = -3/2$. The roots are distinct, and their difference is not a positive integer. Thus, there are two linearly independent solutions, which are of the form $a_0 + a_1 t + \cdots$ and $t^{-3/2}(b_0 + b_1 t + \cdots)$. Thus, one solution approaches a constant as $t \to 0+$ and the other solution becomes infinite like $t^{-3/2}$.

(b) Here $P(t) = 4/t$ and $Q(t) = (2-t)/t^2$. Hence, $t = 0$ is a regular singular point, and $tP(t) = 4$ and $t^2 Q(t) = 2 - t$. Therefore, the indicial equation is given by $m(m-1) + 4m + 2 = 0$; its roots are given by $m_1 = -1$ and $m_2 = -2$. The roots are distinct, and their difference is a positive integer. We can certainly find one solution of the form $y(t) = a_0/t + a_1 + a_2 t \cdots$, which approaches infinity as $t \to 0+$ if $a_0 \neq 0$, and $t^{-3/2}(b_0 + b_1 t + \cdots)$. Thus, one solution approaches a constant as $t \to 0+$ and the other solution becomes infinite like $t^{-3/2}$.

(c) Here $P(t) = 0$ and $Q(t) = (1 - 6t)/t^2$. Hence, $t = 0$ is a regular singular point, and $tP(t) = 0$ and $t^2 Q(t) = 1 - 6t$. Therefore, the indicial equation is given by $m(m-1) + 1 = 0$; its roots are complex and are given by $m = \frac{1}{2} \pm \frac{\sqrt{3}}{2} i$. Thus, there are two linearly independent solutions and since the real part of the roots of the indicial equation is $1/2$, we conclude that both the solutions approach zero as $t \to 0+$.

Exercise 3.6

(a) Rewrite the equation as $y'' + P(t)y' + Q(t)y = 0$ with $P(t) = \dfrac{3t^2}{(t-2)(t+3)^2}$ and $Q(t) = \dfrac{2(t+3)}{(t-2)(t+3)^2}$. We see that P has singularities at $t = 2$ and $t = -3$. However, $(t-2)P(t)$ is analytic at $t = 2$, but $(t+3)P(t)$ is not analytic at $t = -3$. Similarly, Q has singularities at $t = 2$ and $t = -3$, but $(t-2)^2 Q(t)$ is analytic at $t = 2$ and $(t+3)^2 Q(t)$ is analytic at $t = -3$. We conclude that $t = 2$ is a regular singular point and $t = -3$ is an irregular singular point.

(b) Rewrite the equation as $y'' + P(t)y' + Q(t)y = 0$ with $P(t) = \dfrac{\sin t}{t^2}$ and $Q(t) = \dfrac{\cos t}{t^2}$. We see that P has singularity at $t = 0$ but, $tP(t)$ is analytic at $t = 0$. Similarly, Q has singularity at $t = 0$ and $t^2 Q(t)$ is analytic at $t = 0$. We conclude that $t = 0$ is a regular singular point.

(c) Rewrite the equation as $y'' + P(t)y' + Q(t)y = 0$ with $P(t) = \dfrac{\sin t}{(e^t - 1)^2}$ and $Q(t) = \dfrac{3}{(e^t - 1)^2}$. We see that P has singularity at $t = 0$. Observing that $(e^t - 1)/t \to 1$ as $t \to 0$, we see that $tP(t)$ is analytic at $t = 0$. Similarly, Q has singularity at $t = 0$ and $t^2 Q(t)$ is analytic at $t = 0$. We conclude that $t = 0$ is a regular singular point.

(d) Rewrite the equation as $y'' + P(t)y' + Q(t)y = 0$ with $P(t) = 3$ and $Q(t) = \sqrt{t}$. Since P is a constant, it is analytic everywhere. However, $t^k Q$ is not analytic at $t = 0$ for any non-negative integer k, but is analytic at every $t = t_0 > 0$, as follows from the binomial expansion. We conclude that $t = 0$ is an irregular singular point.

Exercise 3.7

(a) We may ignore the constant in the given functions as the equation is linear. We have (ignoring the constant)

$$y_1'(t) = \frac{\cos t}{t^{1/2}} - \frac{\sin t}{2t^{3/2}}, \quad y_1''(t) = -\frac{\sin t}{t^{1/2}} - \frac{\cos t}{t^{3/2}} + \frac{3\sin t}{4t^{5/2}}.$$

Substituting these expressions into the equation, after multiplying by the corresponding powers of t, we verify that y_1 is a solution. Similar computations show that y_2 is also a solution.

(b) Rewriting the equation as $y'' + P(t)y' + Q(t)y = 0$ with $P(t) = 1/t$ and $Q(t) = 1 - 1/(4t^2)$. Therefore,

$$tP(t) = 1 \quad \text{and} \quad t^2 Q(t) = t^2 - 1/4.$$

This immediately gives $p_0 = 1$ and $q_0 = -1/4$. The indicial equation is given by $m(m-1) + p_0 m + q_0 = 0$ or $m^2 - 1/4 = 0$. Thus, the roots are $1/2$ and $-1/2$, and they differ by 1, a positive integer. Nevertheless, we will now see that both the roots give solutions by the method Frobenius, and they are linearly independent.

Consider the root $1/2$ and seek a solution in the form $y(t) = t^{1/2} \sum_{n=0}^{\infty} a_n t^n$. We have

$$y'(t) = \sum_{n=0}^{\infty} (n + 1/2) a_n t^{n-1/2}$$

$$y''(t) = \sum_{n=0}^{\infty} (n + 1/2)(n - 1/2) a_n t^{n-3/2}$$

Substituting these expressions into the equation, we get

$$\sum_{n=0}^{\infty}[(n+1/2)(n-1/2)+(n+1/2)-1/4]a_n t^{n+1/2} + \sum_{n=0}^{\infty} a_n t^{n+5/2} = 0.$$

Simplification leads to

$$2a_1 t^{3/2} + \sum_{n=0}^{\infty}[n(n+1)a_{n+2} + a_n]t^{n+5/2} = 0.$$

Thus, the coefficient a_0 is free, that is, we can choose it arbitrarily and $a_1 = 0$. Recursively, we have $a_{n+2} = -\dfrac{a_n}{n(n+1)}$ for $n = 0, 1, \ldots$. Therefore, $a_{2k+1} = 0$ and $a_{2k} = (-1)^k \dfrac{a_0}{(2k+1)!}$ for $k = 0, 1, 2, \ldots$. The series is therefore given by

$$y(t) = a_0 \sum_{n=0}^{\infty} \frac{t^{2k+1/2}}{(2k+1)!} = \frac{a_0}{\sqrt{t}} \sum_{n=0}^{\infty} \frac{t^{2k+1}}{(2k+1)!} = \frac{a_0}{\sqrt{t}} \sin t.$$

Choosing $a_0 = \sqrt{2/\pi}$ gives the solution y_1 in part (a).

Next, consider the root $-1/2$ and seek a solution in the form $y(t) = t^{-1/2} \sum_{n=0}^{\infty} a_n t^n$. Similar computations now give the equation

$$\sum_{n=0}^{\infty}[(n-1/2)(n-3/2)+(n+1/2)-1/4]a_n t^{n+1/2} + \sum_{n=0}^{\infty} a_n t^{n+3/2} = 0.$$

Simplification leads to

$$\sum_{n=0}^{\infty}[(n-1)na_{n+2} + a_n]t^{n+3/2} = 0.$$

We now see that the coefficients a_0 and $a_1 = 0$ are free. Recursively, we have $a_{n+2} = -\dfrac{a_n}{(n-1)n}$ for $n = 0, 1, \ldots$. Choosing $a_1 = 0$ gives $a_{2k+1} = 0$ and $a_{2k} = (-1)^k \dfrac{a_0}{(2k)!}$ for $k = 0, 1, 2, \ldots$. The series is therefore given by

$$y(t) = a_0 \sum_{n=0}^{\infty} \frac{t^{2k-1/2}}{(2k)!} = \frac{a_0}{\sqrt{t}} \sum_{n=0}^{\infty} \frac{t^{2k}}{(2k)!} = \frac{a_0}{\sqrt{t}} \cos t.$$

Choosing $a_0 = \sqrt{2/\pi}$ gives the solution y_2 in part (a).

This is an example where we find two linearly independent solutions, without any logarithmic term, though the roots of the indicial equation differ by a positive integer, and both the solutions are obtained by the method of Frobenius.

Exercise 3.8

(a) Rewrite the equation as $y'' + P(t)y' + Q(t)y = 0$ with $P(t) = 3/t$ and $Q(t) = 2$. Hence, $t = 0$ is the only (finite) regular singular point of the equation. To find out whether $t = \infty$ is a singular point, we transform the equation in the variable $\tau = 1/t$. We obtain (see equation (3.4.19)):

$$\tau^4 \frac{d^2 y}{d\tau^2} + [2\tau^3 - \tau^2(3\tau)]\frac{dy}{d\tau} + 2y = 0.$$

Thus, $\tau = 0$ is an irregular singular point for this equation, which in turn implies that $t = \infty$ is an irregular singular point for the given equation.

(b) Rewriting the equation as $y'' + \frac{2}{t}y' + \frac{1}{t^5}y = 0$, we see that $t = 0$ is an irregular singular point. Transforming the equation in the variable $\tau = 1/t$, we obtain

$$\tau^4 \frac{d^2 y}{d\tau^2} + [2\tau^3 - \tau^2(2\tau)]\frac{dy}{d\tau} + \tau^5 y = 0.$$

Thus, $\tau = 0$ is an ordinary point for this equation, which in turn implies that $t = \infty$ is an ordinary point for the given equation.

(c) Rewriting the equation as $y'' + \frac{1}{t}y' + [1 - \frac{p^2}{t^2}]y = 0$, we see that $t = 0$ is a regular singular point. Transforming the equation in the variable $\tau = 1/t$, we obtain

$$\tau^4 \frac{d^2 y}{d\tau^2} + [2\tau^3 - \tau^2(\tau)]\frac{dy}{d\tau} + [1 - p^2\tau^2]y = 0.$$

Thus, $\tau = 0$ is an irregular singular point for this equation, which in turn implies that $t = \infty$ is an irregular singular point for Bessel's equation.

4

Regular Sturm–Liouville Theory and Boundary Value Problems

4.1 Introduction

A *regular Sturm–Liouville system*, or simply S–L system, refers to the following boundary value problem (BVP):

$$\mathcal{L}u(t) := -\frac{d}{dt}\left(p(t)\frac{du}{dt}\right) + q(t)u(t) = \lambda \rho(t)u(t),$$

with prescribed boundary conditions at the end points a and b of the given interval $[a,b]$. Here p is a positive C^1 function, q and ρ are continuous functions and $\rho > 0$, all defined on a bounded interval $[a,b]$, and λ is a real parameter. When p vanishes somewhere in the interval or the interval under consideration is unbounded, the problem is termed as *singular*. Singular BVPs are more difficult to deal with. The interested reader may refer to Ref. [15].

It turns out that the non-trivial solutions to the BVP exist only for a discrete set of values of the parameter λ, tending to infinity. The situation may thus be compared with the eigenvalues of a matrix, considered as a linear operator on a finite-dimensional space. The main difference is that we are now working in an infinite dimensional space. In analogy with matrices, the discrete set of the values of the parameter for which the non-trivial solutions exist is called the *eigenvalues* of the BVP, and the corresponding non-trivial solutions are called the *eigenfunctions*.

Again, continuing with the similarity with matrices, we know that any vector in $\mathbb{R}^n$ can be written as a unique linear combination of eigenvectors of a real, symmetric matrix. In this case, the eigenvectors are also orthogonal; we discuss more regarding orthogonality in the next section. Surprisingly, these concepts can be extended to the operator $\mathcal{L}$. In the case of a BVP too, one may express an *arbitrary* function as a *linear combination* of the corresponding eigenfunctions. Since these are (infinite) power series, we need to discuss the convergence and so on, and the tools required come from functional analysis (Hilbert space).

The S–L systems arise in many physical problems, for example, in the following situation. Consider the longitudinal vibrations of an elastic bar of local stiffness $p(\xi)$ and density $\rho(\xi)$. The mean longitudinal displacement $v(\xi, \tau)$, at position ξ and time τ, of the section of such a bar from its equilibrium satisfies the one-dimensional wave equation

$$\rho(\xi)\frac{\partial^2 v}{\partial \tau^2} = \frac{\partial}{\partial \xi}\left[p(\xi)\frac{\partial v}{\partial \xi}\right].$$

If we now seek the solution in the following form of *simple harmonic* vibrations (or the *normal modes* of vibrations) as

$$v(\xi, \tau) = u(\xi)\cos(k(\tau - \tau_0)),$$

we obtain an S–L system for u, changing ξ variable to t variable, with $q \equiv 0$ and $\lambda = k^2$. For a finite bar, say, $\xi \in [a, b]$, the following are some of the natural boundary conditions:

$$u(a) = u(b) = 0 \text{ (rigidly fixed ends)}$$
$$u'(a) = u'(b) = 0 \text{ (free ends)}$$
$$u'(a) + \alpha u(a) = u'(b) + \beta u(b) = 0 \text{ (elastically held ends)}$$
$$u(a) = u(b), u'(a) = u'(b) \text{ (periodic conditions)}.$$

Similarly, if we consider the *vibrating circular membrane*, we obtain the wave equation in two space dimensions. After the introduction of polar coordinates and assuming the radial symmetry of the solution, the equation can be reduced to a one-dimensional wave equation. If we again seek a solution by the separation of variables, as in the case of the elastic bar above, this will lead to a singular S–L system involving the Bessel's equation. For more details, see Ref. [9].

The familiar Fourier series involving sine and cosine functions is an example. Another example is $\mathcal{L}u = -u''$. It is easy to see that non-trivial solutions to the BVP $\mathcal{L}u = \lambda u$, $u(0) = u(\pi) = 0$, exist if and only if $\lambda = n^2$, $n \in \mathbb{Z}$, $n \neq 0$. Thus, the eigenvalues are n^2, n a non-zero integer and the corresponding eigenfunctions are $\sin(nt)$. If, instead, we consider the boundary conditions as $u(-\pi) = u(\pi) = 0$, the eigenvalues are now $n^2/4$, $n \in \mathbb{Z}$, $n \neq 0$, and the eigenfunctions are $\sin(nt/2)$ and $\cos(nt/2)$ for n even and n odd, respectively. A reader familiar with Fourier (sine and cosine) series recognizes that any suitable function satisfying the given boundary conditions can be expressed as an infinite series involving the corresponding eigenfunctions.

The Fourier series arises if we replace the above boundary conditions by the *periodic boundary conditions* $u(-\pi) = u(\pi)$.

From this example, we learn that the form of the boundary conditions plays an important role in the determination of the eigenvalues; it is also important in making the operator $\mathcal{L}$ *self-adjoint* and the orthogonality condition of the eigenfunctions, as we will see in the next section.

4.1.1 Boundary Value Problems

The BVPs for linear and non-linear second-order equations arise in a vast number of practical situations ranging from physics, engineering to biology. For a very good collection of such problems and their detailed descriptions, we refer to Ref. [4].

The analysis, in the linear case, makes use of the existence of two linearly independent solutions. These are used to construct the so-called *Green's function* of the given BVP, which in turn will generate the required solution for the inhomogeneous equation.

The non-linear case is more delicate. We only mention a well-known method called the *shooting method* to prove the existence and uniqueness of solution to BVP; see Ref. [33]. Also, performing a phase space (plane) analysis, wherever possible, may help decide whether a non-trivial solution to the given BVP is possible or not. However, in most of the situations, a suitable numerical scheme needs to be used to obtain a solution. The study of existence and uniqueness of solutions to a BVP is more difficult than that of an IVP, even in the linear case.

4.1.2 Existence and Uniqueness of Green's Function

We now discuss briefly the existence and uniqueness of the Green's function for a BVP of a second-order linear ordinary differential equation (ODE). More generally, we consider the following BVP[1]:

$$\mathcal{L}u \equiv p_0 u'' + p_1 u' + p_2 u = 0, \text{ in } (a,b), \tag{4.1.1}$$

$$A \begin{bmatrix} u(a) \\ u'(a) \end{bmatrix} + B \begin{bmatrix} u(b) \\ u'(b) \end{bmatrix} = \begin{bmatrix} 0 \\ 0 \end{bmatrix}. \tag{4.1.2}$$

Here p_i, $i = 0, 1, 2$ are real valued functions defined on the interval $[a,b]$, $p_0 \in C^2$, $p_0 > 0$; $p_1 \in C^1$, $p_2 \in C^0$ and A, B are 2×2 real non-zero matrices. In general, the boundary conditions at $t = a$ and $t = b$ are coupled.

We say that 0 is *not* an eigenvalue of the BVP (4.1.1) and (4.1.2) if $u \equiv 0$ is the only solution satisfying equations (4.1.1) and (4.1.2). We now construct the Green's function, denoted by $G(t,s)$, for the above BVP under the assumption that 0 is not an eigenvalue. We will discuss the issues that arise when 0 is an eigenvalue in the exercises.

---

[1] The analysis can be carried through for higher-order equation requiring more algebra.

We first observe that the condition 0 is not an eigenvalue is equivalent to the condition $AW(a) + BW(b)$ is non-singular, where W denotes the Wronski matrix of any two linearly independent solutions of equation (4.1.1). For, let u_1 and u_2 be any two linearly independent solutions of equation (4.1.1). Then any solution u of equation (4.1.1) can be written as $u = c_1 u_1 + c_2 u_2$ for some constants c_1 and c_2. If u satisfies the boundary conditions (4.1.2), then $[AW(a) + BW(b)] \begin{bmatrix} c_1 \\ c_2 \end{bmatrix} = \begin{bmatrix} 0 \\ 0 \end{bmatrix}$, where $W(t) = \begin{bmatrix} u_1(t) & u_2(t) \\ u_1'(t) & u_2'(t) \end{bmatrix}$ is the Wronski matrix of u_1 and u_2. The conclusion is now immediate.

The importance of the Green's function lies in obtaining the solution to the inhomogeneous equation

$$Lu = f \quad (4.1.3)$$

satisfying the boundary conditions (4.1.2), by an integration:

$$u(t) = \int_a^b G(t,s) f(s)\, ds, \quad (4.1.4)$$

for every continuous function f defined on $[a, b]$. When 0 is an eigenvalue of equations (4.1.1) and (4.1.2), then there may be some restrictions on f for the existence of a solution to the inhomogeneous equation, and also, there may be non-uniqueness.

Existence of the Green's Function: Assume 0 is not an eigenvalue of equations (4.1.1) and (4.1.2). Fix $a < s < b$. Seek a solution, call it $G(t, s)$, of equation (4.1.1) in the intervals $[a, s]$ and $[s, b]$ satisfying the following:

- The function $t \mapsto G(t, s)$ is continuous in $[a, b]$.

- The first derivative $\dfrac{\partial G}{\partial t}$ has a jump discontinuity[2] at $t = s$, having jump $1/p_0(s)$:

$$\frac{\partial G}{\partial t}(s+, s) - \frac{\partial G}{\partial t}(s-, s) = \frac{1}{p_0(s)}.$$

- The function $t \mapsto G(t, s)$ satisfies the boundary conditions (4.1.2).

Suppose u_1 and u_2 are any two linearly independent solutions of equation (4.1.1) and let

---

[2] This condition is motivated from: For the operator $(pu')' + qu$, where p is a positive C^1 function, the Wronskian of any two linearly independent solutions equals $1/p$ up to a constant multiple.

$$G(t,s) = a_1 u_1(t) + a_2 u_2(t) \text{ for } t \in [a,s],$$
$$G(t,s) = b_1 u_1(t) + b_2 u_2(t) \text{ for } t \in [s,b],$$

for some constants a_i, b_i for $i = 1, 2$. The first two conditions imposed on G give

$$a_1 u_1(s) + a_2 u_2(s) = b_1 u_1(s) + b_2 u_2(s),$$
$$b_1 u_1'(s) + b_2 u_2'(s) = a_1 u_1'(s) + a_2 u_2'(s) + 1/p_0(s),$$

or, $W(s) \begin{bmatrix} c_1 \\ c_2 \end{bmatrix} = \begin{bmatrix} 0 \\ 1/p_0(s) \end{bmatrix}$, where W is the Wronskii matrix of u_1 and u_2 and $c_i = b_i - a_i$, $i = 1, 2$. Thus, c_1 and c_2 are uniquely determined.
Next, from the requirement that G satisfies the boundary conditions (4.1.2), we obtain

$$A \begin{bmatrix} a_1 u_1(a) + a_2 u_2(a) \\ a_1 u_1'(a) + a_2 u_2'(a) \end{bmatrix} + B \begin{bmatrix} b_1 u_1(b) + b_2 u_2(b) \\ b_1 u_1'(b) + b_2 u_2'(b) \end{bmatrix} = \begin{bmatrix} 0 \\ 0 \end{bmatrix}.$$

Rewriting, we get $[AW(a) + BW(b)] \begin{bmatrix} b_1 \\ b_2 \end{bmatrix} = AW(a) \begin{bmatrix} c_1 \\ c_2 \end{bmatrix}$. Using the hypothesis, we now conclude that the constants a_i, b_i for $i = 1, 2$ are uniquely determined and thus G is unique.

Alternative Methods: Though the above-mentioned procedure of constructing the Green's function is quite satisfactory, the algebra may be cumbersome at times. We now describe two alternative methods to construct the Green's function. The first one is this: Suppose u_1 and u_2 are any two linearly independent solutions of equation (4.1.1). Put

$$F(t,s) = \alpha u_1(t) + \beta u_2(t) \pm \frac{u_1(t)u_2(s) - u_1(s)u_2(t)}{2p_0(s)[u_1(s)u_2'(s) - u_1'(s)u_2(s)]}, \qquad (4.1.5)$$

where the positive sign is chosen when $t \in [a,s]$ and the negative sign is chosen when $t \in [s,b]$. The function F is continuous in $[a,b]$, and its first derivative has jump discontinuity at $t = s$ with jump equal to $1/p_0(s)$, and is otherwise continuous in $[a,b]$. Note that the third term in the expression for F is independent of the functions u_1 and u_2 chosen. Thus, the function F is very similar to G. If we can choose the constants α and β so that F satisfies the boundary conditions (4.1.2), then $F = G$ by uniqueness. We will compute F and G in some exercises.

The second method is applicable when the boundary conditions are *decoupled*. Suppose u_1 and u_2 are any two linearly independent solutions of equation (4.1.1) such that u_1 satisfies the boundary condition at $t = a$, u_2 at $t = b$ and the normalization condition $u_1(t)u_2'(t) - u_1'(t)u_2(t) = 1/p_0(t)$. Then the Green's function is given by

$$G(t, s) = \begin{cases} u_1(t)u_2(s) & \text{for } a \leqslant t \leqslant s, \\ u_1(s)u_2(t) & \text{for } s \leqslant t \leqslant b. \end{cases} \quad (4.1.6)$$

Using the Green's function and its properties, we now directly verify that the function u defined (4.1.4) satisfies the inhomogeneous equation $\mathcal{L}u = f$ and the boundary conditions (4.1.2). Since G satisfies the said boundary conditions, it is immediate that u also satisfies the boundary conditions (4.1.2). Since G is uniformly continuous in $[a, b]$, it follows that

$$u'(t) = \int_a^b \frac{\partial G}{\partial t}(t, s) f(s) \, ds.$$

Since now the integrand has discontinuity at $t = s$, we have

$$u''(t) = \frac{d}{dt} \int_a^t \frac{\partial G}{\partial t}(t, s) f(s) \, ds + \frac{d}{dt} \int_t^b \frac{\partial G}{\partial t}(t, s) f(s) \, ds$$

$$= \int_a^t \frac{\partial^2 G}{\partial t^2}(t, s) f(s) \, ds + \int_t^b \frac{\partial^2 G}{\partial t^2}(t, s) f(s) \, ds + \lim_{\varepsilon \to 0} \frac{\partial G}{\partial t}(t, s) f(s) \Big|_{s=t-\varepsilon}^{s=t+\varepsilon}$$

$$= \int_a^b \frac{\partial^2 G}{\partial t^2}(t, s) f(s) \, ds + \frac{f(t)}{p_0(t)}.$$

It follows that

$$\mathcal{L}u = \int_a^b \mathcal{L}G \, f(s) \, ds + f(t) = f(t)$$

as $\mathcal{L}G = 0$.

4.1.3 Green's Function for the Adjoint Equation (Symmetry)

The *adjoint equation* of equation (4.1.1) is given by

$$\mathcal{L}^* u \equiv (p_0 u)'' - (p_1 u)' + p_2 u = 0, \text{ in } (a, b), \quad (4.1.7)$$

$$C \begin{bmatrix} u(a) \\ u'(a) \end{bmatrix} + D \begin{bmatrix} u(b) \\ u'(b) \end{bmatrix} = \begin{bmatrix} 0 \\ 0 \end{bmatrix}, \quad (4.1.8)$$

with C and D being 2×2 real non-zero matrices. If u and v are C^2 functions, they satisfy the *Lagrange identity*:

$$v\mathcal{L}u - u\mathcal{L}^*v = \frac{d}{dt}P(u,v), \tag{4.1.9}$$

where

$$P(u,v) = u[p_1 v - (p_0 v)'] + u'(p_0 v).$$

Upon integration, we obtain the Green's formula:

$$\int_a^b v\mathcal{L}u - u\mathcal{L}^*v = P(u,v)\Big|_{t=a}^{t=b}. \tag{4.1.10}$$

Let $H(t,s)$ be the Green's function for the adjoint problem (4.1.7) and (4.1.8). Let $a < s_1 < s_2 < b$. Applying the Green's formula (4.1.10) with $u = G(t, s_1)$ and $v = H(t, s_2)$, we obtain

$$\int_a^b v\mathcal{L}u - u\mathcal{L}^*v = P(u,v)\Big|_{t=a}^{t=b},$$

with the understanding that the range of integration is regarded as the limiting case, as $\varepsilon \to 0$, of the integrals over the intervals $[a, s_1 - \varepsilon]$, $[s_1 + \varepsilon, s_2 - \varepsilon]$ and $[s_2 + \varepsilon, b]$. In each of these subintervals, we have $\mathcal{L}G = 0$ and $\mathcal{L}^*H = 0$. Therefore,

$$\lim_{\varepsilon \to 0} P(G,H)\Big|_{t=a}^{t=s_1-\varepsilon} + \lim_{\varepsilon \to 0} P(G,H)\Big|_{t=s_1+\varepsilon}^{t=s_2-\varepsilon} + \lim_{\varepsilon \to 0} P(G,H)\Big|_{s_2+\varepsilon}^{t=b} = 0.$$

The boundary conditions give that $P(G,H) = 0$ at the points $t = a$ and $t = b$. Thus, we obtain

$$\lim_{\varepsilon \to 0} P(G,H)\Big|_{t=s_1-\varepsilon}^{t=s_1+\varepsilon} + \lim_{\varepsilon \to 0} P(G,H)\Big|_{t=s_2-\varepsilon}^{s_2+\varepsilon} = 0.$$

But the only discontinuous term in $P(G,H)$ is $p_0\left(H\dfrac{\partial G}{\partial t} - G\dfrac{\partial H}{\partial t}\right)$. Hence, we obtain

$$p_0(s_1)H(s_1,s_2)\lim_{\varepsilon \to 0}\frac{\partial G}{\partial t}\Big|_{t=s_1-\varepsilon}^{t=s_1+\varepsilon} - p_0(s_2)G(s_2,s_1)\lim_{\varepsilon \to 0}\frac{\partial H}{\partial t}\Big|_{t=s_1-\varepsilon}^{t=s_1+\varepsilon} = 0.$$

Since

$$p_0(s_1)\lim_{\varepsilon \to 0}\frac{\partial G}{\partial t}\Big|_{t=s_1-\varepsilon}^{t=s_1+\varepsilon} = p_0(s_2)\lim_{\varepsilon \to 0}\frac{\partial H}{\partial t}\Big|_{t=s_1-\varepsilon}^{t=s_1+\varepsilon} = 1,$$

it follows that $H(s_1, s_2) = G(s_2, s_1)$. This relation is obtained under the assumption that $s_1 < s_2$. We obtain in a similar fashion the same relation when $s_2 < s_1$. We therefore conclude that $H(t,s) = G(s,t)$ for all $s, t \in [a,b]$. Thus, the Green's function for the adjoint problem is $G(s,t)$. Furthermore, if the given system is *self-adjoint*, that is, $\mathcal{L} = \mathcal{L}^*$, then the Green's function is symmetric: $G(t,s) = G(s,t)$. The converse is also true. Namely, if the Green's function for a given system is symmetric, then the system is self-adjoint. This follows from the uniqueness of the Green's function.

If $\mathcal{L}$ is self-adjoint, it is straight forward to see that $\mathcal{L}$ is of the form[3]:

$$\mathcal{L}u = -(pu')' + qu,$$

where p is a positive C^1 function and q a continuous function. Conversely, any such operator is self-adjoint.

4.2 Prüfer Substitution

Consider the BVP for a second-order linear homogeneous equation

$$(P(t)u'(t))' + Q(t)u(t) = 0, \ t \in (a,b), \quad (4.2.1)$$

along with the boundary conditions in equation (4.1.2). We assume that $P > 0$ is differentiable and Q is continuous in the interval $[a,b]$. For the S–L system, we have $P = p$ and $Q = \lambda \rho - q$.

To study the oscillations of the solution u, we introduce a powerful tool known as *Prüfer substitution*:

$$P(t)u'(t) = r(t)\cos\theta(t), \ u(t) = r(t)\sin\theta(t). \quad (4.2.2)$$

We have

$$r^2 = u^2 + P^2(u')^2, \ \theta = \arctan\left(\frac{u}{Pu'}\right). \quad (4.2.3)$$

Here r is called the *amplitude* variable and θ, the *phase* variable. When $P \equiv 1$, this gives the usual polar coordinates in the (u', u) plane. For $r \neq 0$, the correspondence $(Pu', u) = (r, \theta)$ as defined above, is smooth (in fact, *analytic*) with non-vanishing Jacobian. Also, for non-trivial solutions u, we have $r > 0$. For, if $r(t_0) = 0$, then $u(t_0) = 0$ and $P(t_0)u'(t_0) = 0$. Since $P > 0$, it follows that $u(t_0) = 0$ and $u'(t_0) = 0$. By uniqueness, $u \equiv 0$.

---

[3] The negative sign in the second derivative term makes $\mathcal{L}$ a positive operator, so that its eigenvalues are all positive.

The equivalent system of ODEs for r and θ can be easily derived. Using, $\tan\theta = \dfrac{u}{Pu'}$ or $\cot\theta = \dfrac{Pu'}{u}$, we obtain

$$-\csc^2(\theta)\,\theta' = \frac{u(Pu')'}{u^2} - \frac{P(u')^2}{u^2} = -Q - \frac{1}{P}\cot^2\theta.$$

This, after multiplication throughout by $\sin^2\theta$ gives

$$\theta' = Q\sin^2\theta + \frac{1}{P}\cos^2\theta \equiv F(t,\theta). \tag{4.2.4}$$

Similarly, by differentiating the expression $r^2 = u^2 + P^2(u')^2$, with respect to t, we obtain

$$r' = \left(\frac{1}{P} - Q\right) r\sin\theta\cos\theta = \frac{1}{2}\left(\frac{1}{P} - Q\right) r\sin 2\theta. \tag{4.2.5}$$

This system is equivalent to the original system (4.2.1) in the sense that every non-trivial solution of equation (4.2.1) defines a unique solution of the ODEs (4.2.4) and (4.2.5) by the Prüfer substitution. Next, observe that $F(t,\theta)$ is Lipschitz with respect to θ, as

$$\frac{\partial F}{\partial \theta} = Q\sin 2\theta - \frac{1}{P}\sin 2\theta$$

and therefore

$$\left|\frac{\partial F}{\partial \theta}\right| \leqslant \sup_{t\in[a,b]} |Q(t)| + \sup_{t\in[a,b]} \frac{1}{|P(t)|}.$$

Hence, we obtain a unique solution θ defined on $[a,b]$ for any initial value $\theta(a) = \gamma$. Once θ is known, the equation (4.2.5) for r gives

$$r(t) = r(a)\exp\left[(1/2)\int_a^t \left\{\frac{1}{P(s)} - Q(s)\right\}\sin 2\theta(s)\,ds\right],$$

for all $t \in [a,b]$. Each solution of the Prüfer system depends on an initial amplitude $r(a)$ and an initial phase $\gamma = \theta(a)$. Changing $r(a)$ just multiplies the solution u by a constant factor. Thus, the zeros of any solution u can be found by studying only the ODE for the phase θ.

From equation (4.2.3), we see that the zeros of any non-trivial solution u of equation (4.2.1) occur where the phase function θ assumes the values $n\pi$, $n \in \mathbb{Z}$. At these points, $\cos^2\theta = 1$ and $\theta' > 0$, as follows from equation (4.2.4). Geometrically, this means that the curve $(P(t)u'(t), u(t))$, $t \in [a,b]$ in the (Pu', u) plane, corresponding to a solution u, can cross the Pu'-axis at $\theta = n\pi$ only counterclockwise.

The advantage of Prüfer substitution in studying the zeros of the solution u is now evident from equation (4.2.4) satisfied by the phase variable. It is only a first-order equation for θ and *does not* contain r, and the solution exists in $[a,b]$ for any given initial condition.

The general references are [46], [8], [29], [15].

4.3 Exercises

Exercise 4.1

Let u, v satisfy the following equations

$$\frac{d}{dt}\left(P_1(t)\frac{du}{dt}(t)\right) - Q_1(t)u(t) = 0, \qquad (4.3.1)$$

$$\frac{d}{dt}\left(P_2(t)\frac{dv}{dt}(t)\right) - Q_2(t)v(t) = 0, \qquad (4.3.2)$$

in an interval $[a,b] \subset \mathbb{R}$, where $P_1 \geq P_2 > 0$ are continuously differentiable functions and $Q_1 \geq Q_2$ are continuous functions. Derive the *Picone formula*

$$\left[\frac{u}{v}(P_1 u'v - P_2 uv')\right]_a^b = \int_a^b (Q_1 - Q_2)u^2\, dt + \int_a^b (P_1 - P_2)u'^2\, dt \\ + \int_a^b P_2 \frac{u'v - uv'}{v^2}\, dt \qquad (4.3.3)$$

provided that v does not vanish in $[a,b]$. Here, for a function χ, $[\chi]_a^b = \chi(b) - \chi(a)$. Further, deduce the Sturm comparison theorem using the Picone formula. See Ref. [46].

Exercise 4.2

Using the Picone formula derived in Exercise 4.1, prove the following: Suppose u is a non-trivial, common solution of the equations (4.3.1) and (4.3.2), with the assumptions on P_i and Q_i stated therein. If $a < b$ are two consecutive zeros of u, then $Q_1 = Q_2$ in $[a,b]$ and $P_1 = P_2$ in $[a,b]$, except on subintervals of $[a,b]$ where $Q_1 = Q_2 = 0$.

Exercise 4.3

Using the Prüfer substitution, solve the equation $Pu'' + Qu = 0$, where P and Q are positive constants.

Exercise 4.4

Assuming $P:\mathbb{R} \to \mathbb{R}$ is a positive C^1 function, solve the equation

$$(Pu')' \pm \frac{1}{P}u = 0.$$

Exercise 4.5

In the following exercises, a linear second-order differential operator $\mathcal{L}$ is given on an interval I with prescribed boundary conditions. Find the eigenvalues and the (normalized) eigenfunctions of the corresponding Sturm–Liouville problem.

(a) $\mathcal{L}u = u''$, $I = [0,1]$, $u(0) = u(1) = 0$.
(b) $\mathcal{L}u = \alpha u''$, $I = [0,\ell]$, $u(0) = u(\ell) = 0$ ($\alpha > 0$, $\ell > 0$).
(c) $\mathcal{L}u = u''$, $I = [0,1]$, $u'(0) = u'(1) = 0$.

Exercise 4.6

Determine the function F (see equation (4.1.5)) and the Green's function for the following boundary value problems:

(a) $u'' = 0$, $u(0) = u(1) = 0$.
(b) $u'' - a^2 u = 0$, $u(0) = u(1) = 0$.
(c) $u'' + a^2 u = 0$, $u(0) = u(1)$, $u'(0) = u'(1)$ ($a \neq 2n\pi$, $n \in \mathbb{Z}$, $n \neq 0$). What happens if $a = 2n\pi$?
(d) $u'' + u = 0$, $u(0) = u'(\ell) = 0$ ($\ell < \pi/2$). What happens if $\ell = \pi/2$?

Exercise 4.7

Consider the operator $-\dfrac{d}{dt}\left(p(t)\dfrac{d}{dt}\right) + q(t)$ in an interval $[a,b]$, where p is a C^1 positive function and q a continuous function. Verify that the boundary conditions $\alpha_1 u(a) + \alpha_2 u'(a) = 0$ and $\beta_1 u(b) + \beta_2 u'(b) = 0$ associated with this operator, where $\alpha_1^2 + \alpha_2^2 > 0$ and $\beta_1^2 + \beta_2^2 > 0$, can be written in the form

$$\mathbf{A}\begin{bmatrix} u(a) \\ u'(a) \end{bmatrix} + \mathbf{B}\begin{bmatrix} u(b) \\ u'(b) \end{bmatrix} = \begin{bmatrix} 0 \\ 0 \end{bmatrix},$$

where $\mathbf{A}$ and $\mathbf{B}$ are 2×2 matrices such that the block matrix $\begin{bmatrix} \mathbf{A} & \mathbf{B} \end{bmatrix}$ has rank 2 and satisfy the relation $p(a)\det(\mathbf{B}) = p(b)\det(\mathbf{A})$.

Exercise 4.8

In the following exercises, a linear second-order differential operator $\mathcal{L}$ is given on an interval I with prescribed boundary conditions. Find the eigenvalues and the (normalized) eigenfunctions of the corresponding Sturm–Liouville problem:
(a) $\mathcal{L}u = u''$, $I = [0,1]$, $u(0) = u(1)$ and $u'(0) = u'(1)$.
(b) $\mathcal{L}u = u'' - u$, $I = [0,1]$, $u(0) - u'(0) + u(1) = 0$ and $u(0) + u'(0) + 2u'(1) = 0$.

Exercise 4.9

Determine the values of λ for which a Green's function can be constructed for the equation $y'' + \lambda y = f(t)$, with the following prescribed boundary conditions. Further, construct a Green's function for all such values of λ:
(a) $y(0) = y(1) = 0$. (b) $y(0) + y'(0) = 0$, $y(1) = y'(1)$.
(c) $y(0) = y'(0)$, $y(\ell) = 0$ ($\ell > 0$). (d) $y(0) = y'(0)$, $y(\pi) = y'(\pi)$.

Exercise 4.10

Determine the values of λ for which the BVP

$$u'' + 2u' + \lambda u = 0, \quad u(0) = u(\ell) = 0, \quad (\ell > 0),$$

has non-trivial solutions.

Exercise 4.11

Determine the values of λ for which the BVP

$$u'' + \lambda e^u = 0, \quad u(0) = u(1) = 0$$

has non-trivial solutions.

4.4 Solutions

Exercise 4.1

We begin by deriving an identity. Consider

$$\frac{d}{dt}\left[\frac{u}{v}(P_1 u'v - P_2 uv')\right] = \frac{u}{v}\left\{v\frac{d}{dt}(P_1 u') - u\frac{d}{dt}(P_2 v') + (P_1 - P_2)u'v'\right\}$$
$$+ \frac{vu' - uv'}{v^2}(P_1 u'v - P_2 uv')$$

$$= \frac{u}{v}\{Q_1 uv - Q_2 uv + (P_1 - P_2)u'v'\} + P_1(u')^2$$
$$- (P_1 + P_2)uu'\frac{v'}{v} + P_2 u^2 \left(\frac{v'}{v}\right)^2$$
$$= (Q_1 - Q_2)u^2 + (P_1 - P_2)uu'\frac{v'}{v} + P_1(u')^2$$
$$- (P_1 + P_2)uu'\frac{v'}{v} + P_2 u^2 \left(\frac{v'}{v}\right)^2$$
$$= (Q_1 - Q_2)u^2 + (P_1 - P_2)(u')^2$$
$$+ P_2\left((u')^2 - 2uu'\frac{v'}{v} + u^2\left(\frac{v'}{v}\right)^2\right)$$
$$= (Q_1 - Q_2)u^2 + (P_1 - P_2)(u')^2 + P_2\left(u' - u\frac{v'}{v}\right)^2.$$

The equality in the third line follows from using the equations (4.3.1) and (4.3.2), and in the fifth line the term $P_2(u')^2$ is added and subtracted. The Picone formula now follows immediately after an integration of the above identity over $[a,b]$.

We now deduce the Sturm comparison theorem. In the first version of this theorem, it is assumed that $P_1 = P_2$ and $Q_1 \geq Q_2$, but $Q_1 \neq Q_2$ in the relevant interval. Suppose u is a non-trivial solution of equation (4.3.1) and let $a < b$ be two consecutive zeros of u. We must show that any non-trivial solution v of equation (4.3.2) must vanish in (a,b). Here, we prove this result with an additional assumption that Q_1 and Q_2 are not both identically zero in any subinterval $[c,d] \subset [a,b]$.

First assume that $v \neq 0$ in $[a,b]$. Then, the term on the left-hand side of the Picone formula vanishes. We first observe that the same result holds even if we assume either $v(a) = 0$ or $v(b) = 0$ or both. Suppose $v(a) = 0$. In this case, the indeterminate quantity u/v has to be replaced by its limiting value u'/v', which is finite as u' and v' do not vanish at points where u and v vanish. Thus,

$$\lim_{t \to a}\left[\frac{u}{v}(P_1 u'v - P_2 uv')\right] = ((P_1 - P_2)uu')(a) = 0.$$

Therefore, we conclude that the term on the left-hand side of the Picone formula vanishes irrespective of whether v vanishes at a or b, or not.

We now show that the term on the right-hand side of the Picone formula is positive, thus arriving at a contradiction. Each term on the right-hand side is certainly non-negative. If $Q_1 > Q_2$ at a point in (a,b), then, by continuity, the first term is positive. The first two terms on the right-hand side can possibly be zero if $Q_1 = Q_2$ in (a,b) and $P_1 = P_2$ in a subinterval $[c,d]$ and $u' = 0$ outside $[c,d]$. Using equation (4.3.1), we then conclude that $Q_1 = 0$ outside $[c,d]$. But then, Q_2 is also zero outside $[c,d]$, which is a contradiction to our assumption. The third term can be zero if u and v are linearly dependent. In this case, u, also v, satisfies both the equations (4.3.1) and (4.3.2). We conclude that $Q_1 = Q_2$ in $[a,b]$ and $P_1 = P_2$ in $[a,b]$, except on subintervals of $[a,b]$ where $Q_1 = Q_2 = 0$; see Exercise 4.2.

Thus, we conclude that v must vanish in (a,b) if, in addition, Q_1 and Q_2 do not both vanish in any subinterval.

Exercise 4.2

Applying the Picone formula, with $v = u$, we obtain

$$\int_a^b (Q_1 - Q_2) u^2 \, dt = 0 \quad \text{and} \quad \int_a^b (P_1 - P_2) u'^2 \, dt = 0.$$

The first relation immediately gives that $Q_1 = Q_2$ in $[a,b]$, as both the terms in the integrand are non-negative and $u \neq 0$ in (a,b). The second relation holds true if $P_1 = P_2$ in some subinterval(s) of $[a,b]$ and $u' = 0$ in the complement of these subintervals; the case of $u' = 0$ in $[a,b]$ is ruled out as u is a non-trivial solution. If $u' = 0$ in a subinterval, then the given equations imply that $Q_1 = Q_2 = 0$ in that subinterval. This completes the proof.

Exercise 4.3

The Prüfer substitution $Pu' = r\cos\theta$ and $u = r\sin\theta$ gives

$$\theta' = Q \sin^2 \theta + \frac{1}{P} \cos^2 \theta, \tag{4.3.4}$$

$$r' = \frac{1}{2}\left(\frac{1}{P} - Q\right) r \sin(2\theta). \tag{4.3.5}$$

Once we solve equation (4.3.4) for θ, we can readily integrate equation (4.3.5) to obtain r. If $Q = 1/P$, we see that $\theta' = Q$ and $\theta(t) = Qt + \theta_0$, where $\theta_0 = \theta(0)$ is the initial condition. On the other hand, the equation (4.3.5) gives $r(t) = r_0$, where r_0 is the initial condition for r. Thus, the solution u in this case is given by $u(t) = r_0 \sin(Qt + \theta_0)$.

Now assume $Q \neq 1/P$. Put $Q = b^2$ and $1/P = a^2$, where a and b are positive real numbers. We have

$$\theta' = Q \sin^2 \theta + \frac{1}{P} \cos^2 \theta = a^2 \cos^2 \theta + b^2 \sin^2 \theta. \tag{4.3.6}$$

We first observe that the solution θ exists for all t and is an increasing function of t. Further,

$$\min(a^2, b^2) t + \theta_0 \leq \theta(t) \leq \max(a^2, b^2) t + \theta_0,$$

for $t \geq 0$ and

$$\lim_{t \to -\infty} \theta(t) = -\infty \quad \text{and} \quad \lim_{t \to \infty} \theta(t) = \infty.$$

We now obtain an explicit formula for the solution θ of equation (4.3.6). The right-hand side of equation (4.3.6) is a smooth periodic function of θ, of period π. Thus, by a translation of θ by an integer multiple of π, we may assume that the initial value $\theta_0 \in (-\pi/2, \pi/2]$. An integration of the first-order equation (4.3.6) gives $\tan \theta$ as a function of t. As the range of θ is $\mathbb{R}$ and the tan function is not a one-one function, we can readily invert this relation to obtain θ; we need to add a correction every time when θ crosses a value, which is an odd integer multiple of $\pi/2$. Instead, we adopt the following procedure.

First, an integration[4] of equation (4.3.6) and some algebra, we obtain

$$\tan \theta = \frac{a}{b} \tan(abt + \xi), \qquad (4.3.7)$$

where $\xi = \arctan(ba^{-1} \tan \theta_0)$. Differentiating both the sides of equation (4.3.7) with respect to t, we get

$$\theta' = \frac{a^2 b^2 \sec^2(abt + \xi)}{b^2 + a^2 \tan^2(abt + \xi)}. \qquad (4.3.8)$$

If $\theta_0 = \pi/2$, we may replace equations (4.3.7) and (4.3.8) by $\tan \theta = -\frac{a}{b} \cot(abt)$ and $\theta' = \frac{a^2 b^2 \csc^2(abt)}{b^2 + a^2 \cot^2(abt)}$, respectively, as $\xi = \pi/2$ in this case.

Therefore, we have obtained θ' as a function of t, and one more integration gives θ in an explicit form. Observe that the function on the right-hand side of equation (4.3.7) is a smooth periodic function of t, with period $\frac{\pi}{ab}$. Its mean value over one period is seen to be ab. By a result of integral of a periodic functions,[5] it follows that the function $\theta(t) - abt$ is a periodic function of period $\frac{\pi}{ab}$.

The periodic function is given as follows. Let $\theta_0 \in (-\pi/2, \pi/2)$. Define

$$p_0(t) = \begin{cases} \arctan\left[\frac{a}{b} \tan(abt + \xi)\right] & \text{if } 0 \leq t \leq \frac{1}{2ab}(\pi - 2\xi), \\ \pi + \arctan\left[\frac{a}{b} \tan(abt + \xi)\right] & \text{if } \frac{1}{2ab}(\pi - 2\xi) \leq t \leq \frac{\pi}{ab}, \end{cases} \qquad (4.3.9)$$

and $p(t) = p_0(t) - abt$ for $t \in \left[0, \frac{\pi}{ab}\right]$. Note that $p(0) = p_0(0) = \theta_0$. Extend the definition of p to $\mathbb{R}$ as a periodic function of period $\frac{\pi}{ab}$. The solution θ is then given by $\theta(t) = abt + p(t)$ for $t \in \mathbb{R}$. Alternatively, we have $\theta(t) = n\pi + p_0(t - n\pi/(ab))$ for

---

[4]
$$\int \frac{d\theta}{a^2 \cos^2 \theta + b^2 \sin^2 \theta} = (ab)^{-1} \arctan(ba^{-1} \tan \theta).$$

[5] If $f : \mathbb{R} \to \mathbb{R}$ is a period function of period ω and $m_f = \frac{1}{\omega} \int_0^\omega f(t) \, dt$ is the mean of f over one period, then the function $t \mapsto g(t) - m_f t$ is a periodic function of period ω. Here $g(t) = \int_0^t f(s) \, ds$, up to a constant.

$t \in I_n \equiv [n\pi/(ab), (n+1)\pi/(ab)]$, $n \in \mathbb{Z}$. If $\theta_0 = \pi/2$, p_0 is changed to $p_0(t) = \pi - \arctan(a\cot(abt)/b)$ for $t \in [0, \pi/(ab)]$.

Next, we obtain r by an integration of equation (4.3.5). Again assuming $\theta_0 \in (-\pi/2, \pi/2)$, we have

$$(1/2)\left(\frac{1}{P} - Q\right)\sin(2\theta) = \frac{1}{2}(a^2 - b^2)\frac{2\tan\theta}{1+\tan^2\theta}$$
$$= (a^2 - b^2)\frac{ab^{-1}\tan(abt+\xi)}{1 + a^2b^{-2}\tan^2(abt+\xi)}$$
$$= (a^2 - b^2)\frac{ab\sin(abt+\xi)\cos(abt+\xi)}{b^2\cos^2(abt+\xi) + a^2\sin^2(abt+\xi)}$$
$$= ab(a^2 - b^2)\frac{\sin(2(abt+\xi))}{(b^2 - a^2)\cos(2(abt+\xi)) + (a^2+b^2)}.$$

Therefore, the solution r is given by

$$r(t) = r_0 \exp\left[\frac{1}{2}\left(\frac{1}{P} - Q\right)\int_0^t \sin(2\theta(s))\,ds\right]$$
$$= r_0 \exp\left[\int_0^t ab(a^2-b^2)\frac{\sin(2(abs+\xi))}{(b^2-a^2)\cos(2(abs+\xi)) + (a^2+b^2)}\,ds\right]$$
$$= r_0 \exp\left[\frac{1}{2}\log\{(b^2-a^2)\cos(2(abs+\xi)) + (a^2+b^2)\}\Big|_0^t\right]$$
$$= r_0 \frac{((b^2-a^2)\cos(2(abt+\xi)) + (a^2+b^2))^{\frac{1}{2}}}{((b^2-a^2)\cos(2\xi) + (a^2+b^2))^{\frac{1}{2}}}$$
$$= r_0 \frac{(b^2\cos^2(abt+\xi) + a^2\sin^2(abt+\xi))^{\frac{1}{2}}}{(b^2\cos^2\xi + a^2\sin^2\xi)^{\frac{1}{2}}}, \qquad (4.3.10)$$

where $r_0 = r(0)$ is the initial value. If $\theta_0 = \pi/2$, we can obtain the expression for r by a similar computation or by letting $\xi \to \pi/2$ in equation (4.3.10):

$$r(t) = r_0\, a^{-1}(a^2\cos^2(abt) + b^2\sin^2(abt))^{\frac{1}{2}}.$$

To obtain the solution u using the relation $u = r\sin\theta$ involves some more lengthy computations (suffices to consider $t \in [0, \pi/(ab)]$). We obtain $u(t) = R\sin(abt+\psi)$ for some suitable constants $R > 0$ and phase shift ψ. Note that $ab = \sqrt{Q}/\sqrt{P}$, as required.[6]

---

[6] $\sin(\arctan\alpha) = \alpha/\sqrt{1+\alpha^2}$ and $\cos(\arctan\alpha) = 1/\sqrt{1+\alpha^2}$.

Exercise 4.4

First, consider the equation

$$(Pu')' + \frac{1}{P}u = 0.$$

Using the Prüfer substitution $Pu' = r\cos\theta$ and $u = r\sin\theta$, we obtain

$$\theta' = \frac{1}{P}\sin^2\theta + \frac{1}{P}\cos^2\theta = \frac{1}{P},$$

$$r' = \frac{1}{2}\left(\frac{1}{P} - \frac{1}{P}\right)r\sin(2\theta) = 0.$$

Thus, if we put $\chi(t) = \int_0^t (P(s))^{-1}\,ds$, we readily see that

$$r(t) = r_0 \text{ and } \theta(t) = \chi(t) + \theta_0,$$

for all $t \in \mathbb{R}$, where $r_0 = r(0)$ and $\theta_0 = \theta(0)$ are the initial values. Therefore,

$$u(t) = r(t)\sin(\theta(t)) = r_0\sin(\chi(t) + \theta_0).$$

Of course, r_0 and θ_0 are easily obtained in terms of the initial values $u(0)$ and $u'(0)$.

Alternatively, by multiplying the equation $(Pu')' + P^{-1}u = 0$, throughout by Pu' and then integrating the resulting equation, we obtain the relation $(Pu')^2 + u^2 = c^2$, where c is a non-negative constant, expressible in terms of the initial conditions of u and u'. If $c = 0$, then $u \equiv 0$. If $c > 0$, then we have $u' = \pm P^{-1}\sqrt{c^2 - u^2}$, which can be readily solved for u. We have $u(t) = c_1\cos(\chi(t)) + c_2\sin(\chi(t))$, where the constants c_1 and c_2 are determined by the initial conditions.

Similarly, for the equation $(Pu')' - P^{-1}u = 0$, we get $u(t) = c_1\exp(\chi(t)) + c_2\exp(-\chi(t))$, where, again, the constants c_1 and c_2 are determined by the initial conditions.

Exercise 4.5

(a) The general solution of the equation $u'' + \lambda u = 0$, $\lambda \in \mathbb{R}$ is given by

$$u(t) = \begin{cases} c_1\sin(\mu t) + c_2\cos(\mu t) & \text{if } \lambda > 0 \ (\mu = \sqrt{\lambda}) \\ c_1 e^{\mu t} + c_2 e^{-\mu t} & \text{if } \lambda < 0 \ (\mu = \sqrt{-\lambda}) \\ c_1 + c_2 t & \text{if } \lambda = 0, \end{cases}$$

for arbitrary constants c_1 and c_2. It is straight forward to check that if $\lambda \leqslant 0$, the boundary conditions $u(0) = u(1) = 0$ are satisfied only when $c_1 = c_2 = 0$. Hence, only the trivial solution satisfies the boundary conditions. Thus, there are no eigenvalues in the negative real axis $(-\infty, 0]$.

If $\lambda > 0$, then the boundary condition $u(0) = 0$ implies $c_2 = 0$. The other boundary condition $u(1) = 0$ then implies $c_1 \sin(\mu) = 0$. For the solution u to be non-trivial, we may thus choose μ such that $\sin(\mu) = 0$ or $\mu = n\pi$ with $n \in \mathbb{Z}$, $n \neq 0$. Therefore, the eigenvalues are given by $\lambda_n = n^2\pi^2$ for $n = 1, 2, \ldots$. The corresponding normalized eigenfunction is given by $u_n(t) = \sqrt{2} \sin(n\pi t)$. The constant is chosen so that
$$\int_0^1 u_n^2(t)\, dt = 1.$$

(b) We follow the same procedure. The general solution, in this case, of the equation $\alpha u'' + \lambda u = 0$, $\lambda \in \mathbb{R}$ is given by
$$u(t) = \begin{cases} c_1 \sin(\mu t) + c_2 \cos(\mu t) & \text{if } \lambda > 0 \ (\mu = \sqrt{(\lambda/\alpha)}), \\ c_1 e^{\mu t} + c_2 e^{-\mu t} & \text{if } \lambda < 0 \ (\mu = \sqrt{(-\lambda/\alpha)}), \\ c_1 + c_2 t & \text{if } \lambda = 0, \end{cases}$$
for arbitrary constants c_1 and c_2. Again, we find that there are no eigenvalues in $(-\infty, 0]$. If $\lambda > 0$, then the boundary condition $u(0) = 0$ implies that $c_2 = 0$. The other boundary condition $u(\ell) = 0$ implies, for a non-trivial solution, that $\sin(\mu\ell) = 0$ or $\mu = \dfrac{n\pi}{\ell}$ where n is non-zero integer. Thus, the eigenvalues are given by $\lambda_n = \dfrac{n^2 \alpha \pi^2}{\ell^2}$ and the corresponding normalized eigenfunctions are given by $u_n(t) = \sqrt{\dfrac{2}{\ell}} \sin\left(\dfrac{n\pi t}{\ell}\right)$.

Note that it is possible to reduce the exercise in (b) to the one in (a) by a suitable change of variables.

(c) The differential equation is the same as in (a); only the boundary conditions are different. We now have
$$u'(t) = \begin{cases} c_1 \mu \cos(\mu t) - c_2 \mu \sin(\mu t) & \text{if } \lambda > 0 \ (\mu = \sqrt{\lambda}), \\ c_1 \mu e^{\mu t} - c_2 \mu e^{-\mu t} & \text{if } \lambda < 0 \ (\mu = \sqrt{-\lambda}), \\ c_2 & \text{if } \lambda = 0. \end{cases}$$

Again, it is not difficult to see that there are no eigenvalues in $(-\infty, 0]$, as the trivial solution is the only solution satisfying the boundary conditions if $\lambda \leq 0$. If $\lambda > 0$, the boundary condition $u'(0) = 0$ gives $c_1 = 0$, and therefore, a corresponding eigenfunction is a multiple of $\cos(\mu t)$. For this to be a non-trivial solution satisfying the other boundary condition $u'(1) = 0$, we require that $\mu = n\pi$, where n is a non-zero integer. Therefore, the eigenvalues in this case are $\lambda_n = n^2\pi^2$, $n = 1, 2, \ldots$ and the corresponding normalized eigenfunctions are given by $u_n(t) = \sqrt{2} \cos(n\pi t)$.

Exercise 4.6

We use the notations in § 4.1.2.

(a) Here $p_0 = 1$ and

$$A = \begin{bmatrix} 1 & 0 \\ 0 & 0 \end{bmatrix} \text{ and } B = \begin{bmatrix} 0 & 0 \\ 1 & 0 \end{bmatrix}$$

and the boundary conditions are decoupled. If we take the linearly independent solutions 1 and t, then $AW(0) + BW(1) = \begin{bmatrix} 1 & 1 \\ 0 & 1 \end{bmatrix}$, which is non-singular. The functions t and $t-1$ are linearly independent solutions of $u'' = 0$, one satisfying the boundary condition at $t = 0$ and the other at $t = 1$. Therefore, we have $F(t,s) = \alpha + \beta t \pm (s-t)$ (α, β are constants) and

$$G(t,s) = \begin{cases} t(s-1) & \text{for } 0 \leqslant t \leqslant s, \\ s(t-1) & \text{for } s \leqslant t \leqslant 1. \end{cases}$$

(b) We may take $a > 0$. Here $p_0 = 1$ and

$$A = \begin{bmatrix} 1 & 0 \\ 0 & 0 \end{bmatrix} \text{ and } B = \begin{bmatrix} 0 & 0 \\ 1 & 0 \end{bmatrix}$$

and the boundary conditions are decoupled. If we take the linearly independent solutions e^{at} and e^{-at}, then $AW(0) + BW(1) = \begin{bmatrix} 1 & 1 \\ e^a & e^{-a} \end{bmatrix}$, which is non-singular. The functions $u_1(t) = \sinh(at)$ and $u_2(t) = \sinh(a(t-1))$ are linearly independent solutions, the first one satisfying the boundary condition at $t = 0$ and the second one at $t = 1$, respectively. Also $u_1(t)u_2'(t) - u_1'(t)u_2(t) = a \sinh a$. Thus, after normalization, we have $F(t,s) = \alpha \cosh(at) + \beta \sinh(at) \pm \sinh(a(s-t))/(2a)$ (α, β are constants) and

$$G(t,s) = \begin{cases} \dfrac{\sinh(at)\sinh(a(s-1))}{a \sinh a} & \text{for } 0 \leqslant t \leqslant s, \\ \dfrac{\sinh(as)\sinh(a(t-1))}{a \sinh a} & \text{for } s \leqslant t \leqslant 1. \end{cases}$$

(c) We may take $a > 0$ and $a \neq 2n\pi$. Here $p_0 = 1$ and

$$A = \begin{bmatrix} 1 & 0 \\ 0 & 1 \end{bmatrix} \text{ and } B = \begin{bmatrix} -1 & 0 \\ 0 & -1 \end{bmatrix}$$

and the boundary conditions are coupled. If we take the linearly independent solutions $\cos(at)$ and $\sin(at)$, then $AW(0) + BW(1) = \begin{bmatrix} 1 - \cos a & \sin a \\ -a \sin a & a(1 - \cos a) \end{bmatrix}$, which is non-singular. We have

$$F(t,s) = \alpha \cos(at) + \beta \sin(at) \pm (2a)^{-1} \sin(a(s-t))$$
$$= \alpha \cos(at) + \beta \sin(at) + (2a)^{-1} \sin(a|s-t|),$$

where α and β are constants. As the boundary conditions are coupled, we use the function F to construct the Green's function. If F were to satisfy the given boundary conditions, we obtain the following linear algebraic equations for the constants α and β:

$$\alpha \sin(as)/(2a) = \alpha \cos(a) + \beta \sin(a) + \sin(a(1-s))/(2a)$$
$$a\beta - \cos(as)/2 = -a\alpha \sin(a) + a\beta \cos(a) + \cos(a(1-s))/2$$

Solving these equations, we obtain

$$\alpha = \frac{1}{4a}\left[\sin(a(1-s)) - \sin(as) + \cot(a/2)(\cos(a(1-s)) + \cos(as))\right],$$

$$\beta = \frac{1}{4a}\left[-\cot(a/2)(\sin(a(1-s)) - \sin(as)) + \cos(a(1-s)) + \cos(as)\right].$$

Substituting these values in the expression for F, we obtain the Green's function

$$G(t,s) = \frac{1}{2a}\left[\cot(a/2)\cos(a(s-t)) + \sin(a|s-t|)\right].$$

We observe that when $a = 2n\pi$ (n a non-zero integer), the Green's function becomes infinite. In this case, the functions $\cos(2n\pi t)$ and $\sin(2n\pi t)$ are non-trivial solutions of the BVP and are linearly independent. The general solution of the inhomogeneous equation $u'' - 4n^2\pi^2 u = f$ is given by

$$u(t) = c_1 \cos(2n\pi t) + c_2 \sin(2n\pi t) + \frac{1}{2n\pi}\int_0^t \sin(2n\pi(t-s))f(s)\,ds,$$

by using the method of variation of parameters (constants). If this u were to satisfy the given boundary conditions, then f necessarily satisfies the following orthogonal conditions:

$$\int_0^1 f(t)\cos(2n\pi t)\,dt = \int_0^1 f(t)\sin(2n\pi t)\,dt = 0.$$

Furthermore, the solution to the inhomogeneous equation is not unique. Note that the eigenspace corresponding to the eigenvalue 0 is two dimensional.

(d) Here $p_0 = 1$ and $A = \begin{bmatrix} 1 & 0 \\ 0 & 0 \end{bmatrix}$ and $B = \begin{bmatrix} 0 & 0 \\ 0 & 1 \end{bmatrix}$ and the boundary conditions are decoupled. If we take the linearly independent solutions $\sin(t)$ and $\cos(t)$, then

$$AW(0) + BW(\ell) = \begin{bmatrix} 0 & 1 \\ \cos(\ell) & -\sin(\ell) \end{bmatrix},$$

which is non-singular as $\ell < \pi/2$. We have $F(t,s) = \alpha \cos(at) + \beta \sin(at) + \sin(|s-t|)$, where α and β are constants. The functions $u_1(t) = \sin(t)$ and $u_2(t) = \cot(\ell)\cos(t) + \sin(t)$ are linearly independent solutions of the given equation such that u_1 satisfies the boundary condition at $t = 0$ and u_2 at $t = \ell$. Also, $u_1(t)u_2'(t) - u_1'(t)u_2(t) = -\cot(\ell)$. Thus, after normalization, we obtain the Green's function:

$$G(t,s) = \begin{cases} -\sin(t)(\cos(s) + \tan(\ell)\sin(s)) & \text{for } 0 \leqslant t \leqslant s, \\ -\sin(s)(\cos(t) + \tan(\ell)\sin(t)) & \text{for } s \leqslant t \leqslant \ell. \end{cases}$$

We observe that when $\ell = \pi/2$, the Green's function becomes infinite. In this case, the function $\sin(t)$ is a non-trivial solution of the BVP. The general solution of the inhomogeneous equation $u'' + u = f$ is given by

$$u(t) = c_1 \cos(t) + c_2 \sin(t) + \int_0^t \sin(t-s) f(s)\, ds,$$

by using the method of variation of parameters (constants). If this u were to satisfy the boundary conditions $u(0) = 0$, $u'(\pi/2) = 0$, then $c_1 = 0$ and f necessarily satisfies the following orthogonal condition: $\int_0^1 f(t) \sin(t)\, dt = 0$. Furthermore, the solution to the inhomogeneous equation is not unique. Note that the eigenspace corresponding to the eigenvalue 0 is one-dimensional.

Exercise 4.7

The given boundary conditions can be written as

$$\begin{bmatrix} \alpha_1 & \alpha_2 \\ 0 & 0 \end{bmatrix} \begin{bmatrix} u(a) \\ u'(a) \end{bmatrix} + \begin{bmatrix} 0 & 0 \\ \beta_1 & \beta_2 \end{bmatrix} \begin{bmatrix} u(b) \\ u'(b) \end{bmatrix} = \begin{bmatrix} 0 \\ 0 \end{bmatrix}.$$

Thus, we take

$$\mathbf{A} = \begin{bmatrix} \alpha_1 & \alpha_2 \\ 0 & 0 \end{bmatrix} \text{ and } \mathbf{B} = \begin{bmatrix} 0 & 0 \\ \beta_1 & \beta_2 \end{bmatrix}.$$

The conditions imposed on α_1, α_2 and β_1, β_2 ensure that the block matrix

$$[\mathbf{A} \ \mathbf{B}] = \begin{bmatrix} \alpha_1 & \alpha_2 & 0 & 0 \\ 0 & 0 & \beta_1 & \beta_2 \end{bmatrix}$$

has rank 2. Since the determinants of both $\mathbf{A}$ and $\mathbf{B}$ are zero, the condition $p(a)\det(\mathbf{B}) = p(b)\det(\mathbf{A})$ is trivially satisfied.

Exercise 4.8

The procedure is exactly similar to the one in Exercise 4.5. Because of the mixed boundary conditions the algebra is more complicated.

(a) As before, the general solution of the equation $u'' + \lambda u = 0$, $\lambda \in \mathbb{R}$ is given by

$$u(t) = \begin{cases} c_1 \sin(\mu t) + c_2 \cos(\mu t) \text{ if } \lambda > 0 \ (\mu = \sqrt{\lambda}), \\ c_1 e^{\mu t} + c_2 e^{-\mu t} \text{ if } \lambda < 0 \ (\mu = \sqrt{-\lambda}), \\ c_1 + c_2 t \text{ if } \lambda = 0, \end{cases}$$

for arbitrary constants c_1 and c_2. First, consider the case $\lambda = 0$. The boundary conditions reduce to $c_1 = c_1 + c_2$ and $c_2 = c_2$. Thus, the constant function 1 is the normalized eigenfunction corresponding to the zero eigenvalue.

If $\lambda < 0$, the boundary conditions reduce to ($\mu = \sqrt{-\lambda}$)

$$c_1 + c_2 = c_1 e^{\mu} + c_2 e^{-\mu}, \quad c_1 \mu - c_2 \mu = c_1 \mu e^{\mu} - c_2 \mu e^{-\mu}.$$

Rewriting these equations, we obtain

$$(e^{\mu} - 1)c_1 + (e^{-\mu} - 1)c_2 = 0$$
$$\mu(e^{\mu} - 1)c_1 - \mu(e^{-\mu} - 1)c_2 = 0.$$

Since the coefficient matrix $\begin{bmatrix} e^{\mu} - 1 & e^{-\mu} - 1 \\ \mu(e^{\mu} - 1) & -\mu(e^{-\mu} - 1) \end{bmatrix}$ is non-singular, we get $c_1 = c_2 = 0$ and therefore, there are no eigenvalues in the interval $(-\infty, 0)$.

Finally, let $\lambda > 0$. In this case, we obtain the equations ($\mu = \sqrt{\lambda}$)

$$c_2 = c_1 \sin \mu + c_2 \cos \mu, \quad c_1 \mu = c_1 \mu \cos \mu - c_2 \mu \sin \mu,$$

for a solution satisfying the boundary conditions. Rewrite these equations as

$$(\sin \mu)c_1 + (\cos \mu - 1)c_2 = 0$$
$$\mu(\cos \mu - 1)c_1 - \mu(\sin \mu)c_2 = 0.$$

Cancelling the common factor μ from the second equation, we see that the coefficient matrix is $\begin{bmatrix} \sin\mu & \cos\mu - 1 \\ \cos\mu - 1 & -\sin\mu \end{bmatrix}$, whose determinant is equal to $-2(1 - \cos\mu)$. This determinant can thus vanish for $\mu = 2n\pi$, where n is a non-zero integer, and we obtain non-trivial solutions c_1 and c_2. Thus, the eigenvalues in this case are $\lambda_n = 4n^2\pi^2$. Corresponding to the eigenvalue λ_n, we have two *linearly independent* (normalized) eigenfunctions, namely $\sqrt{2}\sin(2n\pi t)$ and $\sqrt{2}\cos(2n\pi t)$ and any other eigenfunction is a linear combination of these two. Thus, the eigenspace is *two*-dimensional. Compare this with Exercise 4.5. We conclude that the eigenvalues of the given Sturm–Liouville problem are $4n^2\pi^2$, $n = 0, 1, 2, \ldots$. The eigenspace corresponding to $n = 0$ is one-dimensional, whereas that corresponding to $n \geq 1$ are two dimensional.

(b) The general solution of the equation $u'' - u + \lambda u = 0$, $\lambda \in \mathbb{R}$ is given by

$$u(t) = \begin{cases} c_1 \sin(\mu t) + c_2 \cos(\mu t) & \text{if } \lambda > 1 \ (\mu = \sqrt{\lambda - 1}), \\ c_1 e^{\mu t} + c_2 e^{-\mu t} & \text{if } \lambda < 1 \ (\mu = \sqrt{1 - \lambda}), \\ c_1 + c_2 t & \text{if } \lambda = 1, \end{cases}$$

for arbitrary constants c_1 and c_2. For the case $\lambda = 1$, the given boundary conditions reduce to $c_1 - c_2 + c_1 + c_2 = 0$ and $c_1 + c_2 + 2c_2 = 0$. It follows that $c_1 = c_2 = 0$, and therefore, $\lambda = 1$ is not an eigenvalue.

If $\lambda < 1$, the boundary conditions reduce to ($\mu = \sqrt{1 - \lambda}$)

$$c_1 + c_2 - c_1\mu + c_2\mu + c_1 e^{\mu} + c_2 e^{-\mu} = 0$$
$$c_1 + c_2 + c_1\mu - c_2\mu + 2c_1\mu e^{\mu} - 2c_2\mu e^{-\mu} = 0.$$

Rewriting these equations, we obtain

$$(e^{\mu} - \mu + 1)c_1 + (e^{-\mu} + \mu + 1)c_2 = 0$$
$$(2\mu e^{\mu} + \mu + 1)c_1 - (2\mu e^{-\mu} + \mu - 1)c_2 = 0.$$

A somewhat lengthy computation shows that the determinant d of the coefficient matrix $\begin{bmatrix} (e^{\mu} - \mu + 1) & (e^{-\mu} + \mu + 1) \\ (2\mu e^{\mu} + \mu + 1) & -(2\mu e^{-\mu} + \mu - 1) \end{bmatrix}$ is given by

$$-d = 8\mu + (2\mu^2 + 3\mu - 1)e^{\mu} - (2\mu^2 - 3\mu - 1)e^{-\mu}.$$

Rewrite this as $-d = -a\mu^2 + b\mu - c$, where

$$a = 2c = 2(e^{\mu} - e^{-\mu}) \text{ and } b = 3(e^{\mu} + e^{-\mu}) + 8.$$

Clearly, $b > 0$ for $\mu > 0$; in fact, for all $\mu \in \mathbb{R}$. Also, both a and c are positive for $\mu > 0$ and both vanish only when $\mu = 0$. Next, the discriminant of the above 'quadratic' in μ is

$$b^2 - 4ac = (3(e^{\mu} + e^{-\mu}) + 8)^2 + 8(e^{\mu} - e^{-\mu})^2$$

is positive for all $\mu > 0$. This shows that $d < 0$ for all $\mu > 0$. We therefore conclude that there are no eigenvalues in the interval $(-\infty, 1)$.

Finally, let $\lambda > 1$. In this case, we obtain the equations ($\mu = \sqrt{\lambda - 1}$) $c_2 - c_1\mu + c_1 \sin\mu + c_2 \cos\mu = 0$ and $c_2 + c_1\mu + 2c_1\mu\cos\mu - 2c_2\mu\sin\mu = 0$, for a solution satisfying the boundary conditions. Rewriting these equations, we obtain

$$(\sin\mu - \mu)c_1 + (\cos\mu + 1)c_2 = 0$$
$$(2\mu\cos\mu + \mu)c_1 + (1 - 2\mu\sin\mu)c_2 = 0.$$

Now the coefficient matrix is $\begin{bmatrix} (\sin\mu - \mu) & (\cos\mu + 1) \\ 2\mu(\cos\mu + 1) & (1 - 2\mu\sin\mu) \end{bmatrix}$. Again, a lengthy computation shows that its determinant equals $(2\mu^2 + 1)\sin\mu - 3\mu\cos\mu - 4\mu$. Thus, we look for $\mu > 0$ satisfying the equation

$$\sin\mu = \frac{\mu}{2\mu^2 + 1}(4 + \cos\mu). \tag{4.3.11}$$

As the term on the right-hand side of equation (4.3.11) is positive for all $\mu > 0$ and tends to 0 as $\mu \to \infty$, we expect the roots of equation (4.3.11) to be close to the integer multiples of π. As it is not possible to write these roots in explicit form, we draw the curves representing the terms on the left and right sides of equation (4.3.11) in the following Figure 4.1. The points where these two curves intersect give us the required roots. We see that the smallest positive root $\mu_0 = \pi/2 + \varepsilon_0$ is close to $\pi/2$. The other roots are given by $\mu_n = n\pi + (-1)^n \varepsilon_n$ for $n = 1, 2, ...$, where $\varepsilon_n > 0$ for all n. Corresponding to each root μ_n, we obtain a normalized eigenfunction by choosing $c_2 = 1$ and $c_1 = \dfrac{1 + \cos\mu}{\mu - \sin\mu}$ ($\mu = \mu_n$) and finding the appropriate normalizing constant. The eigenvalues are given by $\lambda_n = 1 + \mu_n^2$ for $n = 0, 1, 2, ...$

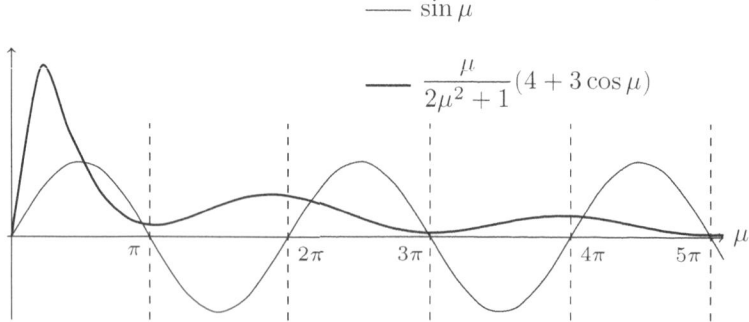

Figure 4.1 | Graphical solution of equation (4.3.11).

Exercise 4.9

It is good to recall Exercise 4.5 here. If there is a non-trivial solution of the homogeneous equation $y'' + \lambda y = 0$ satisfying the given boundary conditions, then λ would be an eigenvalue. In order that the given BVP has a solution, the function f on the right-hand side should satisfy an *orthogonality condition*. On the other hand, if λ is not an eigenvalue, then we look for two linearly independent solutions w_1 and w_2 of the homogeneous equation $y'' + \lambda y = 0$, such that w_1 satisfies the boundary condition at $t = 0$ and w_2 at $t = 1$. We can multiply these functions by suitable constants so that their Wronskian is 1. The required Green's function G is then given by

$$G(t,s) = \begin{cases} w_1(s)w_2(t) & \text{if } 0 \leqslant s \leqslant t, \\ w_1(t)w_2(s) & \text{if } t < s \leqslant 1. \end{cases}$$

We now proceed with details in each of the specified boundary conditions. When the boundary conditions involve both the values of the solution and its derivative, the computations are more and tedious. The general solution of the homogeneous equation $y'' + \lambda y = 0$ is given by

$$y(t) = \begin{cases} c_1 e^{\mu t} + c_2 e^{-\mu t} & \text{if } \lambda < 0, \\ c_1 + c_2 t & \text{if } \lambda = 0, \\ c_1 \cos(\mu t) + c_2 \sin(\mu t) & \text{if } \lambda > 0, \end{cases}$$

for arbitrary constants c_1 and c_2, where $\mu = \sqrt{|\lambda|}$.

Consider the boundary conditions in (a). As observed in Exercise 4.5, λ is not an eigenvalue if $\lambda \leqslant 0$. For $\lambda < 0$, we take

$$w_1(t) = \sinh(\mu t) \quad \text{and} \quad w_2(t) = \frac{\sinh(\mu(1-t))}{\mu \sinh(\mu)}.$$

For $\lambda = 0$, we take $w_1(t) = t$ and $w_2(t) = t - 1$. In the region $\lambda > 0$, there are eigenvalues given by $\lambda_n = n^2 \pi^2$, $n = 1, 2, \ldots$. If $\lambda \neq \lambda_n$ for any $n \geqslant 1$, we take

$$w_1(t) = \sin(\mu t) \quad \text{and} \quad w_2(t) = \frac{\sinh(\mu(t-1))}{\mu \sin(\mu)}.$$

If $\lambda = \lambda_n$ for some $n \geqslant 1$, then the inhomogeneous equation $y'' + \lambda y = f(t)$ has a solution, satisfying the boundary conditions, if and only if[7] f satisfies the orthogonality condition $\int_0^1 f(t) \sin(n\pi t)\, dt = 0$. In this case, the solution is given by $y(t) = c \sin(n\pi t) + \frac{1}{n\pi} \int_0^t f(s) \sin(n\pi(t-s))\, ds$ for arbitrary constant c.

---

[7] Compare this with the finite-dimensional situation. If $\mathbf{A}$ is a symmetric matrix, then the system $\mathbf{Ax} = \mathbf{b}$ has a solution if and only if $\mathbf{b}$ is orthogonal to $\ker \mathbf{A}$.

Next, consider the boundary conditions in (b). Assume $\lambda < 0$. If the general solution $y(t) = c_1 e^{\mu t} + c_2 e^{-\mu t}$, satisfies both the boundary conditions, then we have

$$(1+\mu)c_1 + (1-\mu)c_2 = 0 \text{ and } (1-\mu)e^{\mu}c_1 + (1+\mu)e^{-\mu}c_2 = 0.$$

For c_1 and c_2 to be simultaneously not zero, μ should satisfy the equation $(1+\mu)^2 e^{-\mu} = (1-\mu)^2 e^{\mu}$. Analysing the functions $\mu \mapsto (1+\mu)^2 e^{-\mu}$ and $\mu \mapsto (1-\mu)^2 e^{\mu}$ for $\mu > 0$, we see that there is a unique $\mu^* \in (3/2, 2)$ satisfying this equation. Thus, there is only one eigenvalue in the region $\lambda < 0$, given by $\lambda = -(\mu^*)^2$. For $\mu = 1$, we take $w_1(t) = e^{-t}$ and $w_2(t) = e^t/2$. For $\mu \neq 1, \mu^*$, we take

$$w_1(t) = e^t + \frac{\mu+1}{\mu-1} e^{-t} \text{ and } w_2(t) = c\left[e^t + \frac{\mu-1}{\mu+1} e^{2\mu} e^{-t}\right],$$

for a suitable constant c, so that the Wronskian of w_1 and w_2 is 1. The eigenfunction, up to a constant multiple, for $\mu = \mu^*$ is given by $w(t) = e^t + \frac{\mu^*+1}{\mu^*-1} e^{-t}$. In this case, the solution of the inhomogeneous equation $y'' + \lambda y = f(t)$ can similarly be written down, provided that f satisfies the orthogonality condition $\int_0^1 f(t) w(t)\, dt = 0$. For $\lambda = 0$, we take $w_1(t) = 1 - t$ and $w_2(t) = t$.

Let $\lambda > 0$. Arguing as in the case $\lambda < 0$, we see that there are now infinitely many eigenvalues given as squares of the roots of the equation

$$2\cos\mu + \left(\mu - \frac{1}{\mu}\right)\sin\mu = 0, \mu > 0.$$

Call these roots as μ_n, $n = 1, 2, \ldots$. It is not difficult to see that $\mu_n \approx n\pi$. Suppose $\mu \neq \mu_n$ for any $n \geq 1$. Then, we take $w_1(t) = \mu\cos(\mu t) - \sin(\mu t)$. If $\tan\mu = \mu$, we take $w_2(t) = \frac{1}{\mu^2}\sin(\mu t)$; if $\tan\mu \neq \mu$, we take $w_2(t) = c\left[\cos(\mu t) + \frac{\cos\mu + \mu\sin\mu}{\mu\cos\mu - \sin\mu}\sin(\mu t)\right]$, for suitable constant c.

Finally, consider the case of $\mu = \mu_n$ for some $n \geq 1$. The homogeneous equation $y'' + \lambda y = 0$ has non-trivial solutions, satisfying both the boundary conditions, given by $y(t) = c_1 \cos(\mu t) + c_2 \sin(\mu t)$, where the constants c_1 and c_2 are components of any suitable non-zero vector spanning the kernel of the 2×2 matrix

$$\begin{bmatrix} 1 & \mu_n \\ \cos(\mu_n) + \mu_n \sin(\mu_n) & \sin(\mu_n) - \mu_n \cos(\mu_n) \end{bmatrix}.$$

The case of the inhomogeneous equation $y'' + \lambda y = f(t)$ can be handled similarly, provided that f satisfies the orthogonality condition.

Next consider the boundary conditions in (c), namely $y(0) = y'(0)$ and $y(\ell) = 0$, where $\ell > 0$. Let $\lambda < 0$. If the general solution $y(t) = c_1 e^{\mu t} + c_2 e^{-\mu t}$ satisfies both the boundary conditions, then we have

$$(1 - \mu)c_1 + (1 + \mu)c_2 = 0 \quad \text{and} \quad e^{\mu \ell} c_1 + e^{-\mu \ell} c_2 = 0.$$

The determinant of the coefficient matrix is $(1 - \mu)e^{-\mu \ell} - (1 + \mu)e^{\mu \ell}$ and never vanishes for $\mu > 0$. Thus, there are no eigenvalues in this situation. We take $w_1(t) = e^{\mu t} + \dfrac{\mu - 1}{\mu + 1} e^{-\mu t}$ and $w_2(t) = c \sinh[\mu(\ell - t)]$, for suitable constant c.

For $\lambda = 0$, we take $w_1(t) = 1 + t$ and $w_2(t) = \dfrac{t - \ell}{1 + \ell}$.

Lastly, let $\lambda > 0$. If the general solution $y(t) = c_1 \cos(\mu t) + c_2 \sin(\mu t)$ satisfies both the boundary conditions, then $c_1 = \mu c_2$ and $\cos(\mu \ell) c_1 + \sin(\mu \ell) c_2 = 0$. For y to be non-trivial, μ satisfies the equation $\mu \cos(\mu \ell) + \sin(\mu \ell) = 0$. Thus, there are infinitely many positive values of μ satisfying this equation, and these are the eigenvalues. Denote these roots by μ_n, $n = 1, 2, \ldots$. It is not difficult to see that $\mu_n \approx (2n - 1)\pi/2$ for $n = 1, 2, \ldots$. If $\mu \neq \mu_n$ for any $n \geqslant 1$, we choose $w_1(t) = \cos(\mu t) + \dfrac{1}{\mu} \sin(\mu t)$ and $w_2(t) = c \sin(\mu(t - \ell))$ with $c = [\mu \cos(\mu \ell) + \sin(\mu \ell)]^{-1}$.

On the other hand, if $\mu = \mu_n$ for some $n \geqslant 1$, then the corresponding eigenfunction is given by $w(t) = \cos(\mu_n t) + \dfrac{1}{\mu_n} \sin(\mu_n t)$ and the solution of the inhomogeneous equation $y'' + \lambda y = f(t)$ can be obtained, provided that f satisfies the orthogonality condition $\displaystyle\int_0^\ell f(t) w(t) \, dt = 0$.

Now consider the boundary conditions in (d), namely $y(0) = y'(0)$ and $y(\pi) = y'(\pi)$. The boundary condition at $t = 0$ is the same as the one in (c). Let $\lambda < 0$. If the general solution $y(t) = c_1 e^{\mu t} + c_2 e^{-\mu t}$, satisfies both the boundary conditions, then we have

$$(1 - \mu)c_1 + (1 + \mu)c_2 = 0 \quad \text{and} \quad (1 - \mu)e^{\mu \pi} c_1 + (1 + \mu)e^{-\mu \pi} c_2 = 0.$$

The determinant of the coefficient matrix is $(1 - \mu)^2 e^{-\mu \pi} - (1 + \mu)^2 e^{\mu \pi}$, which never vanishes for $\mu > 0$. Thus, there are no eigenvalues in this situation. We take $w_1(t) = e^{\mu t} + \dfrac{\mu - 1}{\mu + 1} e^{-\mu t}$ and $w_2(t) = c \left[e^{\mu t} + \dfrac{\mu - 1}{\mu + 1} e^{2\mu \pi} e^{-\mu t} \right]$, for suitable constant c.

For $\lambda = 0$, we take $w_1(t) = 1 + t$ and $w_2(t) = (1 - \pi) + t$.

Lastly, let $\lambda > 0$. If the general solution $y(t) = c_1 \cos(\mu t) + c_2 \sin(\mu t)$, satisfies both the boundary conditions, then

$$c_1 = \mu c_2 \quad \text{and} \quad [\cos(\mu \pi) + \mu \sin(\mu \pi)] c_1 + [\sin(\mu \pi) - \mu \cos(\mu \pi)] c_2 = 0.$$

For y to be non-trivial, μ satisfies the equation

$$\mu[\cos(\mu \pi) + \mu \sin(\mu \pi)] + [\sin(\mu \pi) - \mu \cos(\mu \pi)] = 0.$$

The roots of this equation are $\mu_n = n$ for $n = 1, 2, ...$, and these are the eigenvalues. The corresponding eigenfunction is $n\cos(nt) + \sin(nt)$, up to a constant multiple. If $\mu \neq \mu_n$ for any $n \geq 1$, we choose

$$w_1(t) = \cos(\mu t) + \frac{1}{\mu}\sin(\mu t) \quad \text{and}$$

$$w_2(t) = c\left[(\sin(\mu\pi) - \mu\cos(\mu\pi))\cos(\mu t) - (\cos(\mu\pi) + \mu\sin(\mu\pi))\sin(\mu t)\right],$$

for a suitable constant c.

On the other hand, if $\mu = n$ for some $n \geq 1$, then the solution of the inhomogeneous equation $y'' + \lambda y = f(t)$ can be obtained, provided that f satisfies the orthogonality condition $\int_0^\pi f(t)[n\cos(nt) + \sin(nt)]\, dt = 0$.

Exercise 4.10

The characteristic equation is $m^2 + 2m + \lambda = 0$. Its roots are given by $-1 \pm \sqrt{1-\lambda}$. Thus, the general solution of the given ODE is given by

$$u(t) = \begin{cases} c_1 e^{\mu_1 t} + c_2 e^{\mu_2 t} & \text{if } \lambda < 1, \\ e^{-t}(c_1 + c_2 t) & \text{if } \lambda = 1, \\ e^{-t}(c_1 \cos(\alpha t) + c_2 \sin(\alpha t)) & \text{if } \lambda > 1, \end{cases}$$

where $\mu_1 = -1 + \sqrt{1-\lambda}$, $\mu_2 = -1 - \sqrt{1-\lambda}$ ($\lambda < 1$) and $\alpha = \sqrt{\lambda - 1}$ ($\lambda > 1$). For u to satisfy both the boundary conditions, the following relations hold:

For $\lambda < 1$, $c_1 + c_2 = 0$ and $c_1 e^{\mu_1 \ell} + c_2 e^{\mu_2 \ell} = 0$. As $\mu_1 > \mu_2$, we see that $c_1 = c_2 = 0$, and hence there are no non-trivial solutions to the BVP if $\lambda < 1$. For $\lambda = 1$, we have $c_1 = 0$ and $e^{-\ell}\ell c_2 = 0$. Again, we see that $c_1 = c_2 = 0$, and hence there are no non-trivial solutions to the BVP if $\lambda = 1$ as well.

Finally, for $\lambda > 1$, we have $c_1 = 0$ and $e^{-\ell}\sin(\alpha\ell)c_2 = 0$. Again, for c_2 to be non-zero, the condition $\alpha\ell = n\pi$ or $\lambda = 1 + \frac{n^2\pi^2}{\ell^2}$ for $n = 1, 2, ...$, should be satisfied. Thus, there are non-trivial solutions to the BVP for these values of $\lambda > 1$ and the solution is given by $e^{-t}\sin\left(\frac{n\pi t}{\ell}\right)$, up to a non-zero constant multiple.

Exercise 4.11

For $\lambda = 0$, the general solution is $u(t) = c_1 + c_2 t$, and applying the boundary conditions, we get $c_1 = 0$, $c_1 + c_2 = 0$. Hence, there is no non-trivial solution in this case.

For $\lambda > 0$, we utilize the form of the solution we have obtained in Exercise 5.8 in Chapter 5. The function u defined by

$$u(t) = \log k - 2\log[\cosh(\sqrt{k}(\sqrt{\lambda}t + \alpha)/\sqrt{2})],$$

is a solution of the equation $u'' + \lambda e^u = 0$, for any constants $k > 0$ and α real. The boundary conditions are satisfied if

$$\cosh(\sqrt{k}(\sqrt{\lambda} + \alpha)/\sqrt{2}) = \sqrt{k} = \cosh(\sqrt{k}\alpha/\sqrt{2}).$$

This forces $k > 1$ and $\sqrt{\lambda} + \alpha = \pm\alpha$. The positive sign is rejected as it gives $\lambda = 0$ and therefore $\sqrt{\lambda} = -2\alpha$ and

$$u(t) = \log k - 2\log\left[\cosh(\sqrt{k\lambda}(2t-1)/(2\sqrt{2}))\right].$$

The relation between k and λ in order to satisfy the boundary conditions now becomes $\cosh(\sqrt{k\lambda}/(2\sqrt{2})) = \sqrt{k}$. Further simplification yields

$$\sqrt{\lambda} = 2\sqrt{2}\frac{\log\left(\sqrt{k} + \sqrt{k-1}\right)}{\sqrt{k}}.$$

The function of k on the right-hand side for $k > 1$, has a unique positive maximum and tends to 0 as $k \to 0$ and $k \to \infty$. Therefore, there is a critical value λ_{cr} of λ such that the given BVP has no solution for $\lambda > \lambda_{\text{cr}}$, two solutions for $\lambda < \lambda_{\text{cr}}$, and only one solution for $\lambda = \lambda_{\text{cr}}$.

Similarly, for $\lambda < 0$, the function u is defined by any of the following expressions

$$u(t) = \begin{cases} \log k - 2\log|\sinh(\sqrt{k}(\sqrt{|\lambda|}t + \alpha)/\sqrt{2})| & \text{if } k > 0, \\ -2\log[(\sqrt{|\lambda|}t + \sqrt{2}e^{-a/2})/\sqrt{2}] & \text{if } k = 0, \\ \log(-k) - 2\log|\cos(\sqrt{-k}(\sqrt{|\lambda|}t + \alpha)/\sqrt{2})| & \text{if } k < 0, \end{cases}$$

is a solution of the equation $u'' + \lambda e^u = 0$ for any real constants k and α. We should carefully choose any of these expressions so that the domain of u contains the interval $[0,1]$ and then choose appropriate constants to satisfy the boundary conditions. It is not difficult to see that the solutions corresponding to $k \geqslant 0$ do not satisfy both the boundary conditions. This leaves us with a choice of $k < 0$. In this case, we have

$$u(t) = \log(|k|) - 2\log|\cos(\sqrt{|k|}(\sqrt{|\lambda|}t + \alpha)/\sqrt{2})|, \quad k < 0.$$

As $\cos((2n-1)\pi/2) = 0$ for any $n \in \mathbb{Z}$, we should make sure that the interval $[0,1]$ is contained in the set of t values that make the argument of the cosine function lies in the interval $(-\pi/2, \pi/2)$. Assuming this for a moment, the conditions

$$|\cos(\sqrt{|k|}(\sqrt{|\lambda|} + \alpha)/\sqrt{2})| = \sqrt{|k|} = |\cos(\sqrt{|k|}\alpha/\sqrt{2})|$$

need to be satisfied, in order to satisfy the boundary conditions. This forces $k \in [-1, 0)$, and we choose $|\lambda| = -2\alpha$ as in the case of $\lambda > 0$. This gives

$$u(t) = \log(|k|) - 2\log|\cos[\sqrt{|k\lambda|}(2t-1)/(2\sqrt{2})]|.$$

We choose $k \in (-1, 0)$ and $\lambda \in (-2\pi^2, 0)$ to meet the above-mentioned requirement. With these choices, we now obtain the relation

$$\cos[\sqrt{|k\lambda|}/(2\sqrt{2})] = \sqrt{|k|},$$

in order to satisfy the boundary conditions; thus, $k = -1$ is rejected as it implies $\lambda = 0$. This further implies that

$$\sqrt{|\lambda|} = 2\sqrt{2}\frac{\arccos(|k|)}{|k|}.$$

The function $\chi \mapsto \dfrac{\arccos(\chi)}{\chi}$ for $\chi \in (0, 1)$ is a decreasing function and tends to ∞ as $\chi \to 0$. Thus, there is a unique $\chi^* \in (0, 1)$ such that $\dfrac{\arccos(\chi^*)}{\chi^*} = \pi/2$. Thus, if we further restrict k to the interval $(-1, -(\chi^*)^2)$, then for each k in this interval there is a unique $\lambda \in (-2\pi^2, 0)$, satisfying the relation $\sqrt{|\lambda|} = 2\sqrt{2}\dfrac{\arccos(|k|)}{|k|}$, obtaining a non-trivial solution of the BVP. For example, if we choose $k = -1/2$, then $\lambda = -\pi^2$ and the solution is given by

$$u(t) = -\left[\log 2 + 2\log\left(\cos[\pi(2t-1)/4]\right)\right].$$

We remark that it is not difficult to see that we cannot replace the interval $(-\pi/2, \pi/2)$ by $((2n-1)\pi/2, (2n+1)\pi/2)$, with n a non-zero integer, because of the above-mentioned requirement of the set of t values.

5

Qualitative Theory

5.1 Introduction

The stability analysis of *equilibrium points* of an autonomous first-order system

$$\mathbf{x}' = \mathbf{f}(\mathbf{x}), \tag{5.1.1}$$

is an important topic in the qualitative theory of ordinary differential equations (ODE). Here $\mathbf{x} = \mathbf{x}(t) \in \mathbb{R}^n$ is a vector valued unknown function of the independent variable $t \in I$, an interval in $\mathbb{R}$, and $\mathbf{f} \colon \mathbb{R}^n \to \mathbb{R}^n$ is a given vector valued function, which is assumed to be a C^1 or more smooth function. This assumption ensures the uniqueness (local or global) of a solution of the system (5.1.1) with a prescribed initial value at an initial time. The positive integer n is referred to as the *dimension* of the system.

The system (5.1.1) is called an *autonomous* system because the right-side function $\mathbf{f}$ does not depend on t explicitly. When $\mathbf{f}$ depends on t explicitly as well, the system is referred to as *non-autonomous*. For example, the equation $x' = x + t$ (1D or one-dimensional equation) is non-autonomous.

In some situations, we do assume more smoothness on $\mathbf{f}$, so that global existence of a solution is guaranteed; this means existence for all t. Even when a solution does not exist for all t, we can still do the phase space analysis by considering the maximum interval of existence of the solution in question. However, uniqueness plays a crucial role. In what follows, we introduce many concepts, definitions and list many results that are useful in solving the exercises. For proofs and other details, we refer to Ref.[46] or any other book with similar contents.

A solution $\mathbf{x}$ of system (5.1.1) is said to pass through a given $\mathbf{x}_0 \in \mathbb{R}^n$ if $\mathbf{x}(t_0) = \mathbf{x}_0$, for some $t_0 \in \mathbb{R}$.

Definition 5.1. Given a solution $\mathbf{x}$ of system (5.1.1) passing through $\mathbf{x}_0 \in \mathbb{R}^n$ with $\mathbf{x}(t_0) = \mathbf{x}_0$, for some $t_0 \in \mathbb{R}$, the *orbit* through $\mathbf{x}_0$, is the set $\mathcal{O}(\mathbf{x}_0)$ defined by

$\mathcal{O}(\mathbf{x}_0) = \{\mathbf{x}(t) \in \mathbb{R}^n : t \in \mathbb{R}\}$ and the *positive orbit* through $\mathbf{x}_0$, is the set $\mathcal{O}^+(\mathbf{x}_0)$ defined by $\mathcal{O}^+(\mathbf{x}_0) = \{\mathbf{x}(t) \in \mathbb{R}^n : t \geqslant t_0\}$.

Lemma 5.3 shows that any solution passing through $\mathbf{x}_0$ may be used to define $\mathcal{O}(\mathbf{x}_0)$ or $\mathcal{O}^+(\mathbf{x}_0)$ unambiguously. Broadly speaking, the phase space (plane) analysis is about describing all the (positive) orbits of system (5.1.1). The other terminologies used for orbit are *trajectory* and *path*. Geometrically, the orbit $\mathcal{O}(\mathbf{x}_0)$ is the projection onto $\mathbb{R}^n$ of the graph of the solution, namely the set $\{(t, \mathbf{x}(t)) : t \in \mathbb{R}\}$ in $\mathbb{R}^{n+1}$. When we wish to describe an orbit $\mathcal{O}(\mathbf{x}_0)$, which is a curve in $\mathbb{R}^n$, we indicate by an arrow the increasing values of t, as the variable t is suppressed. It is an important property of an autonomous system that any two orbits are either the same or disjoint; see Lemma 5.5.

We will now discuss some important properties of solutions of autonomous systems. In the following results, statements regarding t refer to all $t \in \mathbb{R}$.

Lemma 5.2. If $\mathbf{x}$ is a solution of system (5.1.1), define $\mathbf{x}_c$ by $\mathbf{x}_c(t) = \mathbf{x}(t + c)$ for any fixed c and for all t. Then, $\mathbf{x}_c$ is also a solution of system (5.1.1).

This is an important and very useful property enjoyed by an autonomous system.

Lemma 5.3. If $\mathbf{x}$ and $\mathbf{y}$ are solutions of system (5.1.1) passing through $\mathbf{x}_0 \in \mathbb{R}^n$ with $\mathbf{x}(t_0) = \mathbf{y}(t_1) = \mathbf{x}_0$, for some $t_0, t_1 \in \mathbb{R}$, then $\mathbf{y}(t) = \mathbf{x}(t + t_0 - t_1)$ and $\mathbf{x}(t) = \mathbf{y}(t + t_1 - t_0)$ for all t.

The above lemma shows that $\mathcal{O}(\mathbf{x}_0)$ or $\mathcal{O}^+(\mathbf{x}_0)$ is the same set whether $\mathbf{x}$ or $\mathbf{y}$ is used in its definition.

Corollary 5.4. If $\mathbf{x}_0, \mathbf{x}_1 \in \mathbb{R}^n$ and $\mathbf{x}_1 \in \mathcal{O}(\mathbf{x}_0)$ (respectively, $\mathbf{x}_1 \in \mathcal{O}^+(\mathbf{x}_0)$), then, $\mathcal{O}(\mathbf{x}_0) = \mathcal{O}(\mathbf{x}_1)$ (respectively, $\mathcal{O}^+(\mathbf{x}_0) \supset \mathcal{O}^+(\mathbf{x}_1)$).

Lemma 5.5. Let $\mathbf{x}_0, \mathbf{x}_1 \in \mathbb{R}^n$. Then, either $\mathcal{O}(\mathbf{x}_0) = \mathcal{O}(\mathbf{x}_1)$ or $\mathcal{O}(\mathbf{x}_0) \cap \mathcal{O}(\mathbf{x}_1) = \emptyset$.

Similar statements may be made regarding the positive orbits.

Lemma 5.6. Suppose $\mathbf{x}$ is a solution of system (5.1.1) and that there exist t_0 and $T > 0$ such that $\mathbf{x}(t_0 + T) = \mathbf{x}(t_0)$. Then, $\mathbf{x}(t + T) = \mathbf{x}(t)$ for all t.

Remark 5.7. The solution in Lemma 5.6 is termed as a *periodic solution*. The smallest such a $T > 0$ is called the *period* of $\mathbf{x}$. The orbit of a periodic solution is called a *periodic orbit* or *closed orbit*. If a periodic orbit is *isolated* in the sense that there is no other periodic orbit in its neighbourhood, then the periodic orbit is called a *limit cycle*. For example, the orbits of the second-order equation $x'' + x = 0$ are all periodic orbits, but none of them is a limit cycle. The existence of periodic solutions to system (5.1.1) is an important aspect of the qualitative theory and two important results, namely Poincarè–Bendixson theorem and Leinard's theorem give sufficient conditions for the existence of periodic solutions in $2D$ systems.

We now discuss an important class of solutions to system (5.1.1).

Definition 5.8. A point $\bar{\mathbf{x}} \in \mathbb{R}^n$ is called an *equilibrium point* or *equilibrium solution* of system (5.1.1) if $\mathbf{f}(\bar{\mathbf{x}}) = 0$. An equilibrium point $\bar{\mathbf{x}}$ is *isolated* if there is a neighbourhood of $\bar{\mathbf{x}}$ not containing any other equilibrium point of system (5.1.1). Otherwise, the equilibrium point is *non-isolated*.

Thus, an equilibrium point is precisely a constant solution of system (5.1.1) and for this reason, it is also called a *fixed point, critical point, steady state solution, stationary point* or a *singularity*. The equilibrium solutions of system (5.1.1) are obtained by solving the system of algebraic equations $\mathbf{f}(\mathbf{x}) = 0$; the equilibrium solutions are the common zeros (roots) of the functions f_j, where $\mathbf{f} = \begin{bmatrix} f_1 & \cdots & f_n \end{bmatrix}^T$. For example, for a 2×2 linear system with non-singular coefficient matrix, the origin $(0,0) \in \mathbb{R}^2$ is the only equilibrium point. Here is an important property of equilibrium points.

Lemma 5.9. Suppose $\mathbf{x}$ is a solution of system (5.1.1) and $\lim\limits_{t \to \infty} \mathbf{x}(t) = \boldsymbol{\xi}$ exists. Then, $\boldsymbol{\xi}$ is an equilibrium point. If $\mathbf{x}$ is a non-constant solution of system (5.1.1), then $\mathbf{x}(\tilde{t})$ cannot be an equilibrium point for any $\tilde{t} \in \mathbb{R}$.

5.1.1 Liapunov Stability, Liapunov Function

We confine ourselves to a discussion of the stability of equilibrium solutions of system (5.1.1) in the sense of Liapunov. Further, we consider only isolated equilibrium points.

Definition 5.10 (Liapunov stability). An isolated equilibrium point $\bar{\mathbf{x}} \in \mathbb{R}^n$ of system (5.1.1) is said to be *stable* (*Liapunov stable*) if given $\varepsilon > 0$, there is a $\delta = \delta(\varepsilon) > 0$ such that for any solution $\mathbf{x}$ of system (5.1.1) satisfying $\|\mathbf{x}(t_0) - \bar{\mathbf{x}}\| < \delta$, we have $\|\mathbf{x}(t) - \bar{\mathbf{x}}\| < \varepsilon$ for all $t > t_0$. Otherwise, $\bar{\mathbf{x}}$ is said to be *unstable*.

Often, we take $t_0 = 0$.

Definition 5.11 (Asymptotic stability). An isolated equilibrium point $\bar{\mathbf{x}}$ of system (5.1.1) is said to be *asymptotically stable* if it is stable, and for any solution $\mathbf{x}$ of system (5.1.1), there exists $b > 0$ such that if $\|\mathbf{x}(t_0) - \bar{\mathbf{x}}\| < b$, then $\lim\limits_{t \to \infty} \|\mathbf{x}(t) - \bar{\mathbf{x}}\| = 0$.

Before proceeding further, we again recall from $2D$ linear theory that, in the case of a non-singular coefficient matrix, the only equilibrium point $(0,0)$ is asymptotically stable if all the eigenvalues have negative real parts (complex eigenvalues occur in conjugate pairs); it is stable, but not asymptotically stable if eigenvalues have zero real parts; and unstable in all the other cases. In higher dimensions, similar statements hold true, but the possibilities are more.

5.1.2 Linearization

We now discuss the linearization around an equilibrium point of system (5.1.1). We assume that $\mathbf{f}$ in system (5.1.1) is a C^2 function. If $\bar{\mathbf{x}}$ is an equilibrium point, then by Taylor's formula, we have

$$\mathbf{f}(\bar{\mathbf{x}} + \mathbf{y}) = \mathbf{f}(\bar{\mathbf{x}}) + \mathbf{A}\mathbf{y} + O(\|\mathbf{y}\|^2) = \mathbf{A}\mathbf{y} + O(\|\mathbf{y}\|^2), \tag{5.1.2}$$

where $\mathbf{A} = D\mathbf{f}(\bar{\mathbf{x}}) \equiv \left[\dfrac{\partial f_i}{\partial x_j}(\bar{\mathbf{x}})\right]$ denotes the Jacobian matrix of $\mathbf{f}$ at $\bar{\mathbf{x}}$. Writing $\mathbf{x} = \bar{\mathbf{x}} + \mathbf{y}$ and keeping only linear terms in $\mathbf{y}$, we obtain from equations (5.1.1) and (5.1.2), the following linear system:

$$\mathbf{y}' = \mathbf{A}\mathbf{y}. \tag{5.1.3}$$

System (5.1.3) is referred to as a *linearized system* corresponding to (5.1.1) around the equilibrium point $\bar{\mathbf{x}}$. It is also referred to as the *variation equation* of (5.1.1). Since we are concerned about the stability of $\bar{\mathbf{x}}$, it is quite reasonable to examine the stability of $\mathbf{0}$ for the linearized system (5.1.3). This is termed as *linear stability analysis*. However, we see through some examples that the stability of $\mathbf{0}$ of the linearized system (5.1.3) may or may not imply the stability of $\bar{\mathbf{x}}$ for system (5.1.1). We do have a result in the positive direction, namely:

Theorem 5.12. Suppose the eigenvalues of $\mathbf{A}$ in system (5.1.3) all have negative real parts. Then, the equilibrium point $\bar{\mathbf{x}}$ is asymptotically stable for system (5.1.1).

The above result is a particular case of *Perron's Theorem*, stated below.

Theorem 5.13 (Perron's Theorem). ([15], [46]) Consider the n-dimensional system

$$\mathbf{x}' = \mathbf{A}\mathbf{x} + \mathbf{f}(t, \mathbf{x}), \tag{5.1.4}$$

where $\mathbf{f}$ is a continuous function satisfying

$$\mathbf{f}(t, \mathbf{x}) = o(\|\mathbf{x}\|), \tag{5.1.5}$$

as $\mathbf{x} \to \mathbf{0}$, *uniformly* in t, and the real $n \times n$ matrix $\mathbf{A}$ has eigenvalues, all having negative real parts. Then, a solution $\mathbf{x}$ of system (5.1.4) with sufficiently small $\mathbf{x}(0)$ exists for all $t \geq 0$ and $\mathbf{x}(t)$ tends to $\mathbf{0}$ as $t \to \infty$.

Thus, the identically zero solution of system (5.1.4) is asymptotically stable. The assumption on **A** implies that the zero solution, which is the only equilibrium point, of the linear system $\mathbf{x}' = \mathbf{A}\mathbf{x}$ is asymptotically stable. Therefore, the theorem asserts that the asymptotic stability persists under small non-linear perturbations. The hypothesis may be modified so that we have existence of a solution for small positive time; uniqueness is not an issue here. For related issues, see Ref.[15].

We also remark that the existence of the solution for all $t \geqslant 0$ is not trivial; the reader may wish to compare in this regard the $1D$ equations

$$x' = x^2, \quad x' = -\mu x + x^2, \; \mu > 0.$$

A proof of this theorem using a Liapunov function is presented in the following section.

Definition 5.14. An equilibrium point $\bar{\mathbf{x}}$ of system (5.1.1) is called a *hyperbolic equilibrium point* if all the eigenvalues of $D\mathbf{f}(\bar{\mathbf{x}})$ have non-zero real parts.

We now mention an important theorem, namely Hartman–Grobman theorem, which states that *in a neighbourhood of a hyperbolic equilibrium point, the orbits of a non-linear system and those of its linearized system are linked via a homeomorphism.* For more details, see Ref.[48]. We have already seen (Perron's theorem) that if the linearized system around an equilibrium point has eigenvalues, all with negative real parts, then, the equilibrium point is also asymptotically stable for the non-linear system. Thus, the only case when the orbits of the non-linear system and those of the linearized system around an isolated equilibrium point may not be comparable is when the linearized system has at least one eigenvalue with zero real part, that is, when the equilibrium point is non-hyperbolic. This case can be handled by the construction of an appropriate *Liapunov function*, which is the topic of the next section. The *center manifold theorem*, see Ref.[58], also deals in detail with the case of a non-hyperbolic equilibrium point.

5.1.3 Liapunov Function and Stability

Assume that $\mathbf{0}$ is an isolated equilibrium point of system (5.1.1). Thus, there is an open neighbourhood Ω containing $\mathbf{0}$, which does not contain any other equilibrium point of system (5.1.1). The case of a non-zero equilibrium point can be handled by a translation.

Definition 5.15. A C^1 function $V: \Omega \to \mathbb{R}$ satisfying:

(1) $V(\mathbf{0}) = 0, \quad V(\mathbf{x}) > 0$ for all $\mathbf{x} \in \Omega \setminus \{\mathbf{0}\}$,

(2) $\nabla V \cdot \mathbf{f} = \sum_{j=0}^{n} \dfrac{\partial V}{\partial x_j} f_j \leqslant 0$ in Ω,

is called a *Liapunov function* for system (5.1.1).

Here, ∇V denotes the gradient of V. We have the following stability result:

Theorem 5.16. Assume that $\mathbf{0}$ is an isolated equilibrium point of system (5.1.1) and the system (5.1.1) possesses a Liapunov function V. Then, the origin $\mathbf{0}$ is stable. If, in addition, $\nabla V \cdot \mathbf{f} < 0$ in $\Omega \setminus \{\mathbf{0}\}$, then $\mathbf{0}$ is asymptotically stable.

On similar lines, we have the following instability result.

Theorem 5.17 (Instability Result)**.** Suppose there is a C^1 function $V : \Omega \to \mathbb{R}$ satisfying:

(1) $V(\mathbf{0}) = 0$.

(2) Every sphere centred at $\mathbf{0}$ contains at least one point where V is positive.

(3) $\nabla V \cdot \mathbf{f} > 0$ in $\Omega \setminus \{\mathbf{0}\}$.

Then, $\mathbf{0}$ is unstable for system (5.1.1).

Actually, the condition (2) in the above theorem may be replaced by a weaker assumption. The above theorem follows from the following theorem, due to Chetaev.

Theorem 5.18 (Chetaev)**.** Suppose in any neighbourhood of $\mathbf{0}$, there is a non-empty set in which $V > 0$ and $\nabla V \cdot \mathbf{f} > 0$ in the region $\{V > 0\}$. Then, $\mathbf{0}$ is unstable for system (5.1.1).

We now present a proof of *Perron's theorem* using a Liapunov function. This proof is different from the one given in Refs.[46], [15].

Consider an n-dimensional linear system $\mathbf{x}' = \mathbf{A}\mathbf{x}$, where $\mathbf{A}$ is a real non-singular $n \times n$ matrix. Suppose this system has a Liapunov function $V(\mathbf{x})$, which is given by a quadratic form, that is, $V(\mathbf{x}) = \mathbf{x}^\mathrm{T} \mathbf{B} \mathbf{x}$ for some real, symmetric, positive definite matrix $\mathbf{B}$. Suppose $\mathbf{B}$ can be chosen to further satisfy the condition $\mathbf{A}^\mathrm{T}\mathbf{B} + \mathbf{B}\mathbf{A} = -\mathbf{I}$. It then follows that $\dfrac{d}{dt}(V(\mathbf{x})) \leqslant -\|\mathbf{x}\|^2$ for any solution $\mathbf{x}(t)$ of $\mathbf{x}' = \mathbf{A}\mathbf{x}$. We conclude that the only equilibrium point $\mathbf{0}$ of the system $\mathbf{x}' = \mathbf{A}\mathbf{x}$ is asymptotically stable. We claim that the equilibrium point $\mathbf{0}$ of the non-linear system $\mathbf{x}' = \mathbf{A}\mathbf{x} + \mathbf{f}(t, \mathbf{x})$, where $\mathbf{f}$ is a smooth (vector valued) function satisfying the smallness condition: $\mathbf{f}(\mathbf{x}, t) = o(\mathbf{x})$, *uniformly* in t, is also asymptotically stable, by showing the function $V(\mathbf{x})$ is also a Liapunov function for this non-linear system. The smallness assumption on $\mathbf{f}$ means $\displaystyle\lim_{\|\mathbf{x}\| \to 0} \dfrac{\|\mathbf{f}(t, \mathbf{x})\|}{\|\mathbf{x}\|} = 0$, where the limit is uniformly in t.

To prove the claim, let $\mathbf{x}(t)$ be any solution of the non-linear system. Then,

$$\begin{aligned}\frac{d}{dt}(V(\mathbf{x})) = \frac{d}{dt}\left(\mathbf{x}^T\mathbf{B}\mathbf{x}\right) &= (\mathbf{x}')^T\mathbf{B}\mathbf{x} + \mathbf{x}^T\mathbf{B}\mathbf{x}'\\ &= \left[\mathbf{x}^T\mathbf{A}^T + (\mathbf{f}(t,\mathbf{x}))^T\right]\mathbf{B}\mathbf{x} + \mathbf{x}^T\mathbf{B}\left[\mathbf{A}\mathbf{x} + \mathbf{f}(t,\mathbf{x})\right]\\ &= \mathbf{x}^T\left(\mathbf{A}^T\mathbf{B} + \mathbf{B}\mathbf{A}\right)\mathbf{x} + (\mathbf{f}(t,\mathbf{x}))^T\mathbf{B}\mathbf{x} + \mathbf{x}^T\mathbf{B}(\mathbf{f}(t,\mathbf{x}))\\ &= -\|\mathbf{x}\|^2 + (\mathbf{f}(t,\mathbf{x}))^T\mathbf{B}\mathbf{x} + \mathbf{x}^T\mathbf{B}(\mathbf{f}(t,\mathbf{x})).\end{aligned}$$

Since $\mathbf{B}$ is a constant matrix, the smallness assumption on $\mathbf{f}$ implies that

$$|(\mathbf{f}(t,\mathbf{x}))^T\mathbf{B}\mathbf{x} + \mathbf{x}^T\mathbf{B}(\mathbf{f}(t,\mathbf{x}))| \leqslant (1/2)\|\mathbf{x}\|^2.$$

for small $\|\mathbf{x}\|$. Therefore, $\dfrac{d}{dt}(V(\mathbf{x})) \leqslant -(1/2)\|\mathbf{x}\|^2$ for sufficiently small $\|\mathbf{x}\|$. The claim now follows from the Liapunov theorem.

We now construct the matrix $\mathbf{B}$ that is used in the construction of the Liapunov function above, under the assumption that the eigenvalues of the matrix $\mathbf{A}$ all have negative real parts. Using the Jordan canonical form, the following inequality is established:

$$\|\exp(t\mathbf{A})\| \leqslant K e^{-\sigma t}, \text{ for all } t \geqslant 0,$$

for some positive constants K and σ. Since the eigenvalues of the matrix $\mathbf{A}^T$ are the same as those of $\mathbf{A}$, the above estimate also holds for $\mathbf{A}^T$. Thus, the matrix $\mathbf{B}$ is defined by the integral

$$\mathbf{B} = \int_0^\infty \exp\left(t\mathbf{A}^T\right)\exp(t\mathbf{A})\,dt$$

is well-defined. Clearly, $\mathbf{B}$ is symmetric and positive definite. Put $\mathbf{Y}(t) = \exp\left(t\mathbf{A}^T\right)\exp(t\mathbf{A})$. Then,

$$\frac{d}{dt}\mathbf{Y}(t) = \mathbf{A}^T\exp\left(t\mathbf{A}^T\right)\exp(t\mathbf{A}) + \exp\left(t\mathbf{A}^T\right)\exp(t\mathbf{A})\mathbf{A}.$$

Integrating both sides over $[0,\infty)$, we get $\lim_{t\to\infty}\mathbf{Y}(t) - \mathbf{Y}(0) = \mathbf{A}^T\mathbf{B} + \mathbf{B}\mathbf{A}$, which gives the desired condition $\mathbf{A}^T\mathbf{B} + \mathbf{B}\mathbf{A} = -\mathbf{I}$, as the matrices $\mathbf{A}$ and $\mathbf{A}^T$ have the same eigenvalues, all having negative real parts.

5.1.4 Hamiltonian System

A $2n$-dimensional system of the form

$$\frac{d\mathbf{u}}{dt} = \mathbf{J}H_{\mathbf{u}} \tag{5.1.6}$$

is called a *Hamiltonian system*. Here, $H: \mathbb{R}^{2n} \to \mathbb{R}$ is a C^2 function and is called a *Hamiltonian* or *energy functional*, and $H_{\mathbf{u}} = \nabla_{\mathbf{u}} H$ denotes the gradient of H, and $\mathbf{J}$ is the real $2n \times 2n$ matrix given by

$$\mathbf{J} = \begin{bmatrix} \mathbf{O}_n & \mathbf{I}_n \\ -\mathbf{I}_n & \mathbf{O}_n \end{bmatrix},$$

where $\mathbf{O}_n$ and $\mathbf{I}_n$ are the zero matrix and the identity matrix of order n, respectively. It is more customary to write the system (5.1.6) as follows. Write a (column) vector $\mathbf{u} \in \mathbb{R}^{2n}$ as $\mathbf{u} = (\mathbf{x}, \mathbf{y})$, where $\mathbf{x}$ and $\mathbf{y}$ are vectors in $\mathbb{R}^n$. Considering the Hamiltonian $H = H(\mathbf{u}) = H(\mathbf{x}, \mathbf{y})$ now as a function of $\mathbf{x}$ and $\mathbf{y}$, the system (5.1.6) transforms into

$$\left. \begin{array}{l} \dfrac{d\mathbf{x}}{dt} = H_{\mathbf{y}} = \dfrac{\partial H}{\partial \mathbf{y}} \\ \dfrac{d\mathbf{y}}{dt} = -H_{\mathbf{x}} = -\dfrac{\partial H}{\partial \mathbf{x}} \end{array} \right\} \quad (5.1.7)$$

If $(\mathbf{x}(t), \mathbf{y}(t))$ is an orbit of system (5.1.7), it follows that $H(\mathbf{x}(t), \mathbf{y}(t))$ is a constant. For, $\dfrac{d}{dt}(H(\mathbf{x}(t), \mathbf{y}(t))) = H_{\mathbf{x}} \mathbf{x}' + H_{\mathbf{y}} \mathbf{y}' = 0$, using system (5.1.7). This energy relation is particularly useful for analysing the phase portrait and would, in principle, suffice to describe the orbits of a two-dimensional Hamiltonian system. For the two-dimensional Hamiltonian system

$$x' = \frac{\partial H}{\partial y}, \quad y' = -\frac{\partial H}{\partial x},$$

the Jacobian matrix is given by

$$\begin{bmatrix} H_{xy} & H_{yy} \\ -H_{xx} & -H_{xy} \end{bmatrix} = \begin{bmatrix} 0 & 1 \\ -1 & 0 \end{bmatrix} D^2 H,$$

where $D^2 H$ denotes the Hessian matrix of H. Evaluated at any equilibrium point then gives the linearized system around that equilibrium point.

A function $\mathbf{f}: \mathbb{R}^{2n} \to \mathbb{R}^{2n}$ is called a *canonical transformation* if its Jacobian matrix $\left[\dfrac{\partial \mathbf{f}}{\partial \mathbf{u}} \right]$ is a *symplectic matrix*. This means that the Jacobian matrix satisfies the relation

$$\left[\frac{\partial \mathbf{f}}{\partial \mathbf{u}} \right]^T \mathbf{J} \left[\frac{\partial \mathbf{f}}{\partial \mathbf{u}} \right] = \mathbf{J}, \quad (5.1.8)$$

where the superscript T denotes the transpose of a matrix (also of a column vector). In particular, a canonical transformation is non-singular, and its inverse is also a canonical transformation. This follows from some simple properties of symplectic matrices.

In general, a real $2n \times 2n$ matrix $\mathbf{M}$ is called symplectic matrix if $\mathbf{M}^T \mathbf{J} \mathbf{M} = \mathbf{J}$. It follows that $\mathbf{M}$ is non-singular and $\det(\mathbf{M}) = \pm 1$. In fact, $\det(\mathbf{M}) = 1$, but this does not follow immediately. It further follows that $\mathbf{M}^{-1}$ and $\mathbf{M}^T$ are also symplectic matrices and, $\mathbf{M}$ and $\mathbf{M}^{-1}$ are similar. Also, $\mathbf{J}\mathbf{M} = \mathbf{J}(\mathbf{M}^T)^{-1}$.

5.1.5 Two-Dimensional Potential Flows

An n-dimensional first-order autonomous system $\mathbf{x}' = \mathbf{F}(\mathbf{x})$ is called a *potential flow* if $\mathbf{F} = \nabla u$ for a smooth function u defined in a region of $\mathbb{R}^n$. A particularly simple class of $2D$ potential flows occurs when we take u as the real or imaginary part of an analytic function. These are quite simple to analyse, and their phase portraits (i.e., displaying all the orbits of a given system in the phase plane) are easier to obtain. Such systems are also closely related to steady two-dimensional incompressible fluid flow. These phase portraits also play an important role in the study of dynamical systems in differential geometry; see Ref.[43].

Consider a $2D$ potential flow

$$x' = u_x, \quad y' = u_y, \qquad (5.1.9)$$

where u is the real part of an analytic function $f = u + iv$ defined in a region of $\mathbb{C}$, which we identify with $\mathbb{R}^2$. The functions u and v are harmonic and satisfy the Cauchy–Riemann(CR) equations: $u_x = v_y$ and $u_y = -v_x$; the function v is called the *harmonic conjugate* of u and vice versa. The simplicity of the system (5.1.9) lies in the fact that it is readily integrable and the function v is an integral of system (5.1.9), that is, along any solution $(x(t), y(t))$ of system (5.1.9), v is a constant. For,

$$\frac{d}{dt} v(x(t), y(t)) = v_x x' + v_y y' = v_x u_x + v_y u_y = 0,$$

using CR equations. The system corresponding to v is

$$x' = v_x, \quad y' = v_y. \qquad (5.1.10)$$

Again, using CR equations, we see that the orbits of systems (5.1.9) and (5.1.10) are mutually orthogonal. Sometimes it is useful to use the polar co-ordinates r and θ, especially when u and v can be easily described. If u is expressed in terms of polar co-ordinates r and θ, then system (5.1.9) can be written in polar co-ordinates as

$$r' = u_r, \quad \theta' = r^{-2} u_\theta, \ (r > 0). \qquad (5.1.11)$$

The systems (5.1.9) and (5.1.11) are equivalent in the region $\{r > 0\}$.

In the context of fluid mechanics, a steady-state two-dimensional irrotational flow of an incompressible fluid is characterized by an analytic function $f = u + iv$. The function f is called the *complex potential* or *characteristic function* of the flow; u is called the *potential function* and v the *stream function*. The level curves $u =$ constant are called the *equipotential lines*, and the level curves $v =$ constant are called the *streamlines*. Thus, the orbits or the trajectories of the system (5.1.9) are precisely the streamlines in the context of fluid mechanics. The velocity of the flow is given by $\mathbf{V} = \nabla u$, that is, $\mathbf{V} = (u_x, u_y)$.

The general references are [46], [15], [48], [58], [29], [52], [3], [11], [10], [32], [38], [41], [54].

5.2 Exercises

Exercise 5.1

Find the equilibrium points $x' = y + \sin x$, $y' = x - y\cos x$, and discuss their linear stability.

Exercise 5.2

Describe the orbits of the Lotka–Volterra prey–predator system given by

$$x' = ax - bxy, \quad y' = -cy + dxy,$$

where a, b, c and d are positive real numbers, in the first quadrant of the $x - y$ plane.

Exercise 5.3

Show that the (open) first quadrant in the $x - y$ plane is an invariant set for the system $x' = x^2 + y\sin x$, $y' = -1 + xy + \cos y$.

Exercise 5.4

For the real 2×2 matrix $\mathbf{A} = \begin{bmatrix} a & b \\ c & d \end{bmatrix}$, having eigenvalues with negative real parts, construct the Liapunov function V via a symmetric positive definite matrix $\mathbf{B}$ mentioned in Section 5.1.3.

Exercise 5.5

Solve the first-order equation $y' = c + \sin y$, $c \in \mathbb{R}$.

This simple-looking equation exhibits some interesting qualitative behaviour of solutions and can be thought of as a simple *bifurcation* problem with c as the bifurcation parameter. In deriving an explicit formula for the solution, we need to 'invert' a relation of the form $\tan(y/2) = \cdots$. Since the tan function is not a one-one function, a correction term needs to be added whenever y takes the value of an odd integer multiple[1] of π. This is where the qualitative analysis will guide us towards a correct expression for the solution.

Exercise 5.6

Consider the second-order equation in conservative form:

$$x'' + \frac{d}{dx}U(x) = 0, \qquad (5.2.1)$$

where the potential function $U:\mathbb{R} \to \mathbb{R}$ is a C^2 function. If $U \geqslant 0$, show that any solution of equation (5.2.1) exists globally, that is, defined for all $t \in \mathbb{R}$.

Exercise 5.7

Consider Duffing's equation without damping: $x'' - x + x^3 = 0$. This is an equation in conservative form, and the corresponding conserved quantity is $(1/2)(x')^2 + U(x) = E$, where $U(x) = -x^2/2 + x^4/4$ is the potential function. Here E is a constant, referred to as the energy level of the solution. Since U is bounded below, it follows that any solution of the given equation exists for all t, by Exercise 5.6. Solve the Duffing's equation at the energy level $E = 0$.

Exercise 5.8

Solve the second-order conservative equations:

(1) $x'' + e^x = 0$ and (2) $x'' - e^x = 0$.

Exercise 5.9

Draw the phase portrait of the second-order equation $x'' = (1/2)(x^2 - 1)$.

Exercise 5.10

Solve the second-order equation $-x'' + xx' = 0$.

---

[1]
$$\arctan(\tan a) = \begin{cases} a & \text{if } |a| \leqslant \pi/2, \\ a - n\pi & \text{if } |a| > \pi/2, \end{cases}$$
where $n \in \mathbb{Z}$ is chosen so that $|a - n\pi| \leqslant \pi/2$.

Exercise 5.11

The first-order partial differential equation (PDE) $u_\tau + u u_\xi = \varepsilon u_{\xi\xi}$, where $\varepsilon > 0$, is termed as the *viscous Burgers' equation*. Here u is the unknown function of the real variables ξ and τ. By suitably rescaling the variables, we may assume $\varepsilon = 1$. A *travelling wave solution* of the viscous Burgers' equation is a solution of the form $u(\xi, \tau) = x(t)$ where $t = \xi - c\tau$ for some constant c. Deduce that the function x, after suitable adjustments, satisfies the equation in Exercise 5.10. Further, find a solution x such that $\lim_{t \to -\infty} x(t) = a$ and $\lim_{t \to \infty} x(t) = -a$ for suitable constant $a \ne 0$.

Exercise 5.12

Derive the necessary and sufficient conditions on the real $n \times n$ matrices $\mathbf{A}$, $\mathbf{B}$, $\mathbf{C}$ and $\mathbf{D}$ such that the linear system

$$\frac{d\mathbf{x}}{dt} = \mathbf{A}\mathbf{x} + \mathbf{B}\mathbf{y}, \quad \frac{d\mathbf{y}}{dt} = \mathbf{C}\mathbf{x} + \mathbf{D}\mathbf{y}$$

is a Hamiltonian system.

Exercise 5.13

Suppose $\mathbf{u}$ satisfies a Hamiltonian system $\dfrac{d\mathbf{u}}{dt} = JH_\mathbf{u}$ in an interval in $\mathbb{R}$, for some Hamiltonian H. If $\mathbf{f} \colon \mathbb{R}^{2n} \to \mathbb{R}^{2n}$ is a canonical transformation and $\mathbf{v} = \mathbf{f}(\mathbf{u})$, show that $\mathbf{v}$ also satisfies a Hamiltonian system. Thus, under a canonical transformation a Hamiltonian system changes into another Hamiltonian system.

Exercise 5.14

Using the Poincarè–Bendixson theorem, determine whether the following two-dimensional autonomous systems possess any periodic orbit. If so, identify the region in the phase plane where such a periodic orbit is located and find the same wherever possible:

(1) $x' = x + y - x \exp(x^2 + y^2 - 4)$, $y' = -x + y - y \exp(x^2 + y^2 - 4)$.
(2) $x' = 2x + 3y - x(x^2 + y^2)$, $y' = -3x + 2y - y(x^2 + y^2)$.

Exercise 5.15

Using the Poincarè–Bendixson theorem, show that the second-order equation $x'' + (x^2 + 2(x')^2 - 1)x' + x = 0$ has a non-trivial periodic orbit.

Exercise 5.16

Analyse the system $x' = x$, $y' = -y + x^2$ and show that

(1) $x(y - x^2/3)$ is a constant and
(2) $(y(t) - (x(t))^2/3) = e^{-t}(y(0) - (x(0))^2/3)$,

for any solution and for all $t \in \mathbb{R}$.

Exercise 5.17

Discuss the stability of equilibrium point(s) of the spring-mass-dashpot system $mx'' + cx' + kx = 0$, writing it as a first-order system. Here the constants m, k are positive and $c \geqslant 0$.

Exercise 5.18

Discuss the stability of equilibrium point(s) of the non-linear equation $mx'' + cx' + k\sin x = 0$, where the constants m, k are positive and $c \geqslant 0$. This is a non-linear version of the equation in Exercise 5.17 and with suitable interpretation of the constants, it is the non-linear pendulum equation.

Exercise 5.19

Analyse the stability of the equilibrium point(s) of the following systems by the method of Liapunov:
(a) $x' = -2xy$, $y' = x^2 - y^3$.
(b) $x' = -3x^3 - y$, $y' = x^5 - 2y^3$.

Exercise 5.20

Analyse the phase portrait of the potential flow for the monomial $f(z) = z^n$, where n is a positive integer.

Exercise 5.21

Analyse the phase portrait of the potential flow for the function $f(z) = z^{-n}$, where n is a positive integer.

Exercise 5.22

The *Joukowski transformation* is the analytic function $f(z) = z + \dfrac{1}{z}$, $z \neq 0$. Analyse the phase portrait of the potential flow (5.1.9) for $u = \Re(f)$. Note that f is analytic in the punctured plane $\mathbb{C}\setminus\{0\}$.

The Joukowski transformation plays an important role in the design of airfoils. Such an airfoil, usually called a *Joukowski airfoil*, is the cross-section of a airplane wing and is constructed as follows. Let γ be a circle of radius > 1, with one of its chords as the line segment with end points as -1 and 1, in the complex plane. Let γ' be any other circle, containing γ in its interior, except the point 1, where they are tangent to each other. A Joukowski airfoil is the image of γ' under the Joukowski transformation $z \mapsto (1/2)(z + 1/z)$; the factor $1/2$ is used to map the point 1 to itself. If the point 1 is replaced by any other $a > 0$, then we use the transformation $z \mapsto (1/2)(z + a^2/z)$. The shape and size of an airfoil can be changed by suitably choosing γ and γ'; see Ref.[39]. A typical airfoil is shown in Figure 5.11(a).

Exercise 5.23

Show that the two-dimensional autonomous system

$$x' = y(13 - x^2 - y^2), \ y' = 12 - x(13 - x^2 - y^2)$$

is a Hamiltonian system and analyse its phase portrait.

5.3 Solutions

Exercise 5.1

If (x, y) is an equilibrium point, it is easily seen that $x + (\sin x)(\cos x) = 0$. Thus $x = 0$, and we conclude that $(0, 0)$ is the only equilibrium point. The linearized system is then given by $x' = y + x$, $y' = x - y$. The eigenvalues of the coefficient matrix $\begin{bmatrix} 1 & 1 \\ 1 & -1 \end{bmatrix}$ are $\pm\sqrt{2}$, and hence the equilibrium point is hyperbolic and unstable. By Hartman–Grobman theorem, the equilibrium point is also unstable for the non-linear system.

Exercise 5.2

It is easily checked that all the equilibrium points of the given system are $(0, 0)$ and $(c/d, a/b)$. It is not difficult to do a linear stability analysis at these equilibrium points and see that $(0, 0)$ is unstable and $(c/d, a/b)$ is stable but not asymptotically stable.

It is easy to see that both the positive axes are orbits of the system. Therefore, the (open) first quadrant is an invariant subset, that is, any orbit starting in the first quadrant will remain there for all future times. We will now analyse the orbits in more detail. From the given system, we get

$$\frac{dy}{dx} = \frac{dy/dt}{dx/dt} = \frac{y(-c + dx)}{x(a - by)}.$$

Writing this equation in variable separable form and integrating once, we obtain

$$y^a e^{-by} = K x^{-c} e^{dx}, \tag{5.2.2}$$

where $K = \dfrac{y_0^a e^{-by_0}}{x_0^{-c} e^{dx_0}}$ with $x_0 = x(0)$ and $y_0 = y(0)$ being the initial values, both being positive. It is not possible to express either x or y in terms of the other from (5.2.2). However, we now show that all the points (x, y) satisfying equation (5.2.2) form a simple closed curve in the $x - y$ plane, as follows. Let $z(x) = K x^{-c} e^{dx}$, $x > 0$. Observe that z has a unique minimum at $x = c/d$ with minimum value $z_{\min} = K(c/d)^{-c} e^c$ and $z(x) \to \infty$ as $x \to 0+$ and $x \to \infty$. It follows that for each $A > z_{\min}$, there are precisely two values of x, call them x_1 and x_2, such that $z(x_1) = z(x_2) = A$. Further, observe that $0 < x_1 < c/d < x_2$.

Similarly, the function w defined by $w(y) = y^a e^{-by}$, $y > 0$, has a unique maximum at $y = a/b$ with maximum value $w_{\max} = (a/b)^a e^{-a}$. Also, $w(y) \to 0$ as $y \to 0+$ and $y \to \infty$. It follows that for each $0 < A < w_{\max}$, there are precisely two values of y, call them y_1 and y_2, such that $w(y_1) = w(y_2) = A$ and $0 < y_1 < a/b < y_2$.

Next observe that a pair (x, y) satisfies equation (5.2.2) if and only if $z(x) = w(y)$. Thus, it suffices to find all the pairs (x, y) that satisfy $z(x) = w(y)$. Put $K_0 = \dfrac{(a/b)^a e^{-a}}{(c/d)^{-c} e^c}$, which is the constant corresponding to the initial values $x(0) = c/d$ and $y(0) = a/b$. We notice that $K < K_0$ if $(x_0, y_0) \neq (c/d, a/b)$. This in turn implies that $z_{\min} \leq w_{\max}$ with equality holding only when $K = K_0$. If $x(0) = c/d$ and $y(0) = a/b$, then $K = K_0$ and $z_{\min} = w_{\max}$. Therefore, there is only one pair (x, y), namely $x = c/d$ and $y = a/b$, that satisfies equation (5.2.2). Of course, this is an equilibrium solution.

Suppose then that $(x_0, y_0) \neq (c/d, a/b)$ and both x_0 and y_0 are positive. Then, we have $K < K_0$ and $z_{\min} < w_{\max}$. The equality $z(x) = w(y) = A$, say, for some x, y, is therefore possible only if $A \in [z_{\min}, w_{\max}]$. Choose $X_1 < c/d < X_2$ such that $z(X_1) = z(X_2) = w_{\max}$. It follows that the pairs $(X_i, a/b)$, $i = 1, 2$ satisfy equation (5.2.2). Similarly, choosing $Y_1 < a/b < Y_2$ such that $w(Y_1) = w(Y_2) = z_{\min}$, we see that the pairs $(c/d, Y_i)$, $i = 1, 2$ satisfy equation (5.2.2). Next, let $z_{\min} < A < w_{\max}$. Then, there exist x_1, x_2, and y_1, y_2, such that $z(x_1) = z(x_2) = A = w(y_1) = w(y_2)$; see Figure 5.1. It follows that the four pairs (x_i, y_j), $i, j = 1, 2$ satisfy equation (5.2.2). Note that all these four points (x_i, y_j), $i, j = 1, 2$ lie within the rectangle $[X_1, X_2] \times [Y_1, Y_2]$. Further, it is easy to see that as $A \to z_{\min}$, the points $(x_i, y_j) \to (c/d, Y_j)$ for $i, j = 1, 2$, respectively. Similarly, as $A \to w_{\max}$, the points $(x_i, y_j) \to (X_i, a/b)$ for $i, j = 1, 2$, respectively. These observations show that the pairs (x, y) satisfying equation (5.2.2) form a simple closed curve in the $x - y$ plane, lying in the compact rectangle $[X_1, X_2] \times [Y_1, Y_2]$. A qualitative picture of an orbit is depicted in Figure 5.2, where the arrows indicate the increasing values of t, which are determined by the signs of x' and y' using the given equations.

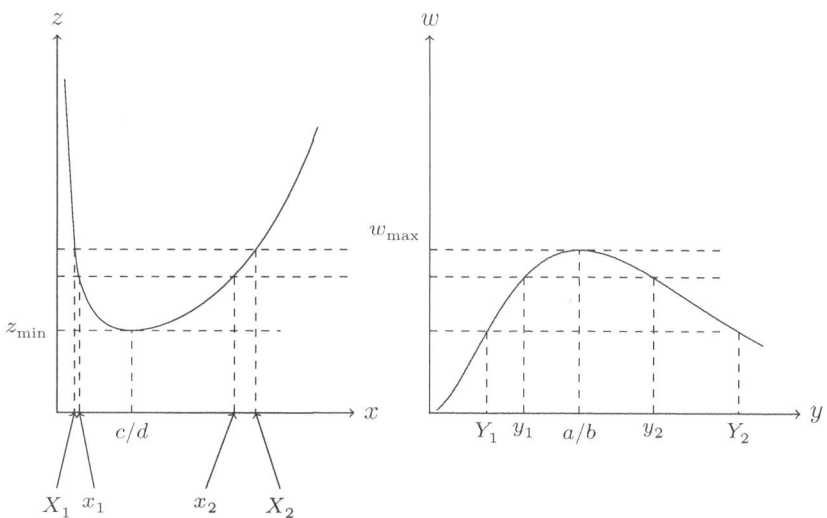

Figure 5.1 | Functions z and w in Exercise 5.2.

Orbits of the predator-prey model

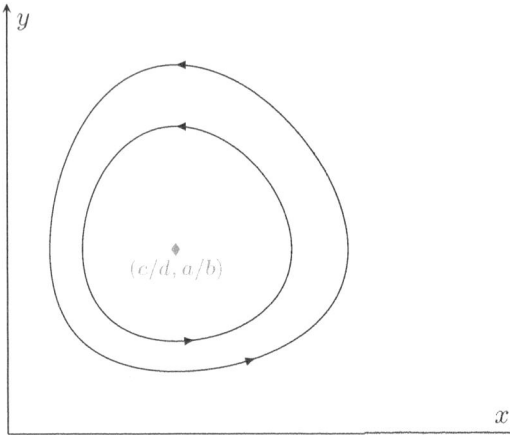

Figure 5.2 | Typical orbits in Exercise 5.2.

We conclude that for $(x_0, y_0) \neq (c/d, a/b)$ and both x_0, y_0 are positive, the solutions x and y are periodic, with period $T > 0$, say. It is possible to get some information about T by solving the system numerically; of course, T depends on x_0, y_0. However, we have the following interesting fact regarding the *averages* of x and y over a period. Consider the first equation $x' = ax - bxy = x(a - by)$. Dividing throughout by x and integrating over $[0, T]$, we have

$$\int_0^T \frac{x'}{x}\,dt = \int_0^T (a - by)\,dt = aT - b\int_0^T y(t)\,dt.$$

The integral on the left-hand side is zero because x is periodic. Therefore, $\dfrac{1}{T}\int_0^T y(t)\,dt = \dfrac{a}{b}$.

Similarly, $\dfrac{1}{T}\int_0^T x(t)\,dt = \dfrac{c}{d}$.

Exercise 5.3

Clearly $(0, 0)$ is an equilibrium point. If $y \equiv 0$, then $x' = x^2$. Thus, the positive x-axis is an orbit. Similarly, the positive y-axis is a union of orbits consisting of the equilibrium points $(0, (4n+1)\pi/2)$, $n = 0, 1, \ldots$ and the open intervals between them. As the given system is an autonomous system, this shows that the (open) first quadrant is an invariant set.

Exercise 5.4

Suppose λ and μ are the eigenvalues of $\mathbf{A}$. It is easy to check that λ and μ have negative real parts if and only if $a + d < 0$ and $ad - bc > 0$. The advantage we have here is that the sum, difference and product of λ and μ are easily expressed in terms of the matrix entities a, b, c and d.

The matrix $\mathbf{B}$ is given by

$$\mathbf{B} = \int_0^\infty \exp(t\mathbf{A}^T)\exp(t\mathbf{A})\,dt.$$

Since $\mathbf{A}$ is a 2×2 matrix, it is not difficult to see that

$$e^{\mathbf{A}} = \begin{cases} \dfrac{e^\lambda - e^\mu}{\lambda - \mu}\mathbf{A} + \dfrac{\lambda e^\mu - \mu e^\lambda}{\lambda - \mu}\mathbf{I} & \text{if } \lambda \neq \mu, \\ e^\lambda \mathbf{A} + (1-\lambda)e^\lambda \mathbf{I} & \text{if } \lambda = \mu. \end{cases}$$

Therefore,

$$e^{t\mathbf{A}} = \begin{cases} \dfrac{e^{t\lambda} - e^{t\mu}}{\lambda - \mu}\mathbf{A} + \dfrac{\lambda e^{t\mu} - \mu e^{t\lambda}}{\lambda - \mu}\mathbf{I} & \text{if } \lambda \neq \mu, \\ e^{t\lambda}\mathbf{A} + (1-t\lambda)e^{t\lambda}\mathbf{I} & \text{if } \lambda = \mu, \end{cases}$$

and a similar expression for $e^{t\mathbf{A}^T}$. A bit lengthy calculation shows that

$$e^{t\mathbf{A}^T}e^{t\mathbf{A}} = \frac{1}{(\lambda - \mu)^2}\Big[\big(e^{2t\lambda} + e^{2t\lambda} - e^{2t(\lambda+\mu)}\big)\mathbf{A}^T\mathbf{A}$$
$$+ \big((\lambda+\mu)e^{(\lambda+\mu)t} - \mu e^{2t\lambda} - \lambda e^{2t\mu}\big)(\mathbf{A}^T + \mathbf{A})$$
$$+ \big(\lambda^2 e^{2\mu t} + \mu e^{2\lambda t} - 2\lambda\mu e^{(\lambda+\mu)t}\big)\mathbf{I}\Big],$$

if $\lambda \neq \mu$; similar expression holds when $\lambda = \mu$. Using the fact that $\displaystyle\int_0^\infty e^{\eta t}\,dt = -1/\eta$ for any η with $\Re(\eta) < 0$, we can compute the integrals of all the terms in the above expression. Note that the terms involving λ and μ can all be expressed in terms of a, b, c, d. For example, we have

$$\lambda + \mu = a + d,\ \lambda\mu = ad - bc \text{ and } (\lambda - \mu)^2 = (a-d)^2 + 4bc.$$

We introduce some notations: $p = -(a+d)$, $q = ad - bc$ and $D = pq$. By our assumption on $\mathbf{A}$, both p and q are positive. Some more computations and simplifications lead to the following. If $\mathbf{B} = \begin{bmatrix} b_{11} & b_{12} \\ b_{12} & b_{22} \end{bmatrix}$, then we have

$$b_{11} = \frac{c^2 + d^2 + q}{D},\ b_{12} = -\frac{ac + bd}{D},\ b_{22} = \frac{a^2 + b^2 + q}{D}.$$

Denoting a (column) vector in $\mathbb{R}^2$ by $\mathbf{x} = \begin{bmatrix} x \\ y \end{bmatrix}$, the Liapunov function is given by

$$V(\mathbf{x}) = \mathbf{x}^T\mathbf{B}\mathbf{x} = b_{11}x^2 + 2b_{12}xy + b_{22}y^2,$$

where b_{ij} are as above. We remark that this expression coincides with the one mentioned in Ref.[52], though the topic of matrix exponential is not discussed there.

Exercise 5.5

We may assume that $c \geq 0$ by changing y to $-y$ if $c < 0$. Since the function $c + \sin y$ is differentiable with a bounded derivative, it follows that the solution of the IVP $y' = c + \sin y$, $y(0) = y_0$ exists for all t and is unique for any given $y_0 \in \mathbb{R}$. The limits $\lim_{t \to \pm\infty} y(t)$ depend on the values of c and y_0. Four distinct cases arise:

- **Case $c = 0$**: In this case, the equilibrium points of the ODE are $n\pi$, $n \in \mathbb{Z}$. If the initial condition $y(0) = y_0$ is not an equilibrium point, then we can find an integer m such that $y_0 = m\pi + x_0$ with $0 < x_0 < \pi$. With the substitution $y = m\pi + x$, the function x satisfies the IVP $x' = (-1)^m \sin x$, $x(0) = x_0$. It follows that the unique solution x satisfies the condition that $x(t) \in (0, \pi)$ for all $t \in \mathbb{R}$. Since $\sin x$ and $\tan(x/2)$ are positive in the interval $(0, \pi)$, an integration of this equation yields the following relation:

$$\tan(x/2) = \tan(x_0/2) \exp((-1)^m t) = \exp(\pm(t + \alpha)),$$

where $\pm = (-1)^m$ and $\alpha = \pm \log(\tan(x_0/2))$. From this, it follows that

$$x(t) = 2 \arctan\left[\exp(\pm(t + \alpha))\right].$$

Thus,

$$\lim_{t \to \infty} x(t) = \pi \text{ and } \lim_{t \to -\infty} x(t) = 0,$$

if m is even. The limits are interchanged if m is odd. The solution y is obtained by adding the constant $m\pi$ to x. Further, we see that the equilibrium points $(2m+1)\pi$ are asymptotically stable, whereas the other equilibrium points $2m\pi$ are unstable. See Figure 5.3.

Figure 5.3 | Orbits of Exercise 5.5: $c = 0$.

- **Case $0 < c < 1$**: Choose $\alpha_0 \in (0, \pi/2)$ such that $\sin \alpha_0 = c$. The equilibrium points are then given by $2n\pi - \alpha_0$ and $(2n+1)\pi + \alpha_0$ for $n \in \mathbb{Z}$. If the initial condition $y(0) = y_0$ is not an equilibrium point, we can choose an even integer $2k$ such that $y_0 - 2k\pi$ lies in $(-\alpha_0, \pi + \alpha_0)$ or $(\pi + \alpha_0, 2\pi - \alpha_0)$. The substitution $y = 2k\pi + x$ shows that $x' = c + \sin x$, $x(0) = x_0 \equiv y_0 - 2k\pi$, which lies in $(-\alpha_0, \pi + \alpha_0)$ or $(\pi + \alpha_0, 2\pi - \alpha_0)$.

Put $b = \sqrt{1 - c^2} > 0$. The relations

$$\alpha_0 = 2 \arctan\left(\frac{1-b}{c}\right) \text{ and } \pi - \alpha_0 = 2 \arctan\left(\frac{1+b}{c}\right)$$

are easily derived and will be used in what follows in verifying the initial condition and/or limiting values of the solution as $t \to \pm\infty$. First consider the case $x_0 \in (-\alpha_0, \pi + \alpha_0)$. It follows that the solution x is an increasing function and satisfies the condition $x(t) \in (-\alpha_0, \pi + \alpha_0)$ for all t. As such, it attains the value π at some t and makes the solution formula little more complicated.

Since $c + \sin x > 0$ in $(-\alpha_0, \pi + \alpha_0)$, it follows that

$$\int \frac{dx}{c + \sin x} = \frac{1}{b} \log \left[\frac{c \tan(x/2) + 1 - b}{c \tan(x/2) + 1 + b} \right].$$

Thus, we have

$$\frac{1}{b} \log \left[\frac{c \tan(x/2) + 1 - b}{c \tan(x/2) + 1 + b} \right] = t + \gamma,$$

where the constant of integration γ is given by

$$\gamma = \frac{1}{b} \log \left[\frac{c \tan(x_0/2) + 1 - b}{c \tan(x_0/2) + 1 + b} \right].$$

Note that $\gamma \leqslant 0$ and $= 0$ if $x_0 = \pi$. Performing some more algebra,[2] we obtain

$$c \tan(x/2) + 1 = -b \coth(b(t + \gamma)/2).$$

As the right-hand side term of this expression becomes infinite when $t = -\gamma$, care should be exercised in taking inverse to obtain the solution x. We have

$$x(t) = \begin{cases} -2 \arctan \left[c^{-1} (1 + b \coth(b(t + \gamma)/2)) \right] & \text{if } t < -\gamma, \\ \pi & \text{if } t = -\gamma, \\ 2\pi - 2 \arctan \left[c^{-1} (1 + b \coth(b(t + \gamma)/2)) \right] & \text{if } t > -\gamma. \end{cases}$$

We further observe that

$$\lim_{t \to -\infty} x(t) = -2 \arctan \left(\frac{1 - b}{c} \right) = -\alpha_0$$

and

$$\lim_{t \to \infty} x(t) = 2\pi - 2 \arctan \left(\frac{1 + b}{c} \right) = \pi + \alpha_0.$$

The expression for the solution y is obtained by adding $k\pi$ to the solution x.

---

[2] Let $a > 0$. If $|x| > a$, then the relation $\log \left(\frac{x-a}{x+a} \right) = y$ implies that $x = -a \coth(y/2)$. If $|x| < a$, then the relation $\log \left(\frac{a+x}{a-x} \right) = y$ implies that $x = a \tanh(y/2)$.

Next, consider the case $x_0 \in (\pi + \alpha_0, 2\pi - \alpha_0)$. In this case, the solution $x(t)$ lies in the interval $(\pi + \alpha_0, 2\pi - \alpha_0)$ for all $t \in \mathbb{R}$. As $c + \sin x < 0$ in $(\pi + \alpha_0, 2\pi - \alpha_0)$, the solution x is a decreasing function of t and we have

$$\int \frac{dx}{c + \sin x} = -\frac{1}{b} \log \left[\frac{b + (c\tan(x/2) + 1)}{b - (c\tan(x/2) + 1)} \right].$$

This in turn gives

$$c\tan(x/2) + 1 = -b\tanh(b(t+\gamma)/2),$$

where, now,

$$\gamma = -\frac{1}{b} \log \left[\frac{b + (c\tan(x_0/2) + 1)}{b - (c\tan(x_0/2) + 1)} \right].$$

As the term $\tanh(b(t+\gamma)/2)$ remains finite for all t, the solution x is given by

$$x(t) = 2\pi - 2\arctan\left[c^{-1}\left(1 + b\tanh(b(t+\gamma)/2)\right)\right].$$

We further observe that

$$\lim_{t \to -\infty} x(t) = 2\pi - 2\arctan\left(\frac{1-b}{c}\right) = 2\pi - \alpha_0$$

and

$$\lim_{t \to \infty} x(t) = 2\pi - 2\arctan\left(\frac{1+b}{c}\right) = \pi + \alpha_0,$$

as required. The expression for the solution y is obtained again by adding $2k\pi$ to x. We also conclude that the equilibrium points $(2k+1)\pi + \alpha_0$ are asymptotically stable and the rest are unstable. See Figure 5.4.

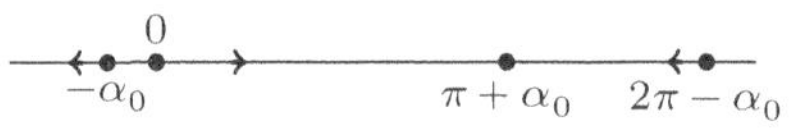

Figure 5.4 | Orbits of Exercise 5.5: $0 < c < 1$.

- **Case $c = 1$:** In this case, the equilibrium points are given by $(4k-1)\pi/2$ with $k \in \mathbb{Z}$. Since $1 + \sin y \geq 0$ for all y, any solution is non-decreasing and therefore all the equilibrium points are unstable. We may assume that the initial value $y_0 \in (-\pi/2, 3\pi/2)$, by using a simple translation, similar to the above two cases. We then conclude that the solution y exists for all t, which is unique and increasing, and

$$\lim_{t \to -\infty} y(t) = -\pi/2 \text{ and } \lim_{t \to \infty} y(t) = 3\pi/2.$$

Since $1 + \sin y > 0$ in $(-\pi/2, 3\pi/2)$, we have

$$\int \frac{dy}{1 + \sin y} = -2(1 + \tan(y/2))^{-1}.$$

Qualitative Theory | 161

Therefore, an integration of the given equation $y' = 1 + \sin y$ gives the relation

$$-2(1 + \tan(y/2))^{-1} = t + \gamma,$$

where $\gamma = -2(1 + \tan(y_0/2))^{-1}$. Note that $\gamma \leqslant 0$ and $= 0$ if $y_0 = \pi$. As t varies over $\mathbb{R}$, so does $t + \gamma$. So care should be exercised in obtaining the solution y by inverting the tan function. We have

$$y(t) = \begin{cases} -2\arctan\left(\dfrac{t+\gamma+2}{t+\gamma}\right) & \text{if } t < -\gamma, \\ \pi & \text{if } t = -\gamma, \\ 2\pi - 2\arctan\left(\dfrac{t+\gamma+2}{t+\gamma}\right) & \text{if } t > -\gamma. \end{cases}$$

See Figure 5.5.

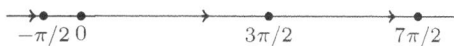

Figure 5.5 | Orbits of Exercise 5.5: $c = 1$.

- **Case $c > 1$:** This case is more technical and subtle. In this case, there are no equilibrium points. We first observe that the solution y exists uniquely for all t and is strictly increasing. Further, it satisfies the inequality

$$(c-1)t + y_0 \leqslant y(t) \leqslant (c+1)t + y_0,$$

for all $t \geqslant 0$, where $y_0 = y(0)$ and $\lim_{t \to \pm\infty} y(t) = \pm\infty$. Thus, every solution is unbounded and has linear growth. We now proceed to find an explicit expression for the solution.

Since a translation of y by $2m\pi$ is also a solution of the equation, we may assume that $y_0 \in [-\pi, \pi]$. As $c > 1$, an integration of the equation $y' = c + \sin y$ yields

$$\int \frac{dy}{c + \sin y} = \frac{2}{a} \arctan\left[a^{-1}(c\tan(y/2) + 1)\right] = t + \gamma,$$

where $a = \sqrt{c^2 - 1} > 0$ and γ is a constant of integration. A first look at this relation makes one feel that the main task of obtaining y as a function of t is not possible as the value of the middle term is confined to a finite interval for all y, whereas the last term assumes any real value as t varies over $\mathbb{R}$. However, a judicious procedure of integration combined with the qualitative properties of the solution y shows a way as follows.

Choose $t_0 < t_1$ such that $y(t_0) = -\pi$ and $y(t_1) = \pi$. Note that $t_0 \leqslant 0$ and $t_1 \geqslant 0$ by our choice of y_0. Integrating the equation $y' = c + \sin y$ over the interval $[t_0, t_1]$, we get

$$\int_{-\pi}^{\pi} \frac{dy}{c + \sin y} = \int_{t_0}^{t_1} dt = t_1 - t_0.$$

Evaluating the integral yields $t_1 - t_0 = 2\pi/a$. It follows that the function $y : [t_0, t_1] \to [-\pi, \pi]$ is a strictly increasing and bijective function. This enables us to obtain an expression for the solution $y(t)$ for $t \in [t_0, t_1]$, by an integration of $y' = c + \sin y$ over the interval $[t_0, t]$. Performing a few computations, we get

$$y(t) = -2 \arctan \left[\frac{a + \tan(a(t-t_0)/2)}{c \tan(a(t-t_0)/2)} \right], \tag{5.2.3}$$

for $t \in [t_0, t_1]$. We now use this information to obtain $y(t)$ for all t. Choose a sequence $t_0 < t_1 < \cdots < t_n < \cdots$ such that $y(t_n) = (2n-1)\pi$ for $n = 0, 1, 2, \ldots$. Now integrate $y' = c + \sin y$ over the interval $[t_n, t_{n+1}]$ to obtain

$$\int_{(2n-1)\pi}^{(2n+1)\pi} \frac{dy}{c + \sin y} = \int_{t_n}^{t_{n+1}} dt = t_{n+1} - t_n.$$

But

$$\int_{(2n-1)\pi}^{(2n+1)\pi} \frac{dy}{c + \sin y} = \int_{-\pi}^{\pi} \frac{dy}{c + \sin y} = 2\pi/a.$$

Therefore, $t_{n+1} - t_n = 2\pi/a$, and hence $t_n = t_0 + 2n\pi/a$ for $n = 1, 2, \ldots$.

Now let $t > t_0$; the arguments are similar for $t < t_0$. Choose $n \in \mathbb{N}$ so that $t \in [t_n, t_{n+1})$. Now integrate $y' = c + \sin y$ over the interval $[t_n, t]$ to obtain

$$\int_{(2n-1)\pi}^{y^*} \frac{dy}{c + \sin y} = \int_{t_n}^{t} dt = t - t_n,$$

where $y^* = y(t)$. But, we have

$$\int_{(2n-1)\pi}^{y^*} \frac{dy}{c + \sin y} = \int_{-\pi}^{y^* - 2n\pi} \frac{dy}{c + \sin y}.$$

As $y^* - 2n\pi \in [-\pi, \pi)$, there is a $\tau \in [t_0, t_1)$ such that

$$\int_{-\pi}^{y^* - 2n\pi} \frac{dy}{c + \sin y} = \int_{t_0}^{\tau} dt = \tau - t_0,$$

and $y^* - 2n\pi = y(\tau)$. Comparison of the above integrals also gives the relation $t - t_n = \tau - t_0$ or $\tau = t - \dfrac{2n\pi}{a}$. Thus, we obtain $y(t) = y^* = 2n\pi + y(t - (2n\pi)/a)$, where the term $y(t - (2n\pi)/a)$ can be evaluated using equation (5.2.3). Now, we can write $2n\pi = a((2n\pi)/a) = at - a(t - (2n\pi)/a)$. Therefore, $y(t) = at + p(t - (2n\pi)/a)$, where $p(t) = y(t) - at$. It is not difficult to see that the function p is a periodic function of period $2\pi/a$. Thus, the solution y is expressed as the sum of a linear term and a periodic function. The linear term has the slope $a = \sqrt{c^2 - 1}$, which, incidentally, is the geometric mean of $c + 1$ and $c - 1$.

Exercise 5.6

Suppose x is a solution of equation (5.2.1) in an interval $|t| \leqslant \alpha$ for some $\alpha > 0$. Then, $(1/2)(x'(t))^2 + U(x(t)) = E$ for all $t \in [-\alpha, \alpha]$, where E is a constant; we refer to this as *energy level E*. Since, by assumption, $U \geqslant 0$, we have $E \geqslant 0$; $E = 0$ only if x is a constant function c, where $U(c) = 0$. Further, it follows that $|x'(t)| \leqslant \sqrt{2E}$. Using $x(t) - x(0) = \int_0^t x'(s)\,ds$, $t > 0$, we then conclude that

$$|x'(t)| \leqslant \sqrt{2E} \text{ and } |x(t) - x(0)| \leqslant \sqrt{2E}\,|t| \qquad (5.2.4)$$

for $t \in [-\alpha, \alpha]$. Fix a $T > 0$ and R be the rectangle $\{|x - x(0)| \leqslant \sqrt{2E}\,T,\ |x'| \leqslant \sqrt{2E}\}$ in the $x - x'$ phase plane. If an orbit of equation (5.2.1) starts in the interior of the rectangle R, the estimates (5.2.4) show that the orbit can leave the rectangle R only through the sides $|x - x(0)| = \sqrt{2E}\,|T|$. Thus, the solution can be extended up to $t = \pm T$. Since T is arbitrary, it follows that the solution can be extended indefinitely.

Remark: Since adding a constant to the potential function does not change the equation, it suffices to assume that U is bounded below. We have the following counter example. Let $U(x) = -x^4/4$, which is not bounded below. A solution of the corresponding equation, namely $x'' - x^3 = 0$ is given by $x(t) = \sqrt{2}\,(t - 1)^{-1}$, which does not exist for all t.

Exercise 5.7

Let $x(0) = a$ and $x'(0) = b$ be the initial conditions. Then, $(1/2)b^2 - a^2/2 + a^4/4 = 0$, as we are assuming $E = 0$. We have $(1/2)(x')^2 - x^2/2 + x^4/4 = 0$, which in turn implies that $|x| \leqslant \sqrt{2}$. Thus, the solution is bounded. Before proceeding with an integration of this first-order equation, some qualitative analysis will be helpful. For a solution x, if there is a t_0 such that $x(t_0) = 0$, then $x'(t_0) = 0$ as well. Thus, by uniqueness, $x \equiv 0$. Therefore, a solution with $a > 0$ (respectively $a < 0$) will be positive (respectively negative) for all t. An integration[3] of the equation $(1/2)(x')^2 - x^2/2 + x^4/4 = 0$ gives

$$\frac{1}{\sqrt{2}} \log\left[\frac{|x|}{\sqrt{2} + \sqrt{2 - x^2}}\right] = \pm \frac{1}{\sqrt{2}}(t + \alpha),$$

---

[3] For $k > 0$, $\int \frac{dx}{x\sqrt{k^2 - x^2}} = \frac{1}{k} \log\left(\frac{|x|}{k + \sqrt{k^2 - x^2}}\right)$.

where α is a constant of integration. If $a > 0$, then $|x| = x$ and if $a < 0$ then $|x| = -x$. Hence, taking exponential on both sides of the above equation (it does not matter which sign we pick up on the right-hand side) and doing some algebraic manipulations, we obtain the solution as

$$x(t) = \pm \frac{2\sqrt{2}}{e^{(t+\alpha)} + e^{-(t+\alpha)}} = \pm\sqrt{2}\,\text{sech}(t+\alpha),$$

where the sign to be chosen is that of a and α is a constant of integration, which can be determined in terms of a and b. In particular, if $b = 0$, which implies that $a = \pm\sqrt{2}$, we have $\alpha = 0$ and the solution is given by $x(t) = \pm\sqrt{2}\,\text{sech}\,t$. Observe that $\lim_{t\to\pm\infty} x(t) = 0$. If we plot the points $(x(t), x'(t))$ for all t in the $x - x'$ phase plane, we obtain two curves, one in the $x > 0$ region and another in the $x < 0$ region. These curves do not depend on α and are called the *homoclinic orbits* of the Duffing's equation at the zero level: $E = 0$. See Figure 5.6(a). Typical solution curves are depicted in Figure 5.6(b).

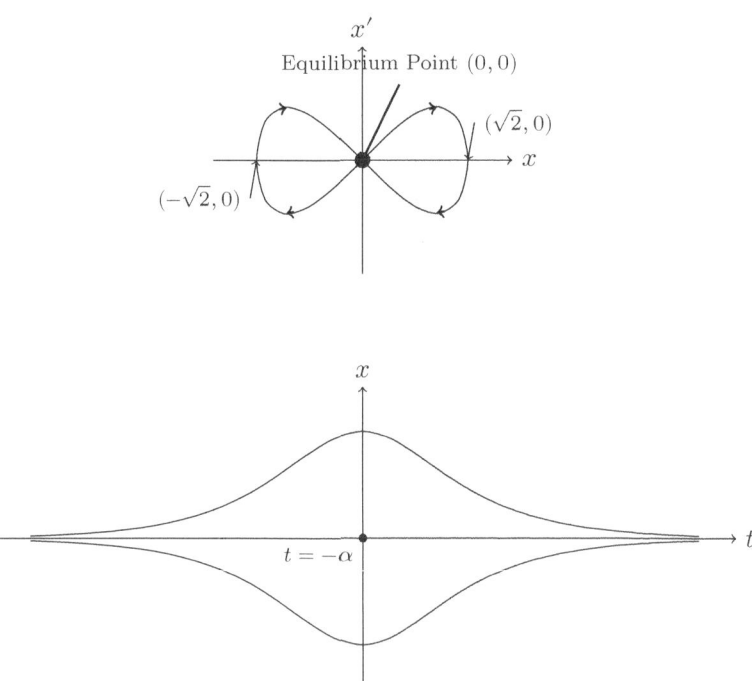

Figure 5.6 | Homoclinic orbits and solution curves of Exercise 5.7.

Exercise 5.8

(1) For the given equation $x'' + e^x = 0$, we take the potential function U as $U(x) = e^x$, which is bounded below. Hence, by Exercise 5.6, any solution of this equation is defined for all t. In the present case, the conservation equation is $\frac{1}{2}(x')^2 + e^x = k$, where the

constant $k = b^2/2 + e^a > 0$, with $x(0) = a$, $x'(0) = b$. Hence $x(t) \leq \log k$ for all t. An integration of the conservation equation gives

$$\frac{1}{\sqrt{k}} \log\left(\frac{\sqrt{k} - \sqrt{k - e^x}}{\sqrt{k} + \sqrt{k - e^x}}\right) = \pm\sqrt{2}(t + \alpha),$$

where α is a constant of integration. After taking exponential on both sides of this equation and simplifying, we obtain

$$x(t) = \log k - 2\log(\cosh(\sqrt{2k}\,(t+\alpha)/2)).$$

The constant of integration α gets uniquely determined in terms of a and b. Observe that $\lim_{t \to \pm\infty} x(t) = -\infty$, so the solution is not bounded below. The above procedure equally applies to the equation $x'' + \lambda e^x = 0$, where $\lambda > 0$.

(2) Here the potential function is given by $U(x) = -e^x$ which is not bounded below. As such, a solution may not be defined for all t. The conservation equation is $(1/2)(x')^2 - e^x = k$, where the constant $k = b^2/2 - e^a$, with $x(0) = a$, $x'(0) = b$. As k can be any real number, the integration of the conservation equation yields different expressions depending on the sign of k. We have

$$\frac{1}{\sqrt{k}} \log\left(\frac{\sqrt{k + e^x} - \sqrt{k}}{\sqrt{k + e^x} + \sqrt{k}}\right) = \pm\sqrt{2}(t+\alpha), \quad \text{if } k > 0,$$

$$-2e^{-x/2} = \pm\sqrt{2}(t+\alpha), \quad \text{if } k = 0,$$

$$\frac{2}{\sqrt{-k}} \arctan\left(\frac{\sqrt{k+e^x}}{\sqrt{-k}}\right) = \pm\sqrt{2}(t+\alpha), \quad \text{if } k < 0,$$

where α is a constant of integration. Several simplifications finally give the solution as

$$x(t) = \begin{cases} \log k - 2\log|\sinh(\sqrt{k}\,(t+\alpha)/\sqrt{2})| & \text{if } k > 0, \\ -2\log((t + \sqrt{2}e^{-a/2})/\sqrt{2}) & \text{if } k = 0, \\ \log(-k) - 2\log|\cos(\sqrt{-k}\,(t+\alpha)/\sqrt{2})| & \text{if } k < 0. \end{cases}$$

We notice that none of the above solutions is defined for all t. The constant of integration gets uniquely determined in terms of a and b.

Exercise 5.9

We take the potential function as $U(x) = x - x^3/3$ and the conserved quantity is $(x')^2 + U(x) = E$. Observe that U has three real roots $0, \pm\sqrt{3}$; it has a (local) maximum at $x = 1$, with maximum value $2/3$, and a minimum at $x = -1$, with minimum value $-2/3$. The points $(\pm 1, 0)$ are the only equilibrium points of the equation. By the linear stability analysis, it is not difficult to see that the equilibrium point $(-1, 0)$ is stable, but not asymptotically stable, whereas, the equilibrium point $(1, 0)$ is unstable. We need to consider the constant E in different regions in order to draw the orbits. For this purpose, first let us understand the potential function by drawing its graph. Looking at the graph of the potential function U as depicted in Figure 5.7(a), we consider the following energy levels E:

Case $E < -2/3$: In this case, we have $x' = \pm\sqrt{E - U(x)}$ and the solution x is restricted to the interval $x \geqslant x_*$, where $U(x_*) = E$. Observe that $x_* > 2$. A typical orbit in this case is depicted in Figure 5.7(b).

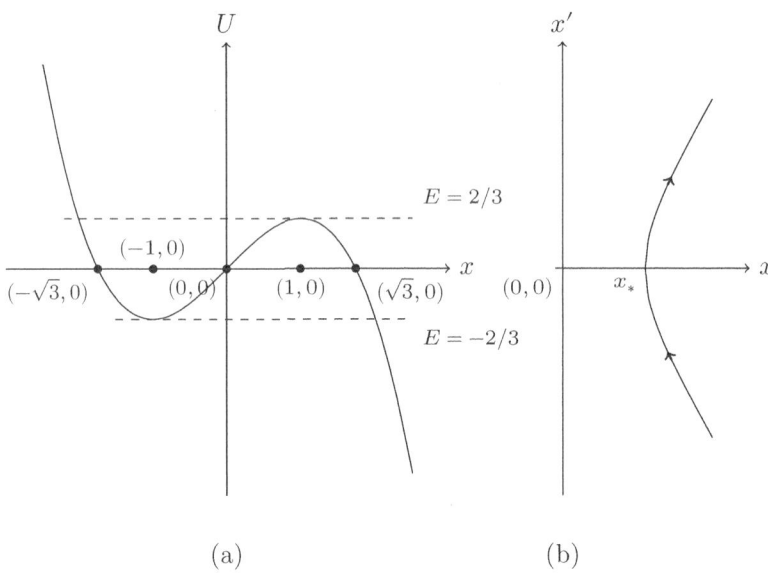

Figure 5.7 | Potential function (a) and an orbit for $E < -2/3$ (b) in Exercise 5.9.

Case $E = -2/3$: In this case, we have $x' = \pm\sqrt{E - U(x)}$, and the solution x is restricted to the interval $x \geqslant 2$, as $U(2) = -2/3$. The orbit in this case is exactly similar to case $E < -2/3$, except that we have in addition the trivial orbit at the equilibrium point $(-1, 0)$. It is possible to write down the solution in explicit form in the region $x \geqslant 2$. As the procedure is the same as in Exercises 5.7 and 5.8, we simply write down the solution: $x(t) = 3\sec^2((t+\alpha)/2) - 1$, where α is a constant of integration, which is uniquely determined in terms of $x(0) = a \geqslant 2$ and $x'(0) = b$. Note that the solution is not for all t.

Case $-2/3 < E < 2/3$: In this case, there are three points x_1, x_2 and x_3 such that $U(x_i) = E$ and $-2 < x_1 < -1 < x_2 < 1 < x_3$. The orbit confined to the region $x \geqslant x_3$ is similar to the case of $E < -2/3$, and the orbit confined to the region $x_1 \leqslant x \leqslant x_2$ is a periodic orbit surrounding the equilibrium point $(-1, 0)$. A typical orbit is depicted in Figure 5.8(a).

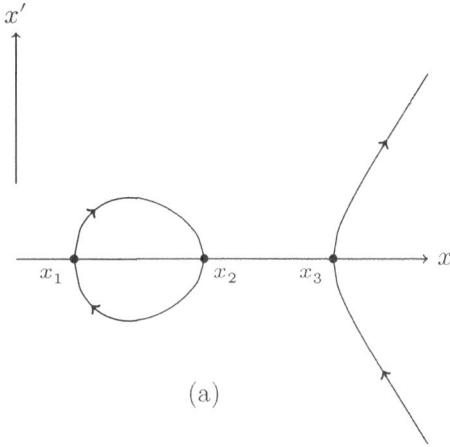

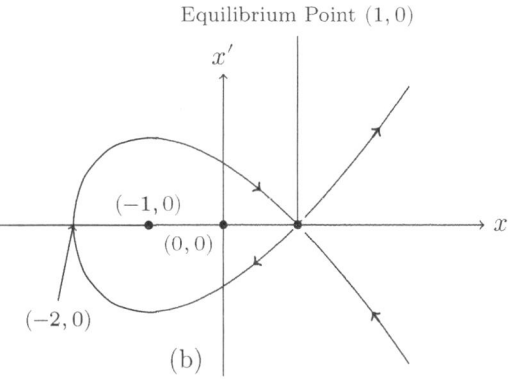

Figure 5.8 | Typical orbits in Exercise 5.9: (a) $-2/3 < E < 2/3$; (b) $E = 2/3$

Case $E = 2/3$: In this case, we have $U(-2) = U(1) = 2/3$, and the $(1, 0)$ is an equilibrium point. The orbit confined to the region $x > 2$ has two branches, one in the region $x' > 0$ and another in $x' < 0$. On the other hand, the orbit confined to the region $-2 \leqslant x < 1$ is a homoclinic orbit approaching the equilibrium point $(1, 0)$ as $t \to \pm\infty$. In this case, it is also possible to obtain the solutions in explicit forms, and the procedure is the same as in Exercises 5.7 and 5.8. We simply write down the solutions: For $x(0) = a \in [-2, 1)$ and $x'(0) = b$, the solution is given by

$$x(t) = 1 - 3\text{sech}^2((t+\alpha)/2),$$

where α is a constant of integration, which is uniquely determined in terms of a and b. For $x(0) = a > 1$ and $x'(0) = b \neq 0$ (implied by the energy relation), the solution is given by

$$x(t) = 1 + \frac{3}{\sinh^2((t+\alpha)/2)},$$

where α is a constant of integration, which is uniquely determined in terms of a and b. Note that in the latter case, the solution is not defined for all t. The orbits are schematically shown in Figure 5.8(b).

Case $E > 2/3$: There is a unique $x_* < -2$ such that $U(x_*) = E$ and the solution is confined to the region $x \geqslant x_*$. The orbit is very similar to the one when $E < -2/3$.

Exercise 5.10

Writing the given equation as a first-order system of two equations, we see that all the points on the x-axis in the $x - x'$ phase plane are equilibrium points and are *non-isolated*. It follows that any solution x with $x'(0) > 0$ (respectively $x'(0) < 0$) satisfies $x'(t) > 0$ (respectively $x'(t) < 0$) for all t in the interval of existence. Thus, any non-constant solution is either strictly increasing or strictly decreasing.

The given equation can be written as $-\frac{d}{dt}(x') + \frac{1}{2}\frac{d}{dt}(x^2) = 0$. An integration then yields $-x' + x^2/2 = c$, where c is a constant, which is determined by the initial conditions. This first-order equation in the variable separable form can readily be integrated. However, the following qualitative behaviour of the solution helps to integrate in the correct way.

First, consider the case of $x'(0) > 0$. Then, the solution is strictly increasing. We observe that the constant c can be any real number, and the following three alternatives hold:

- $c = -a^2/2$, $a > 0$ and we need to integrate the equation $2x' = x^2 + a^2$. Therefore, we get $\frac{1}{a}\arctan\left(\frac{x}{a}\right) = \frac{1}{2}(t + k)$, where k is a constant of integration.

- $c = 0$, but $x(0) \neq 0$, as $x'(0) > 0$. The equation to be integrated is $2x' = x^2$ and we get $-x^{-1} = \frac{1}{2}(t+k)$, where k is a constant of integration and $k \neq 0$.

- $c = a^2/2$, $a > 0$, and we need to integrate the equation $2x' = x^2 - a^2$. As $x' > 0$, it follows that the solution satisfies $|x| > a$. Thus, an integration yields $\frac{1}{2a}\log\left(\frac{x-a}{x+a}\right) = \frac{1}{2}(t+k)$, where k is a constant of integration.

After some simplifications in each of these alternatives, we get the following expressions for the solution:
$$x(t) = \begin{cases} a\tan(a(t+k)/2) & \text{if } c = -a^2/2, a > 0, \\ -2(t+k)^{-1} & (k \neq 0) \text{ if } c = 0, \\ -a\coth(a(t+k)/2) & \text{if } c = a^2/2, a > 0. \end{cases}$$

Note that in each case the solution is not defined for all t.

Now consider the case when $x'(0) < 0$. Then $x'(t) < 0$ for all t, and the solution is strictly decreasing. In this case, the constant c in the first-order equation $-x' + x^2/2 = c$ should be positive, and we put $c = a^2/2$ with $a > 0$. We need to now integrate the equation $2x' = x^2 - a^2$. It follows that the solution satisfies $|x| < a$. An integration and some computations as above, the solution is given by $x(t) = -a \tanh(a(t + k)/2)$. Hence the solution is defined for all $t \in \mathbb{R}$ and we have $\lim_{t \to \pm\infty} x(t) = \mp a$. In all the cases, the constant of integration is uniquely determined in terms of $x(0)$ and $x'(0)$.

Exercise 5.11

For the travelling wave solution, we have $u_\xi = x'$, $u_{\xi\xi} = x''$ and $u_\tau = -cx'$, where $'$ denotes differentiation with respect to the t variable. Assuming $\varepsilon = 1$ and substituting these expressions into the viscous Burgers' equation, we get $-cx' + xx' = x''$. Replacing x by $x + c$ yields $xx' = x''$, which is precisely the equation in Exercise 5.10.

To answer the second question, note that the requirement on the solution is that it exists for all t. The analysis in Exercise 5.10 implies that this is possible only if $x'(0) < 0$ and the solution is strictly decreasing. This further leads to the condition $a > 0$, and the required solution is given by $x(t) = -a \tanh(at/2)$, where the constant of integration is so chosen that $x(0) = 0$.

Exercise 5.12

Assume the given system is a Hamiltonian system. Then, there is C^2 function $H = H(\mathbf{x}, \mathbf{y})$ such that $\dfrac{\partial H}{\partial \mathbf{y}} = \mathbf{Ax} + \mathbf{By}$ and $\dfrac{\partial H}{\partial \mathbf{x}} = -\mathbf{Cx} - \mathbf{Dy}$. Using the fact that the mixed partial derivatives of the second order of a C^2 function are pairwise equal, we infer that the matrices $\mathbf{B}$ and $\mathbf{C}$ are symmetric and $\mathbf{D} = -\mathbf{A}^T$.

Conversely, suppose the matrices $\mathbf{B}$ and $\mathbf{C}$ are symmetric and $\mathbf{D} = -\mathbf{A}^T$. Consider the function H defined by

$$H(\mathbf{x}, \mathbf{y}) = (\mathbf{Ax}) \cdot \mathbf{y} + (1/2)(\mathbf{By}) \cdot \mathbf{y} - (1/2)(\mathbf{Cx}) \cdot \mathbf{x}$$
$$= \mathbf{y}^T \mathbf{Ax} + (1/2)\mathbf{y}^T \mathbf{By} - (1/2)\mathbf{x}^T \mathbf{Cx}.$$

A straight-forward computation[4] shows that

$$\frac{\partial H}{\partial \mathbf{y}} = \mathbf{Ax} + \mathbf{By} \text{ and } \frac{\partial H}{\partial \mathbf{x}} = -\mathbf{Cx} - \mathbf{Dy}.$$

Thus, H is a required Hamiltonian.

---

[4] If $\mathbf{E}$ is a square matrix of order n, then

$$\frac{\partial}{\partial \mathbf{x}}(\mathbf{y}^T \mathbf{Ex}) = \mathbf{E}^T \mathbf{y} \text{ and } \frac{\partial}{\partial \mathbf{y}}(\mathbf{y}^T \mathbf{Ex}) = \mathbf{Ex}.$$

Exercise 5.13

By chain rule, we have $\dfrac{d\mathbf{v}}{dt} = \left[\dfrac{\partial \mathbf{f}}{\partial \mathbf{u}}\right] \dfrac{d\mathbf{u}}{dt}$. Therefore,

$$\dfrac{d\mathbf{v}}{dt} = \left[\dfrac{\partial \mathbf{f}}{\partial \mathbf{u}}\right] \dfrac{d\mathbf{u}}{dt} = \left[\dfrac{\partial \mathbf{f}}{\partial \mathbf{u}}\right] JH_\mathbf{u} = J\left[\dfrac{\partial \mathbf{f}}{\partial \mathbf{u}}\right]^{-T} H_\mathbf{u},$$

where we have used the symplectic property of $\left[\dfrac{\partial \mathbf{f}}{\partial \mathbf{u}}\right]$ and $\left[\dfrac{\partial \mathbf{f}}{\partial \mathbf{u}}\right]^{-T}$ denotes the inverse of the matrix $\left[\dfrac{\partial \mathbf{f}}{\partial \mathbf{u}}\right]^{T}$. Let $K(\mathbf{v}) = H(\mathbf{u}) = H(\mathbf{f}^{-1}(\mathbf{v}))$. Using the chain rule once more, we get $K_\mathbf{v} = \left[\dfrac{\partial \mathbf{f}}{\partial \mathbf{u}}\right]^{-T} H_\mathbf{u}$. Therefore, $\mathbf{v}$ satisfies the Hamiltonian system $\dfrac{d\mathbf{v}}{dt} = K_\mathbf{v}$. This completes the proof. ∎

Exercise 5.14

(1) The given system has only one equilibrium point at $(0, 0)$. In the polar co-ordinates $x = r\cos\theta$ and $y = r\sin\theta$, the system in (1) reduces to $rr' = xx' + yy' = x^2 + y^2 - (x^2 + y^2)\exp(x^2 + y^2 - 4)$, or $r' = r(1 - \exp(r^2 - 4))$; and $\theta' = r^{-2}(xy' - yx')$, or $\theta' = -1$. This immediately implies that $\theta(t) = a - t$, where a is a constant.

Next, looking at the equation satisfied by r, we observe that r is decreasing in the region $r > 2$ and increasing in the region $r < 2$. Thus, if an orbit of this system either starts or enters the annulus $\{3/2 \leqslant r \leqslant 5/2\}$, then that orbit remains in this annulus for all future times. Since this annulus is compact and does not contain any equilibrium point of the given system, the Poincaré–Bendixson theorem asserts the existence of a periodic orbit situated within this annulus. A closer look at the equation satisfied by r, we see that $r = 2$ is a constant solution of this equation. Thus, $x(t) = 2\cos(a - t)$ and $y(t) = 2\sin(a - t)$ is a periodic orbit of the given system. This is a circle of radius 2, traversed in the counterclockwise direction for increasing values of t.

(2) Again, introducing polar co-ordinates as in (1), the given system transforms into:

$$r' = r(2 - r^2), \quad \theta' = -3.$$

This immediately gives $\theta(t) = a - 3t$. The equation satisfied by r shows that $r = \sqrt{2}$ is a constant solution. Thus, $x(t) = \sqrt{2}\cos(a - 3t)$ and $y(t) = \sqrt{2}\sin(a - 3t)$ is a periodic orbit of the given system.

Since we can integrate the equation satisfied by r explicitly, we can say more. Put $r_0 = r(0)$. Then, if $r_0 < \sqrt{2}$, we have $r(t) = \dfrac{\sqrt{2}\,c\,e^{2t}}{\sqrt{1 + c^2 e^{4t}}}$, where the positive constant c is expressed in terms of r_0. It follows that $r(t) < \sqrt{2}$ for all $t > 0$ and $r(t) \to \sqrt{2}$ as $t \to \infty$. Thus, any orbit that starts in the region $r < \sqrt{2}$ remains there for all future times and *spirals* towards the circle $r = \sqrt{2}$ as $t \to \infty$. Similarly, if $r_0 > \sqrt{2}$, we have

$r(t) = \dfrac{\sqrt{2}\, e^{2t}}{\sqrt{e^{4t} - c^2}}$, where the positive constant c is expressed in terms of r_0. Again, we see that $r(t) > \sqrt{2}$ for all $t > 0$ and $r(t) \to \sqrt{2}$ as $t \to \infty$. Therefore, any orbit that starts in the region $r > \sqrt{2}$ remains there for all future times and *spirals* towards the circle $r = \sqrt{2}$ as $t \to \infty$. In such a scenario, the periodic orbit given by $x(t) = \sqrt{2}\cos(a - 3t)$ and $y(t) = \sqrt{2}\sin(a - 3t)$ is termed as a *limit cycle*. A schematic depiction of this situation is shown Figure 5.9.

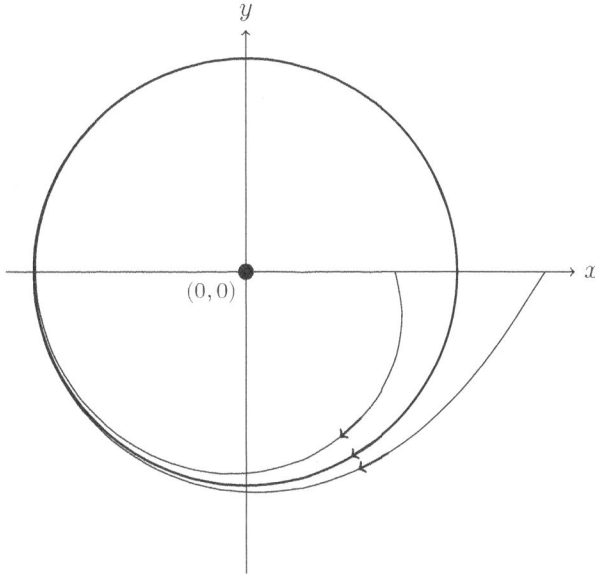

Figure 5.9 | Limit cycle in Exercise 5.14.

Exercise 5.15

First write the given equation as a first-order system by introducing the variable $y = x'$: $x' = y$, $y' = -x + y(1 - x^2 - 2y^2)$. Clearly, the origin $(0,0)$ is the only equilibrium point of this system. Using the polar co-ordinates $x = r\cos\theta$ and $y = r\sin\theta$, we have $r^2 = x^2 + y^2$ and therefore

$$r' = (x/r)x' + (y/r)y' = (x/r)y + (y/r)(-x + y(1 - x^2 - 2y^2))$$
$$= (y^2/r)(1 - x^2 - 2y^2).$$

Observe that $1 - x^2 - 2y^2 = 1 - r^2 - y^2 < 0$ for $r > 1$ and $1 - x^2 - 2y^2 \geqslant 1 - 2r^2 > 0$ for $r < 1/\sqrt{2}$. Hence r is decreasing in the region $r > 1$ and increasing in the region $r < 1/\sqrt{2}$. This implies that if an orbit starts in the annulus $\{1/\sqrt{2} \leqslant r \leqslant 1\}$ at time $t = t_0$, it will remain there for all $t \geqslant t_0$. Since this annulus is compact and does not contain the equilibrium point, the Poincarè–Bendixson theorem guarantees the existence of a periodic orbit in this annulus.

Exercise 5.16

The linearized system $x' = x$, $y' = -y$ is decoupled and we see that both the axes are invariant. For the equilibrium point $(0,0)$, the y-axis is a stable manifold, denoted by $W^s(0,0)$. This means that if an orbit starts on the y-axis, it remains there for all t and tends to $(0,0)$ as $t \to \infty$. On the other hand, though x-axis is also invariant, it is an unstable manifold, denoted by $W^u(0,0)$. Let us analyse the situation under a non-linear perturbation.

The y-axis is still invariant and is a stable manifold. The x-axis is no longer invariant. A straight-forward computation shows that

$$\frac{d}{dt}[x(y - x^2/3)] = x(y - x^2/3) + x(-y + x^2 - (2/3)x^2) = 0,$$

proving (1). Showing the validity of (2) is also simple. It further shows that the parabola $y = x^2/3$ is invariant. It is also easily checked that this parabola is an unstable manifold for the given non-linear system. Therefore, we observe that the unstable manifold of the linear system gets deformed for the non-linear system, whereas, the stable manifold remains unchanged.

All the observations can also be seen from the explicit expressions for the solution of the non-linear system, which are given by

$$x(t) = x(0)e^t, \quad y(t) = e^{-t}[y(0) + (1/3)(x(0))^2(e^{3t} - 1)].$$

Exercise 5.17

Put $y = x'$. Then the given equation can be written as a first-order system $\mathbf{x}' = \mathbf{A}\mathbf{x}$, where

$$\mathbf{x} = \begin{bmatrix} x \\ y \end{bmatrix} \quad \text{and} \quad \mathbf{A} = \begin{bmatrix} 0 & 1 \\ -k/m & -c/m \end{bmatrix}.$$

It is easily checked that the origin $(0,0)$ is the only equilibrium point of this system. The eigenvalues of the matrix $\mathbf{A}$ are given by $-(c/2m) \pm \sqrt{(c/2m)^2 - (k/m)}$. If $c = 0$, the eigenvalues are purely imaginary, and hence $(0,0)$ is stable but not asymptotically stable. In this case, all the orbits in the $x - y$ plane are periodic. On the other hand, if $c > 0$, the eigenvalues are either real and negative or have negative real parts. Thus, $(0,0)$ is asymptotically stable in this case. The orbits spiral around the origin and tend to 0 as $t \to \infty$.

The same conclusion may also be arrived at using the method of Liapunov. Let $V(x,y) = (1/2)(my^2 + kx^2)$, which is the sum of the kinetic energy and the potential energy. Observe that $V \geqslant 0$, $V(0,0) = 0$, and $V(x,y) > 0$ for all $(x,y) \neq (0,0)$. Further, along any trajectory $(x(t), y(t))$ of this system, we have

$$\frac{d}{dt}[V(x(t), y(t))] = \frac{\partial V}{\partial x}x' + \frac{\partial V}{\partial y}y'$$

$$= \nabla V \cdot (\mathbf{A}\mathbf{x}) = \begin{bmatrix} kx & my \end{bmatrix} \begin{bmatrix} y \\ -(1/m)(kx + cy) \end{bmatrix}$$

$$= kxy - kxy - cy^2 = -cy^2 \leqslant 0.$$

Thus, the energy functional V is indeed serves as a Liapunov function, and we conclude that $(0,0)$ is stable if $c = 0$ (in which case the orbits are described by the curves $V = $ constant). However, to deduce the asymptotically stability of $(0,0)$ when $c > 0$, we need to modify the function V, which is done in the following Exercise 5.18.

Exercise 5.18

Introducing the variable $y = x'$, the given equation can be written as a first-order non-linear system:

$$\begin{bmatrix} x' \\ y' \end{bmatrix} = \begin{bmatrix} y \\ -(c/m)y - (k/m)\sin x \end{bmatrix}.$$

It is easy to see that the equilibrium points of this system are given by $(n\pi, 0)$ for $n \in \mathbb{Z}$. It is not difficult to write down the linearized system around an equilibrium point $(n\pi, 0)$: It is given by $\mathbf{Z}' = \mathbf{A}\mathbf{Z}$, where

$$\mathbf{Z} = \begin{bmatrix} X \\ Y \end{bmatrix} \quad \text{and} \quad \mathbf{A} = \begin{bmatrix} 0 & 1 \\ (-1)^{n+1}(k/m) & -c/m \end{bmatrix},$$

with $X = x - n\pi$ and $Y = y$. The eigenvalues of the matrix $\mathbf{A}$ are given by $-(c/2m) \pm \sqrt{(c/2m)^2 - (-1)^n(k/m)}$. Thus, if n is an odd integer, there is a positive eigenvalue, implying that the equilibrium point $(n\pi, 0)$ with n an odd integer is unstable. By Hartman–Grobman theorem, this instability persists for the non-linear system as well.

On the other hand, if n is an even integer, the eigenvalues are purely imaginary or have negative real parts according to $c = 0$ or $c > 0$, respectively. Thus, the equilibrium points $(n\pi, 0)$ with n an even integer are stable or asymptotically stable. For $c > 0$, we use the method of Liapunov to show that the equilibrium point $(n\pi, 0)$ with n an even integer is asymptotically stable for the non-linear system as well.

Assume $n = 0$; the case of other equilibrium points is similar. We modify the Liapunov function used in Exercise 5.17 to take advantage of the condition $c > 0$. Let $V(x,y) = (1/2)[(k^2 + c^2 + km)x^2 + 2cmxy + m(m+k)y^2]$. It is straight forward to check that $V(0,0) = 0$ and $V(x,y) > 0$ for all $(x,y) \neq (0,0)$. If $(x(t), y(t))$ is any trajectory of the linear system $x' = y$, $y' = -(k/m)x - (c/m)y$, we have

$$\frac{d}{dt}[V(x(t), y(t))] = \frac{\partial V}{\partial x}x' + \frac{\partial V}{\partial y}y'$$
$$= [(k^2 + c^2 + km)x + cmy]y +$$
$$[cmx + m(m+k)y][-(k/m)x - (c/m)y]$$
$$= -ck(x^2 + y^2) < 0 \quad \text{for all } (x,y) \neq (0,0).$$

This shows that V is indeed a Liapunov function for the linear system and $(0,0)$ is asymptotically stable. We now show that V is also a Liapunov function for the non-linear system $x' = y$, $y' = -(k/m)\sin x - (c/m)y$ as well, thereby proving the asymptotic stability of $(0,0)$. Along any trajectory $(x(t), y(t))$ of the non-linear system, we have

$$\frac{d}{dt}[V(x(t), y(t))] = \frac{\partial V}{\partial x}x' + \frac{\partial V}{\partial y}y'$$
$$= [(k^2 + c^2 + km)x + cmy]y +$$
$$[cmx + m(m+k)y][-(k/m)\sin x - (c/m)y]$$
$$= -ck(x^2 + y^2) + k[cx + (k+m)y](x - \sin x).$$

Now we have $\lim_{(x^2+y^2)\to 0} \dfrac{x - \sin x}{\sqrt{x^2 + y^2}} = 0$. Therefore, as $c > 0$, we conclude that $|cx + (k+m)y|(x - \sin x)| \leq (c/2)(x^2 + y^2)$ for sufficiently small $x^2 + y^2$. Thus,

$$\frac{d}{dt}[V(x(t), y(t))] \leq -(ck/2)(x^2 + y^2),$$

for sufficiently small $x^2 + y^2$. This proves the assertion that $(0,0)$ is also asymptotically stable for the non-linear system.

What is described above is a particular case of a general result discussed in detail in Section 5.1.3.

Exercise 5.19

We look for a Liapunov function of the form $V(x, y) = ax^{2m} + by^{2n}$, where the positive constants a and b and the positive integers m and n to be chosen. Clearly, $V(0,0) = 0$ and $V(x,y) > 0$ for all $(x,y) \neq (0,0)$. The systems in (a) and (b) have only one equilibrium point at $(0,0)$. Along any orbit $(x(t), y(t))$ of the system in (a), we have

$$\frac{d}{dt}[V(x(t), y(t))] = \frac{\partial V}{\partial x}x' + \frac{\partial V}{\partial y}y'$$
$$= 2max^{2m-1}(-2xy) + 2nby^{2n-1}(x^2 - y^3)$$
$$= [-4max^{2m}y + 2nbx^2y^{2n-1}] - 2nby^{2n+2}.$$

Thus, if we choose $m = n = 1$, $a = 1$ and $b = 2$, we see that $V(x, y) = x^2 + 2y^2$ and

$$\frac{d}{dt}[V(x(t), y(t))] = -4y^4 \leqslant 0.$$

Hence, $(0,0)$ is stable for the system in (a). We now show that $(0,0)$ is indeed asymptotically stable for the system in (a). First note that both the positive and negative y-axes are orbits of the system and approach the origin as $t \to \infty$ ($x \equiv 0$ and $y(t) = y_0[1 + 2(t - t_0)y_0^2]^{-1/2}$ are particular solutions). Therefore, any orbit starting in either of the regions $\{x > 0\}$ and $\{x < 0\}$ will remain there for all future times. For definiteness, consider an orbit $(x(t), y(t))$ starting in the region $\{x > 0\}$ at time $t = 0$. From the relation $\frac{d}{dt}[x^2(t) + 2y^2(t)] = -4y^4$, we infer that $\lim_{t\to\infty}[x^2(t) + 2y^2(t)]$ exists; in particular both $x(t)$ and $y(t)$ are bounded. The second equation in the system then implies that y' is also bounded; in particular, $y(t)$ is Lipschitz continuous and hence uniformly continuous on $[0, \infty)$. Further,

$$x^2(t) + 2y^2(t) = c - 4\int_0^t y^4(s)\,ds,$$

for all $t \geqslant 0$, where $c > 0$. In particular, $x^2(t) \leqslant c - 4\int_0^t y^4(s)\,ds$ for all $t \geqslant 0$. If there is a $t^* < \infty$ such that $\int_0^{t^*} y^4(s)\,ds = c/4$, then $x(t_1) = 0$ for some $t_1 \leqslant t^*$, which cannot happen. Thus, we conclude that $\int_0^\infty y^4(t)\,dt \leqslant c/4 < \infty$. It is not difficult to see now that[5] $\lim_{t\to\infty} y(t) = 0$. It then follows that, using the fact that $\lim_{t\to\infty}[x^2(t) + 2y^2(t)]$ exists, the orbit $(x(t), y(t))$ has a limit as $t \to \infty$, which ought to be an equilibrium point. Since $(0,0)$ is the only equilibrium point of the system, we conclude that $\lim_{t\to\infty}(x(t), y(t)) = (0,0)$, thereby proving that $(0,0)$ is asymptotically stable.

Similarly, for any trajectory $(x(t), y(t))$ of the system in (b), we have

$$\frac{d}{dt}[V(x(t), y(t))] = \frac{\partial V}{\partial x}x' + \frac{\partial V}{\partial y}y'$$
$$= 2max^{2m-1}(-3x^3 - y) + 2nby^{2n-1}(x^5 - 2y^3)$$
$$= -6max^{2m+2} - 2max^{2m-1}y + 2nbx^5y^{2n-1}] - 4nby^{2n+2}.$$

---

[5] Suppose $f:\mathbb{R}\to\mathbb{R}$ is uniformly continuous and $\int_{-\infty}^\infty |f(t)|^p\,dt < \infty$ for some $p > 0$. Then $f(t) \to 0$ as $t \to \pm\infty$.

To get rid of the terms involving the product xy, we choose $m = 3$, $n = 1$ and then a, b such that $6a = 10b$. Thus, we take $V(x,y) = 5x^6 + 3y^2$ and obtain that $\dfrac{d}{dt}[V(x(t), y(t))] = -(90x^8 + 12y^4)$. Thus, $(0,0)$ is asymptotically stable for the system in (b).

Exercise 5.20

We write $z^n = r^n \cos n\theta + ir^n \sin n\theta$ in polar co-ordinates so that $u(r, \theta) = r^n \cos n\theta$ in system (5.1.11). It follows that any solution of system (5.1.9) or system (5.1.11) satisfies the relation $r^n \sin n\theta = k$, a constant. Taking $k = 0$, we see that the phase plane $\mathbb{R}^2$ is divided into $2n$ regions, separated by the lines $\theta = \theta_j = \dfrac{j\pi}{n}$ for $j = 0, 1, \ldots, 2n - 1$. Each of these lines itself is an orbit of system (5.1.9), hence invariant under system (5.1.9). So are the regions between any two consecutive lines among these $2n$ lines. The only equilibrium point $(0, 0)$ is easily seen to be of saddle type; this means that some orbits converge to the origin and some other move away from the origin, as $t \to \infty$, or the right end point of the maximum interval of existence I of the solution. The phase portrait is shown in Figure 5.10 (top) for $n = 4$.

Exercise 5.21

The function $f(z) = z^{-n}$ is analytic in the punctured plane $\mathbb{C}\setminus\{0\}$ and in polar co-ordinates $z^{-n} = r^{-n} \cos n\theta - ir^{-n} \sin n\theta$, $z \neq 0$. It follows that any solution of system (5.1.9) or equivalently that of system (5.1.11), with $u = r^{-n} \cos n\theta$ satisfies the relation $r^{-n} \sin n\theta = k$, a constant. Again, the phase plane $\mathbb{R}^2$, excluding the origin, is divided by the lines $\theta = \theta_j = \dfrac{j\pi}{n}$ for $j = 0, 1, \ldots, 2n - 1$. Each of these lines is an orbit of system (5.1.9), hence invariant under system (5.1.9). So are the regions between any two consecutive lines among these $2n$ lines. This system has no equilibrium point. The phase portrait is shown in Figure 5.10 (bottom) for $n = 4$.

Exercise 5.22

Here $u(x,y) = x\left(1 + \dfrac{1}{x^2 + y^2}\right)$ and $v(x,y) = y\left(1 - \dfrac{1}{x^2 + y^2}\right)$. We have

$$\frac{\partial u}{\partial x} = 1 + \frac{y^2 - x^2}{(x^2 + y^2)^2} \quad \text{and} \quad \frac{\partial u}{\partial y} = -\frac{2xy}{(x^2 + y^2)^2}.$$

Now it can be easily checked that system (5.1.9) has only two equilibrium points $(\pm 1, 0)$ and both are unstable. The orbits are the level curves $v = k$, where k is a constant. Corresponding to $k = 0$, we obtain the two equilibrium points, and we see that the x-axis is a union of orbits, namely the intervals $(-\infty, -1)$, $(-1, 0)$, $(0, 1)$ and $(1, \infty)$, besides the two equilibrium points $(\pm 1, 0)$. Also, again corresponding to $k = 0$ and $y \neq 0$, there are two heteroclinic orbits connecting the two equilibrium points, namely the upper unit semicircle and lower unit semicircle. These are depicted in Figure 5.11(b).

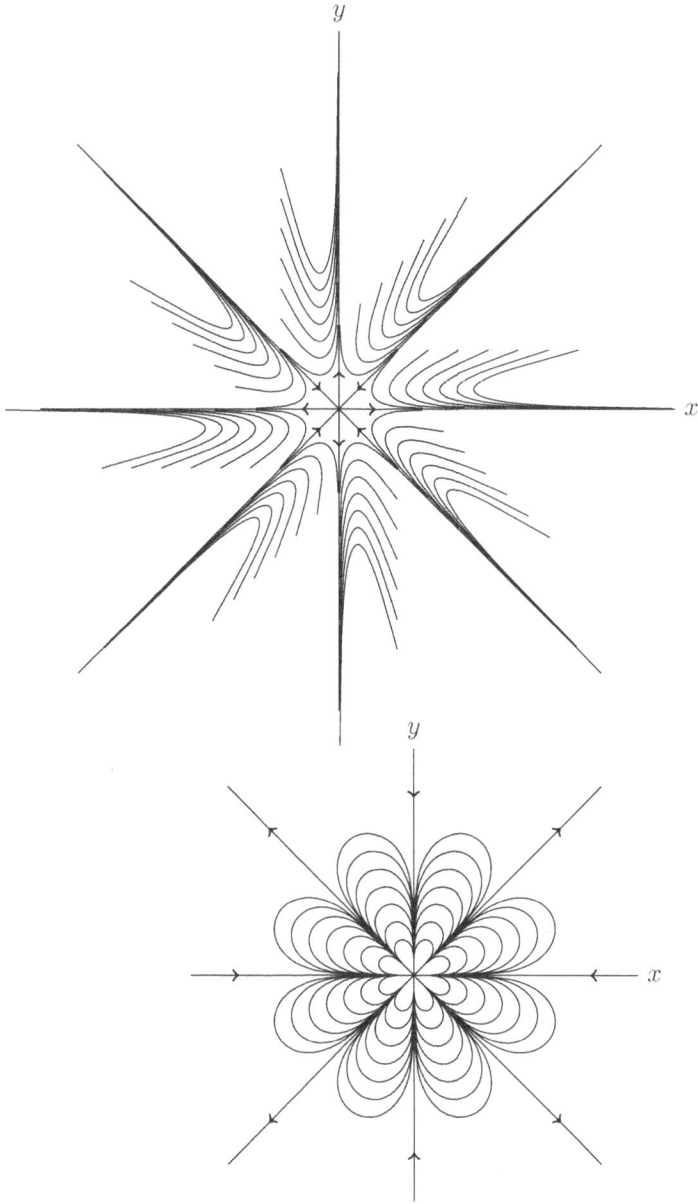

Figure 5.10 | Phase portrait of potential flow for $\Re(z^4)$ (*top*) and $\Re(z^{-4})$ (*bottom*).

Next consider the case $k \neq 0$. Simplifying the relation $v = k$, we obtain

$$x^2 = \frac{y}{y-k}[1 - y(y-k)], \ (x,y) \neq (0,0).$$

For $k > 0$, we see that $x^2 + y^2 > 1$ if $y > 0$ and $x^2 + y^2 < 1$ if $y < 0$. Thus, we see that $y \in (k, (1/2)(k + \sqrt{k^2 + 4})]$ if $y > 0$ and $y \in [(1/2)(k - \sqrt{k^2 + 4}), 0)$ if $y < 0$. Similar statements hold for $k < 0$. With these observations, it is now straight forward to describe the orbits geometrically. These are shown in Figure 5.11(b).

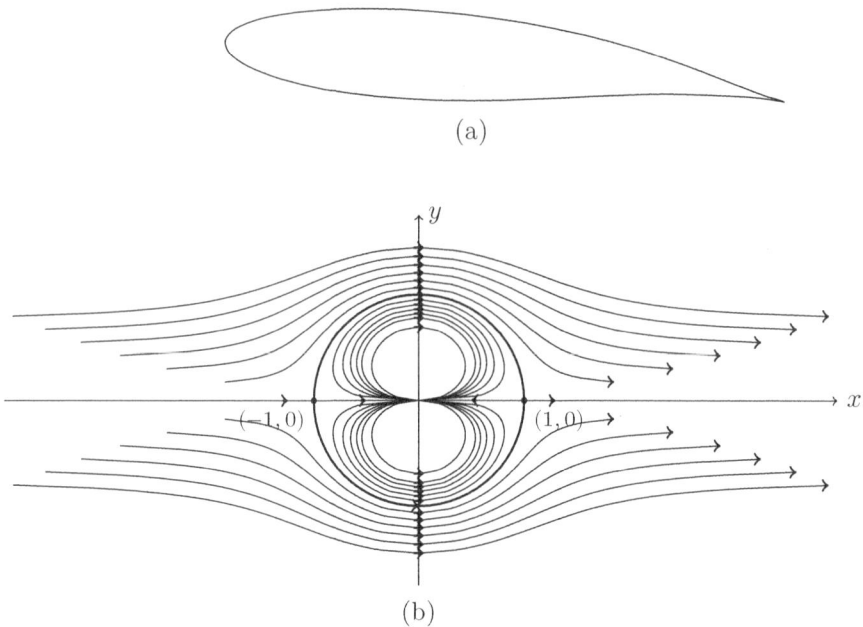

Figure 5.11 | (a) Joukowski airfoil and (b) phase portrait for Joukowski function.

Exercise 5.23

We need to search for a function $H(x, y)$ such that

$$\frac{\partial H}{\partial y} = y(13 - x^2 - y^2) \text{ and } \frac{\partial H}{\partial x} = -12 + x(13 - x^2 - y^2).$$

The first relation is satisfied if we take $H(x, y) = \frac{y^2}{2}(13 - x^2) - \frac{y^4}{4} + f(x)$, where f is an arbitrary function of x. Similarly, the second relation is satisfied if $H(x, y) = -12x + \frac{x^2}{2}(13 - y^2) - \frac{x^4}{4} + g(y)$, where g is an arbitrary function of y. Thus, we take $f(x) = -12x + \frac{13}{2}x^2 - \frac{x^4}{4}$ and $g(y) = \frac{13}{2}y^2 - \frac{y^4}{4}$. After some simplification, we arrive at $H(x, y) = -(1/4)(13 - x^2 - y^2)^2 - 12x$ as a Hamiltonian for the given system; of course, any two Hamiltonians differ by a constant. Thus, the orbits are the level curves of H, that is, the curves $H(x,y) = -E$, describe the orbits of the given system for all energy levels $E \in \mathbb{R}$, for which the set $\{(x,y) \in \mathbb{R}^2\}$ satisfying $H(x,y) = -E$ is non-empty.

Before we move further, let us look for the equilibrium points of the system and study their linear stability. The equilibrium points are the solutions of the simultaneous equations

$$y(13 - x^2 - y^2) = 0 \text{ and } 12 - x(13 - x^2 - y^2) = 0.$$

We find that $(-4, 0)$, $(1, 0)$ and $(3, 0)$ are the three equilibrium points. The linearized systems at each of these equilibrium points are given by

at $(-4, 0)$: $X' = -3Y$, $Y' = 35X$, with $X = x + 4$ and $Y = y$.
at $(1, 0)$: $X' = 12Y$, $Y' = -11X$, with $X = x - 1$ and $Y = y$.
at $(3, 0)$: $X' = 4Y$, $Y' = 5X$, with $X = x - 3$ and $Y = y$.

We find that the equilibrium points $(-4, 0)$ and $(1, 0)$ are centres, while $(3, 0)$ is a saddle point in the linear analysis. By Hartman–Grobman theorem, the equilibrium point $(3, 0)$ is unstable for the non-linear system as well.

We will now analyse the phase portrait of the non-linear system. As observed above, the orbits of the system are the level curves of the Hamiltonian H. The equation for the level curves may be written as

$$(1/4)\left(13 - x^2 - y^2\right)^2 + 12x = E.$$

Equivalently,

$$y^2 = f_{\pm}(x) - x^2 \equiv 13 \pm 2\sqrt{E - 12x} - x^2, \tag{5.2.5}$$

provided that $x \leq E/12$ and $f_{\pm}(x) \geq x^2$. This leads to the task of determining the interval(s) in which $f_{\pm}(x) \geq x^2$. We can then use equation (5.2.5) to sketch the orbits. We observe that the orbits are symmetric with respect to the x-axis, that is, if (x, y) lies on an orbit, so does $(x, -y)$. Since we are looking for the level curves of a quartic, we find that the orbits are bounded. The boundedness of the orbits can also be seen using the polar co-ordinates. In polar co-ordinates $x = r\cos\theta$, $y = r\sin\theta$, the level curve equation can be written as $(13 - r^2)^2 + 48r\cos\theta = 4E$. As $|\cos\theta| \leq 1$, we see that the inequality $|4E - (13 - r^2)^2| \leq 48r$ holds. This in turn proves the boundedness of the orbits.

We use the graphs of the functions $f_{\pm}$ and the parabola x^2 to determine the interval(s) in which $f_{\pm}(x) \geq x^2$. Note that, for $x \leq E/12$, the function f_+ is a decreasing function and the function f_- is an increasing function. Furthermore, $f_-(x) \leq f_+(x)$, $f_{\pm}(E/12) = 13$ and $f_{\pm}(x) \to \pm\infty$ as $x \to -\infty$. Thus, $f_{\pm}$ can intersect the parabola x^2 at most at two points, depending on the energy level E. In further analysis, consideration of the auxiliary energy function defined by $h(x) = -H(x, 0) = (1/4)(x^2 - 13)^2 + 12x$, $x \in \mathbb{R}$, helps. Note that the function h has a global minimum at $x = -4$, with $h(-4) = -183/4$, a local minimum at $x = 3$, with $h(3) = 40$ and a local maximum at $x = 1$ with $h(1) = 48$. A sketch of the function h is depicted in Figure 5.12(a).

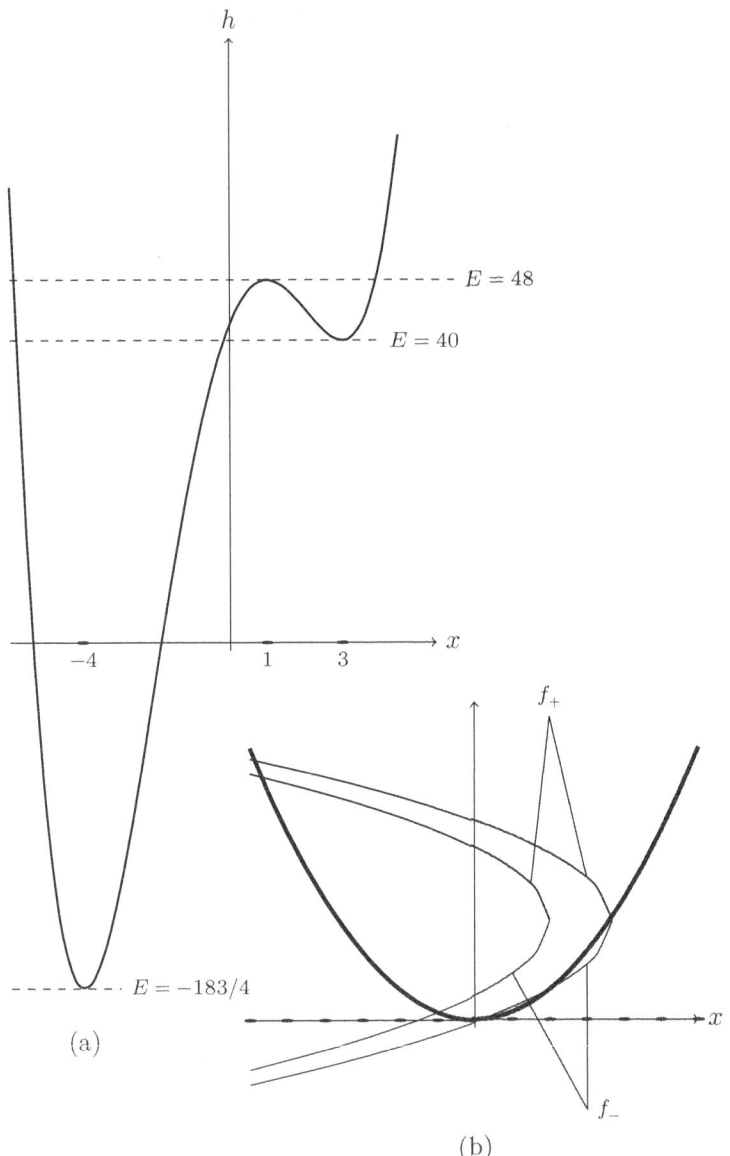

Figure 5.12 | The auxiliary energy function h and the graphs of $f_\pm$ with the parabola x^2.

For an energy level E, observe that $h(x) - E = (1/4)(f_+(x) - x^2)(f_-(x) - x^2)$. This relation helps us in determining the interval(s) in which $f_\pm(x) \geqslant x^2$. For example, suppose $x_1 < x_2$ are two consecutive roots of $h(x) = E$ and $h(x) < 0$ in (x_1, x_2). Then it can be checked that $f_-(x) \geqslant x^2$ in the interval $[x_2, E/12]$ and $f_+(x) \geqslant x^2$ in the interval $[x_1, E/12]$. Similar analysis can be done for the other cases. The qualitative behaviour of the functions $f_\pm$ and the parabola x^2 is depicted in Figure 5.12(b) for some chosen values of E. We now proceed with details for all the possible energy levels E.

<u>$E < -183/4$</u>: As the value $-183/4$ is the global minimum value of h, we find that $f_\pm(x) < x^2$ for all $x \leqslant E/12$. Therefore, these energy levels do not contribute to the phase portrait.

<u>$E = -183/4$</u>: We have $h(4) = -183/4$ and we find that $f_-(x) < x^2$ for all $x \leqslant E/12$ and f_+ and the parabola x^2 are tangent to each other at $x = -4$. Therefore, this energy level contributes only the equilibrium point $(-4, 0)$ to the phase portrait.

<u>$-183/4 < E < 40$</u>: In this case, the equation $h(x) = E$ has two real roots $x_1 < x_2 < 0$ and $h(x) < 0$ in (x_1, x_2); see Figure 5.12. We find that $f_-(x) \geqslant x^2$ in the interval $[x_2, E/12]$ and $f_+(x) \geqslant x^2$ in the interval $[x_1, E/12]$. Therefore, each energy level in this range contributes, using both $f_\pm$, to a periodic (closed) orbit surrounding only the equilibrium point $(-4, 0)$ to the phase portrait.

<u>$E = 40$</u>: In this case, $h(3) = E = 40$, and remember that $(3, 0)$ is an equilibrium point of the given system, and it is an unstable equilibrium point of saddle type, as the linear analysis shows. An orbit can therefore approach this point only as $t \to \infty$ or $t \to -\infty$ or $t \to \pm\infty$. The equation $h(x) = 40$ is equivalent to $(x-3)^2(x^2 + 6x + 1) = 0$, whose roots $3, 3, -3 \pm 2\sqrt{2}$. The function f_- with domain $[-3 + 2\sqrt{2}, 3)$ produces a homoclinic orbit at the equilibrium point $(3, 0)$. The function f_- with the domain $(3, 10/3]$ ($E/12 = 40/12 = 10/3$) and the function f_+ with the domain $[-3 - 2\sqrt{2}, 10/3]$ *together* produce another homoclinic orbit at the equilibrium point $(3, 0)$.

<u>$40 < E < 48$</u>: In this case, the equation $h(x) = E$ has four distinct real roots. Denote them by $x_1 < x_2 < x_3 < x_4$. Observe that $x_2 < 1 < x_3$ and $x_4 > 3$. We find that $f_-(x) \geqslant x^2$ in the interval $[x_2, x_3]$ and $f_+(x) \geqslant x^2$ in the interval $[x_1, x_4]$. Therefore, each energy level in this range contributes, using f_-, a periodic orbit surrounding only the equilibrium point $(1, 0)$, and a periodic orbit, using f_+, surrounding all the three equilibrium points.

<u>$E = 48$</u>: The equation $h(x) = E = 48$ is equivalent to $(x-1)^2(x^2 + 2x - 23) = 0$, whose roots are $1, 1, -1 \pm 2\sqrt{6}$. We find that f_- is tangential to the parabola x^2 at $x = 1$ and gives the equilibrium point $(1, 0)$. The function f_+ with domain $[-1 - 2\sqrt{6}, -1 + 2\sqrt{6}]$ produces a periodic orbit surrounding all the three equilibrium points.

<u>$E > 48$</u>: In this case, the equation $h(x) = E$ has only two distinct real roots. Denote them by $x_1 < x_2$. Note that $x_2 > 3$. We find that $f_-(x) < x^2$ for all $x \leqslant E/12$, thus contributing nothing to the phase portrait. On the other hand, $f_+(x) \geqslant x^2$ in the interval $[x_1, x_2]$ and produces a periodic orbit surrounding all the three equilibrium points.

Some orbits are depicted in Figure 5.13.

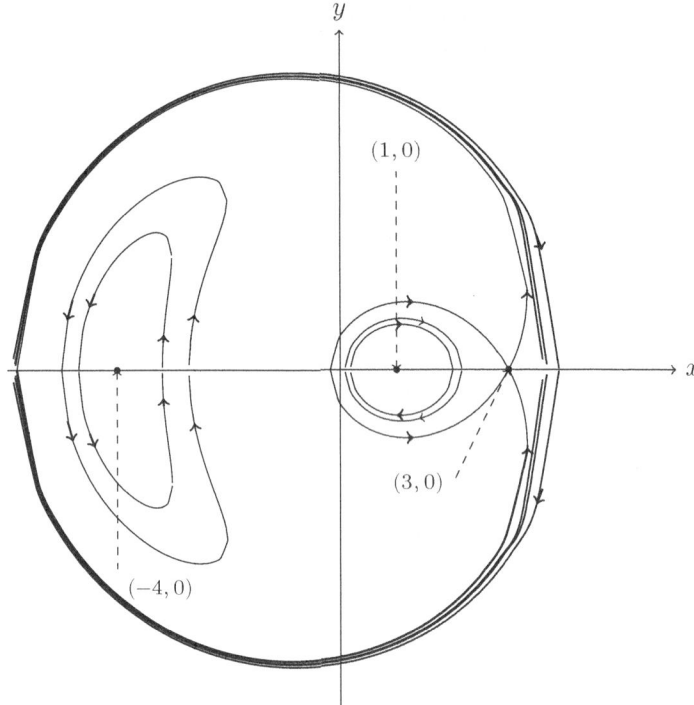

Figure 5.13 | Phase portrait of the Hamiltonian system in Exercise 5.23.

The directions of the orbits, that is, the direction of increasing t is easily determined by using the given equations. For example, $x(t)$ is increasing in t, in the region of the plane where y and $(13 - x^2 - y^2)$ have the same sign. We observe that there are periodic orbits surrounding the equilibrium points $(-4, 0)$ and $(1, 0)$. There is a homoclinic orbit connecting the equilibrium point $(3, 0)$ as $t \to \pm\infty$. Thus, all the equilibrium points have the same qualitative behaviour for the non-linear system, as compared to the corresponding linearized systems.

PART TWO
PARTIAL DIFFERENTIAL EQUATIONS

6

First-Order Partial Differential Equations

6.1 Method of Characteristics

A general first-order partial differential equation (PDE) in n independent variables $x = (x_1, \cdots, x_n)$ can be written as $F(x, u, p) = 0$. Here, $u = u(x)$ is the unknown function to be determined and $p = p(x) = D_x u(x) = \left(\dfrac{\partial u}{\partial x_1}, \cdots, \dfrac{\partial u}{\partial x_n} \right)$ is its gradient. The gradient of a function $u = u(x)$ is also denoted by $Du = D_x u = \nabla u = \nabla_x u$, to emphasize the variables with respect to which the derivatives are taken; and, F is a given real valued C^k ($k \geqslant 2$) function defined on $\Omega \times \mathbb{R} \times \mathbb{R}^n$, where Ω is a domain in $\mathbb{R}^n$. In the two-dimensional case, we also use the notation (x, y) to denote the independent variables and $p = \dfrac{\partial u}{\partial x}, q = \dfrac{\partial u}{\partial y}$ for the first-order derivatives. Sometimes, the variable u is replaced z, as more of a convention. We consider a *Cauchy problem* or an *initial value problem* (IVP) associated with the given equation $F = 0$:

$$F(x, u, p) = 0 \text{ in } \Omega \text{ and } u = u_0 \text{ on } S, \tag{6.1.1}$$

where S is a smooth $(n-1)$ dimensional surface sitting inside Ω. If S has the parametric representation $S = \{h(s) = (h_1(s), \ldots, h_n(s)) : s \in V\}$, where V is a region in $\mathbb{R}^{n-1}$ and $h_1, \ldots, h_n$ are smooth real valued functions defined on V, then the initial condition is written as $u(h(s)) = u_0(s)$ for $s \in V$, where u_0 is a given function defined on V. In general, the function F is also required to satisfy a *compatibility condition* on S.

We attempt to analyse the IVP (6.1.1) through the *method of characteristics*. The characteristic curves $x = x(t)$ associated with the equation $F = 0$ are the solutions of the first-order system of ordinary differential equation (ODE) $\dfrac{dx}{dt} = D_p F$. In general, since F also depends on u and p, this system alone is insufficient to solve for $x(t)$.

To overcome this situation, we augment this ODE system with a system satisfied by p and an equation satisfied by u. Together, they constitute *characteristic equations*, which is a system of $2n+1$ ODE given by

$$\frac{dx}{dt} = D_p F, \quad \frac{du}{dt} = p \cdot D_p F, \quad \frac{dp}{dt} = -D_x F - p F_u.$$

In the case of two variables, these equations are commonly written as

$$\begin{cases} \dfrac{dx}{dt} = F_p, \ \dfrac{dy}{dt} = F_q, \ \dfrac{dz}{dt} = pF_p + qF_q \\ \dfrac{dp}{dt} = -F_x - pF_z, \ \dfrac{dq}{dt} = -F_y - qF_z. \end{cases}$$

These are also symbolically written as

$$\frac{dx}{F_p} = \frac{dy}{F_q} = \frac{dz}{pF_p + qF_q} = \frac{dp}{-F_x - pF_z} = \frac{dq}{-F_y - qF_z}.$$

We remark that the variable t used in the characteristic equations is for the sake of parametrizing the characteristic curve. When convenient, we can use one of the variables in x itself as this parameter, especially when dealing with equations in two variables. Also, t is used as one of the independent variables in many equations of physical interest, designated as the *time variable*. This is the situation in case of a Hamilton–Jacobi equation (HJE) and a conservation law, discussed subsequently.

The initial values given in equation (6.1.1) together with the *strip condition* provide the initial conditions for the characteristic equations. The strip condition is the condition that binds both the given equation and the initial surface and initial conditions. It is stated as

$$\frac{\partial u}{\partial s_i} = \sum_{j=1}^{n} \frac{\partial u}{\partial x_j} \frac{\partial (h_j(s))}{\partial s_i} = \sum_{j=1}^{n} p_j \frac{\partial h_j}{\partial s_i}, \tag{6.1.2}$$

for $i = 1, 2, \ldots, n-1$ and appropriate functions $p_0 = (p_{01}, \ldots, p_{0n})$ are to be chosen to satisfy

$$F(h(s), u_0(s), p_0(h(s))) = 0, \text{ for } s \in V. \tag{6.1.3}$$

See equation (6.1.1). The above system of equations provides initial conditions for p in the characteristic equations and should be thought as a requirement on F and S and the initial functions, in order that a given IVP has a solution. This may lead to non-uniqueness of the solutions or non-existence of solutions. For example, in two variables, the equation $p^2 + q^2 - 1 = 0$ has more than one solution and the equation $p^2 + q^2 + 1 = 0$ has no solution.

If F is linear in p or both u and p, some simplifications are possible. For example, in the case of two variables and when F is linear in u and p, that is, $F = au_x + bu_y$, say, the characteristic equations are

$$\frac{dx}{dt} = a, \quad \frac{dy}{dt} = b,$$

where a and b are smooth functions of x, y only. We notice that whether F is linear or not, the characteristic equations are non-linear, in general. Thus, we can only claim for a local existence result and, in principle, the characteristics can be solved uniquely, of course, under suitable conditions on F, to obtain not only the characteristic curves, but also the solution as well. This is basically the general method of characteristics to solve IVP for first-order equations. As we observed earlier, if the given equation is linear, then the above system can be decoupled to obtain a system for $x(t)$ alone. Further, we obtain an equation for $u(t)$ along the curves $x(t)$. In particular, the characteristic curves are the plane curves in two-dimensional case. Such decoupling is not true in general. However, if the equation is quasi-linear, then a decoupled $n+1$ dimensional system can be obtained for $x(t)$ and $u(t)$ without $p(t)$. Thus, in two-dimensional case, the characteristic curves are space curves.

Further, in case of linear or quasi-linear equations, the strip condition is replaced by a simpler *transversality condition*. For simplicity, we will state this for an equation in two variables. Consider the equation $au_x + bu_y = cu$, where the coefficients a, b and c are functions of x, y or functions of x, y and u (quasi-linear case). The transversality condition is the following:

$$a(x_0(s), y_0(s), u_0(s))\frac{dy_0}{ds}(s) - b(x_0(s), y_0(s), u_0(s))\frac{dx_0}{ds}(s) \neq 0, \qquad (6.1.4)$$

for all $s \in I$. Here, s is the parameter and the functions x_0, y_0, defined on an interval I, defining the initial curve; the function u_0 is the given initial function. The transversality condition says that any characteristic curve of the given equation is never a tangent to the initial curve at any point on it. We will see in the exercises that we cannot define a solution at any point where the transversality condition fails. Also, the transversality condition implies that the IVP we are considering is a *non-characteristic* Cauchy problem. The *characteristic Cauchy problem*, in which the initial condition is prescribed on a characteristic curve, is more difficult to analyse and may lead to non-existence or non-uniqueness of solutions. We will see some examples later.

The solutions of the characteristic equations are denoted by $x(t, s)$, $u(t, s)$ and $p(t, s)$, where the initial conditions are given by $x(0, s) = x_0(s)$ and so on. The strategy then for solving a first-order equation is the following. First, write down the characteristic equations. Using the initial curve or surface and data provided on the same, determine the initial conditions for the characteristic equations and solve them. Using this information about the characteristics, determine the solution of the given IVP in an appropriate domain.

The following notions will be helpful to solve some of the exercises. A relation of the type $g(\alpha, \beta) = 0$, where α, β are known functions of x, y and z (in the two-dimensional case) and g is an arbitrary function, providing a solution to the given first-order PDE is called a *general solution*. Thus, for any choice of g, we obtain a solution. A *complete integral* is a family of solutions consisting of two arbitrary constants a and b say, which can be given by a relation of the form $f(x, y, u, a, b) = 0$. Thus, for any choice of constants a and b, we obtain a solution.

6.2 Hamilton–Jacobi Equation

The general HJE or more generally *Hamilton–Jacobi–Bellman equation* (HJB) from optimal control theory is a first-order non-linear equation of the form

$$u_t + H(x, u, Du) = 0,$$

where $x \in \mathbb{R}^n$ and $t > 0$. Here, $u : \mathbb{R}^n \times [0, \infty) \to \mathbb{R}$ is the unknown function and $Du = \nabla_x u$ is its gradient. The non-linear function H is called a *Hamiltonian*, a terminology borrowed from Hamiltonian mechanics. The unknown u can be thought of as the minimum value of a cost functional or energy functional as the case may be. In optimization, u is also known as the *value function*. In mechanics, the Hamiltonian H is the total energy of the system.

Minimization Problem and Hopf–Lax Formula: Consider the following minimization problem: Find $\overline{w} \in \mathcal{A}_t$ such that

$$J(\overline{w}) = \min_{\mathcal{A}_t} J(w),$$

where the functional $J(w)$ is defined by

$$J(w) = \int_0^t L(w'(s))\,ds + g(w(0)).$$

The Lagrangian $L : \mathbb{R}^n \to \mathbb{R}$ is a given function and $\mathcal{A}_t$ denotes the class of admissible functions defined by

$$\mathcal{A}_t = \{w \in C^2([0, t]; \mathbb{R}^n) : w(0) = y, w(t) = x\},$$

for fixed $t > 0$ and $x, y \in \mathbb{R}^n$. Thus, $\mathcal{A}_t$ consists of all *trajectories* connecting the two points y and x in $\mathbb{R}^n$. In the minimization problem, we fix x, t and vary y. Finally,

$g : \mathbb{R}^n \to \mathbb{R}$ is a given function, which eventually will be the initial data for IVP. The minimizing term can be interpreted as follows. The integral term is the running cost and $g(w(0))$ is the initial cost. The solution $\overline{w}$, when exists, is called the *optimal solution* and $J(\overline{w})$, the *minimum* or *optimal cost*. The minimum cost, $J(\overline{w})$, depends on x and t that we denote by $u(x,t)$, called the *value function*. That is,

$$u(x,t) := \inf_{\mathcal{A}_t} \left\{ \int_0^t L(w'(s)) ds + g(w(0)) \right\}.$$

The above minimum is taken over a class of functions which, in general, is infinite dimensional. Nevertheless, it can be transformed to a minimization problem over the Euclidean space, under certain assumptions on L. This is essentially the *Hopf–Lax formula*, which will be stated in what follows. We assume that the Lagrangian L satisfies the following:

- The mapping $q \mapsto L(q)$ is continuous and convex.
- The mapping L is *coercive* in the sense that $\lim_{|q| \to \infty} \dfrac{L(q)}{|q|} = \infty$.

The second condition means that L has super-linear growth, that is, L roughly behaves like $|q|^{1+\varepsilon}$ for some $\varepsilon > 0$ and for large $|q|$. Under the above assumptions the minimum value u satisfies the *Hopf–Lax formula*

$$u(x,t) \equiv \inf_{y \in \mathbb{R}^n} \left\{ tL\left(\frac{x-y}{t}\right) + g(y) \right\}. \tag{6.2.1}$$

Euler–Lagrange Equations: Consider a more general Lagrangian $L : \mathbb{R}^n \times \mathbb{R}^n \to \mathbb{R}$ and recall the minimization problem described above with the functional given by

$$J(w) = \int_0^t L(w(s), w'(s)) ds.$$

We write $L = L(x, q)$ and use the notation $D_q L := \left(L_{q_1}, \ldots L_{q_n} \right)$ and $D_x L := \left(L_{x_1}, \ldots L_{x_n} \right)$. Let $\overline{w}$ be a minimal solution. Then, $\overline{w}$ satisfies the *Euler–Lagrange* (E–L) equations

$$\left(-\frac{d}{ds} D_q L + D_x L \right)(\overline{w}(s), \overline{w}'(s)) = 0$$

which is a system of n second-order ODEs.

Legendre Transformation: We now give a connection between L and H via the so-called *Legendre transformation*. We begin with the description of the Hamiltonian

H in terms of the Lagrangian L. Let $q = q(p)$ solves $p = D_q L(q)$. Observe that the latter equation corresponds to a critical point of the functional $F(q) := p \cdot q - L(q)$, $q \in \mathbb{R}^n$. In other words, $q = q(p)$ is an extremal point of the functional F. Now define H by $H(p) = F(q(p))$. This motivates the following definition.

Definition 6.1 (Legendre transformation). The Legendre transformation of L, denoted by L^*, is defined by

$$L^*(p) = \sup_{q \in \mathbb{R}^n} \{p \cdot q - L(q)\} = \sup_{q \in \mathbb{R}^n} F(q), \ p \in \mathbb{R}^n.$$

Thus, under the assumption that L is convex and the assumptions of continuity and coercivity, we can prove that $H \equiv L^*$ is also convex, continuous and coercive. Further, $H^* = L$. Now, let g be a continuous function defined on $\mathbb{R}^n$ and u be the value function given by the Lagrangian L. Then u is differentiable a.e. and solves the IVP

$$u_t + H(Du) = 0 \text{ a.e. in } \mathbb{R}^n \times (0, \infty) \text{ and } u = g \text{ in } \mathbb{R}^n.$$

In general, a given HJE can have more than one Lipschitz solution. To obtain the uniqueness of this solution, an additional assumption of *semi-concavity* needs to be imposed; the details can be found in Ref. [45].

6.3 Conservation Laws

The IVP for a conservation law

$$\left. \begin{array}{l} u_t + f(u)_x = 0, \ x \in \mathbb{R}, \ t > 0 \\ u(x, 0) = u_0(x) \ x \in \mathbb{R} \end{array} \right\} \quad (6.3.1)$$

is solved by the method of characteristics. In general, smooth (C^1) solution exists only in a region of the upper-half plane $t > 0$, provided that f and u_0 are smooth. The characteristic passing through a point (x, t) in the upper-half plane is given by

$$x = x_0 + f'(u_0(x_0))t, \quad (6.3.2)$$

x_0 being the point where the characteristic meets the x-axis ($t = 0$). Provided that the equation (6.3.2) can be solved for x_0, the solution of equation (6.3.1) is given by

$$u(x, t) = u_0(x_0), \quad (6.3.3)$$

at the point (x, t). The curve given by equation (6.3.2) is referred to as the *backward characteristic* through the point (x, t). Implicit in the above arguments is that a unique backward characteristic can be drawn through (x, t) so that the value $u(x, t)$ is unambiguously defined.

We remark that the above procedure easily extends to a single conservation law in several variables. Consider the following conservation law in n variables:

$$\frac{\partial u}{\partial t} + a_1(u)\frac{\partial u}{\partial x_1} + \cdots + a_n(u)\frac{\partial u}{\partial x_n} = 0, \ x \in \mathbb{R}^n, \ t > 0$$
$$u(x, 0) = u_0(x), \ x \in \mathbb{R}^n.$$
(6.3.4)

Here, the coefficients a_i are C^1 functions defined in an interval in $\mathbb{R}$ and the initial function u_0 is also a C^1 function defined in $\mathbb{R}^n$. The characteristics corresponding to equation (6.3.4) are given by the system of ODE:

$$\frac{dx_i}{dt} = a_i(u), \ 1 \leqslant i \leqslant n.$$
(6.3.5)

It is easily checked that the solution u is a constant along the characteristics:

$$\frac{d}{dt}u(x_1(t), \ldots, x_n(t), t) = \frac{\partial u}{\partial t} + \sum_{i=1}^{n} \frac{dx_i}{dt}\frac{\partial u}{\partial x_i} = \frac{\partial u}{\partial t} + \sum_{i=1}^{n} a_i(u)\frac{\partial u}{\partial x_i} = 0.$$

Thus, the characteristic curves passing through a point (x, t) in the upper-half space are given by

$$x_i = x_{0i} + a_i(u_0(x_0))t, \ 1 \leqslant i \leqslant n,$$
(6.3.6)

where $x_0 \in \mathbb{R}^n$ is the point the characteristics meet the space $t = 0$. Provided that the system (6.3.6) can be solved for x_0, the solution of equation (6.3.4) is given by

$$u(x, t) = u_0(x_0),$$
(6.3.7)

at the point (x, t). A sufficient condition for solving the system (6.3.6) for x_0 in terms of x and t is provided by the implicit function theorem. This condition is that matrix $I + tD$, where I is the identity matrix and D is the matrix whose $(i, j)^{th}$ element is $\frac{da_i}{du}(u_0(x_0))\frac{\partial u_0}{\partial x_j}(x_0)$, is non-singular. This is certainly true for $t > 0$ sufficiently small.

Back to the one-dimensional case. If for some point (x, t), $t > 0$, more than one backward characteristics can be drawn, then we cannot define $u(x, t)$ unambiguously. We have to then resort to a generalized solution by appropriately defining a discontinuity curve, a *shock*. It may also happen that no point in a region in the upper-half plane is connected to a characteristic. This leads to the notion of a *rarefaction wave*, a different type of solution. Depending on the initial condition u_0, a generalized solution of equation (6.3.1) may exhibit a combination of a smooth regime, a rarefaction regime and a shock regime.

Across a line of discontinuity $x = s(t)$, the solution need to satisfy the *Rankine–Hugoniot condition* (or *jump condition*)

$$\frac{ds}{dt} = \frac{f(u_r) - f(u_\ell)}{u_r - u_\ell}, \tag{6.3.8}$$

where u_r and u_ℓ are the limiting values of u, as we approach the curve $x = s(t)$ from right and left, respectively. If we allow a discontinuous function to be a solution of the IVP equation (6.3.1), then we may face non-uniqueness of solutions. To overcome this situation, an additional condition, called *entropy condition*, is introduced:

$$f'(u_r) < \sigma < f'(u_\ell), \tag{6.3.9}$$

where $\sigma = \dfrac{ds}{dt}$ is the speed of the discontinuity curve and u_ℓ, u_r are as before. Geometrically, the entropy condition (6.3.9) states that the characteristics approaching the discontinuity curve $x = s(t)$ from both the sides *impinge* onto it. A *generalized solution* of equation (6.3.1) is, by definition, a bounded measurable function $u(x, t)$ satisfying the following integral relation:

$$\int_0^\infty \int_{-\infty}^\infty (u\varphi_t + f(u)\varphi_x) \, dx \, dt + \int_{-\infty}^\infty u_0(x)\varphi(x, 0) \, dx = 0, \tag{6.3.10}$$

for all smooth functions φ having compact support in $\mathbb{R} \times [0, \infty)$. It is not difficult to see that if u is a generalized solution and a C^1 function in a region in the upper-half plane $t > 0$, then u actually satisfies the equation (6.3.1) point-wise in that region.

A generalized solution of equation (6.3.1) is an *entropy solution* if it satisfies equations (6.3.8) and (6.3.9) across a discontinuity curve. Such a discontinuity curve is called a *shock*. It can be shown that a generalized solution of the IVP (6.3.1), possessing only shocks as discontinuities, is unique. In this scenario, it is not difficult to see that every point (x, t), $t > 0$ in the upper-half plane is connected to a point x_0 on the line $t = 0$, by a backward characteristic. This is used in the derivation of the Lax–Oleinik formula, discussed next. We note that if $f \in C^2$, with $f'' > 0$, then the condition $f'(u_r) < f'(u_\ell)$ is equivalent to $u_r < u_\ell$. There are other uniqueness results as well.

For convex conservation laws, that is, f is a uniformly convex function, the *Lax–Oleinik formula* gives a formula for the solution of IVP:

$$u(x, t) = g\left(\frac{x-y}{t}\right), \; t > 0, \tag{6.3.11}$$

where $y = y(x,t)$ is the minimizer of:

$$\min_{y \in \mathbb{R}} \left\{ tL\left(\frac{x-y}{t}\right) + w_0(y) \right\}$$

with $g = (f')^{-1}$, L the Legendre transform of f and $w_0(x) = \int_0^x u_0(\eta)\, d\eta$.

The general references are [45], [23], [56], [36], [49], [44] and [5].

6.4 Exercises

General First-Order Equations

The exercises in this section are based on the method of characteristics. Each exercise contains an equation and the initial data and the task is to find the solution, wherever possible, by sketching the characteristics and describing the region where a (smooth) solution is defined and also, verifying the transversality condition or the strip condition.

Exercise 6.1

Solve IVP: $(x+2)u_x + 2yu_y = 2u$, $u(-1,y) = \sqrt{|y|}$.

Exercise 6.2

Consider the PDE $xu_x + yu_y = 2u$ in the region $x > 0, y > 0$. Find solutions satisfying the following initial conditions:

(1) $u = 1$ on the hyperbola $xy = 1$.
(2) $u = 1$ on the circle $x^2 + y^2 = 1$.

Is it possible to prescribe the initial condition on the curve $y = e^x$? Justify.

Exercise 6.3

Solve the IVP $u_x + u_y = 1$ with the initial condition $u = \phi(x)$ on the line $y = 2x$, where ϕ is a given C^1 function.

Exercise 6.4

Find the general solution of $au_x + bu_y + cu = 0$, where a, b and c are constants.

Exercise 6.5

Solve the PDE $(x+1)^2 u_x + (y-1)^2 u_y = (x+y)u$ with the initial condition $u(x,0) = -(1+x)$ for $-1 < x < \infty$.

Exercise 6.6

Solve the following Cauchy problems in the indicated regions:

(1) $xu_x + yu_y = ku$, $x \in \mathbb{R}$, $y \geqslant \alpha > 0$; $u(x, \alpha) = F(x)$, where k and α are fixed constants and F is a given smooth function.

(2) $(x+2)u_x + 2yu_y = \alpha u$; $u(-1, y) = \sqrt{|y|}$, where $\alpha \in \mathbb{R}$.

(3) $yu_x - xu_y = 0$; $u(x, 0) = x^2$.

(4) $x^2 u_x - y^2 u_y = 0$; $u(1, y) = F(y)$.

Exercise 6.7

Find the characteristic curves of the following PDE:

(1) $(x^2 - y^2 + 1)u_x + 2xyu_y = 0$.

(2) $2xyu_x - (x^2 + y^2)u_y = 0$.

Exercise 6.8

Solve the following IVP:

(1) $u_t + (x \cos t)u_x = 0$, $u(x, 0) = \dfrac{1}{1+x^2}$, $x \in \mathbb{R}$, $t > 0$.

(2) $u_t + x^2 u_x = 0$, $u(x, 0) = \phi(x)$, $x \in \mathbb{R}$, $t > 0$.

(3) $u_t + \dfrac{1}{1+|x|} u_x = 0$, $u(x, 0) = \phi(x)$, $x \in \mathbb{R}$, $t > 0$.

(4) $u_t + (x+t)u_x + t(x+1)u = 0$, $u(x, 0) = \phi(x)$, $x \in \mathbb{R}$, $t > 0$.

Exercise 6.9

Discuss the following Cauchy problems for quasi-linear equations:

(1) $uu_x + u_y = 0$, $u(x, 0) = x$.

(2) $uu_x + u_y = 1$, $u(x, x) = x/2$, $x \in [0, 1]$.

Exercise 6.10

Sketch the characteristic curves of the PDE $uu_x + u_y = 0$ with the following initial conditions[1]:

(1) $u(x, 0) = \begin{cases} 0 & \text{if } x < 0, \\ 1 & \text{if } x \geqslant 0. \end{cases}$

---

[1] The initial conditions given here are not continuous functions. Nevertheless, there is no difficulty in sketching the characteristics in the smooth regions. A little care is needed only at the point of discontinuity.

(2) $u(x,0) = \begin{cases} 1 & \text{if } x < 0, \\ 0 & \text{if } x \geq 0. \end{cases}$

(3) $u(x,0) = \begin{cases} 0 & \text{if } x < 0, \\ 1 & \text{if } x > 1. \end{cases}$

and, $u(x,0)$ is smooth and increasing in $[0,1]$ satisfying $u(0,0) = 0$ and $u(1,0) = 1$.

Exercise 6.11

Solve the IVP $p^2 + q^2 = x^2 + y^2$, $u(x,0) = a + bx^2$, where a and b are real constants.

Exercise 6.12

Solve the following non-linear PDE with prescribed initial conditions:

(1) $u_y = u_x^3$, $u(x,0) = 2x^{3/2}$, $x > 0$.

(2) $xu_x + yu_y = (1/2)\left(u_x^2 + u_y^2\right)$, $u(x,0) = a + bx^2$, where a and b are real constants.

Exercise 6.13

Find the integral surface of the equation $xu_x^2 + yu_y = u$ passing through the line $y = 1$, $x + z = 0$.

Exercise 6.14

Solve the IVP for the Burgers' equation $u_t + uu_x = 0$, with the initial condition $u_0(x) = u(x,0) = x^3$, in the upper-half plane $t > 0$.

Exercise 6.15

Solve the IVP: $u_x + u^2 u_y = 0$, $u(0,y) = y$, $y \in \mathbb{R}, x > 0$.

Exercise 6.16

Solve the IVP: $u_x + u_y = 1$ satisfying the condition that $u = 1$ on the unit circle $x^2 + y^2 = 1$.

Exercise 6.17

Show that there are two solutions given by $u(x,y) = \pm(1/\sqrt{2})(x+y-1)$ of the equation $p^2 + q^2 = 1$ satisfying the initial condition $u(x,y) = 0$ on the line $x + y = 1$.

Exercise 6.18

For the PDE $x^2 u_x + y^2 u_y = (x+y)u$, show that there are general solutions of the following forms:

(i) $F\left(\dfrac{x-y}{u}, \dfrac{xy}{u}\right) = 0$,

(ii) $u = xyf\left(\dfrac{x-y}{u}\right)$,

(iii) $u = xyg\left(\dfrac{x-y}{xy}\right)$,

where F, f and g are arbitrary functions.

Exercise 6.19

Find a complete integral of the equation $xu_x^2 + yu_y = u$ and then find an integral surface of which the line $y = 1, x + z = 0$ is a generator.

Exercise 6.20

Show that the PDE $2xu + u_y^2 = x(xu_x + yu_y)$ has a complete integral $z = -a^2x + axy + bx^2$, where a and b are arbitrary constants. Also, obtain a solution that is not in this class by constructing an envelope of a family of curves.

Exercise 6.21

Find a complete integral of the PDE $u_x - u_y = x^2 + y^2$. Hence or otherwise, solve the IVP $u_x - u_y = x^2 + y^2$, $u(x,x) = x^2$, for all $x \in \mathbb{R}$.

Hamilton–Jacobi Equation

Exercise 6.22

Obtain the coupled $2n$-dimensional system of characteristic equations for the HJE $u_t + H(x, Du) = 0$, where $Du = \left(\dfrac{\partial u}{\partial x_1}, \cdots, \dfrac{\partial u}{\partial x_n}\right)$ is the spatial gradient and H is a Hamiltonian.

Exercise 6.23

The following minimization problem comes from the calculus of variation. Fix $x \in (-1, 1)$ and $t \in (0, 1)$. Let $\mathfrak{Lip}$ denote the set of all the Lipschitz functions defined on the interval $[0, 1]$, taking values in $[-1, 1]$ and satisfying the condition[2] $x(t) = x$. The *exit time* of $x(\cdot) \in \mathfrak{Lip}$, denoted $\tau \in (t, 1]$ is the following: either $\tau = 1$ and $|x(s)| \leq 1$ for all $s \in [t, \tau]$; or $|x(\tau)| = 1$ and $|x(s)| < 1$ for all $s \in [t, \tau)$. Consider the Lagrangian $L(v) := L(t, x, v) = 1 + v^2/4$ and the corresponding minimization problem:

$$\min\left\{\int_t^\tau (1 + (x'(s))^2/4)\, ds : x(\cdot) \in \mathfrak{Lip}\right\}.$$

---

[2] The same notation x is used for the function and its value at t and should cause no confusion.

The *value function* is defined by

$$u(t,x) = \min\left\{\int_t^\tau L(t,x(s),x'(s))\,ds : x(\cdot) \in \mathfrak{Lip}\right\}.$$

Show that u is given by $u(t,x) = \min_v\{(\tau-t)L(t,x,v)\}$, where the minimization is over all the real numbers v. Further, show that

$$v^* = \begin{cases} 2 & \text{if } x \geq t, \\ 0 & \text{if } |x| < t, \\ -2 & \text{if } x \leq -t, \end{cases}$$

is a minimizing solution and the corresponding value function is given by

$$u(t,x) = \begin{cases} 1-|x| & \text{if } |x| \geq t, \\ 1-t & \text{if } |x| \leq t. \end{cases}$$

Find the region, where u is differentiable and show that u satisfies the following HJE wherever it is differentiable:

$$-u_t(t,x) + (u_x(t,x))^2 - 1 = 0,$$

and satisfies conditions

$$u(t,1) = u(t,-1) = 0, t \in [0,1]; u(1,x) = 0, x \in [-1,1].$$

Exercise 6.24

(a) Consider the Lagrangian defined by $L(q) = 1 + \frac{1}{4}|q|^2$ for $q \in \mathbb{R}^n$. Find the corresponding Hamiltonian using the Legendre transformation.

(b) Consider the value function corresponding to the Lagrangian in (a):

$$u(x,t) = \min\left[\int_0^t \left(1 + \frac{1}{4}w'(s)^2\right)ds + \frac{1}{2}w(0)^2\right],$$

where the minimum is taken over all smooth trajectories w satisfying the condition $w(t) = x$. Using the Hopf–Lax formula, find u explicitly, write down the HJE and the initial condition it satisfies.

Exercise 6.25

Brachistochrone Problem: This is a well-known and famous problem due to Johann Bernoulli. A frictionless bead located in a vertical plane at a point $A(x_0, y_0)$ slides along a wire under the force of gravity alone whose other end is fixed in the vertical plane at

$B(x_f, y_f)$. The problem is to find the shape of the curve (wire) so that the bead slides from A to B in shortest possible time (Greek: *brachistos* means *shortest* and *chronos* means *time*). Derive the minimizing functional and the E–L equations, and solve the same.

Exercise 6.26

Catenary: Consider a chain with uniform mass density of given length hanging freely (under the gravitational force) between two fixed points. What is the shape of the chain? Formulate the problem and solve it. This problem was posed by Galileo in 1630 and his anticipated solution that the shape is a parabola was proved wrong (see Figure 6.4[b]).

Exercise 6.27

Consider the functional $J(y) = \int_0^1 y(y')^2 dx$, defined for C^2 functions $y = y(x)$, $x \in [0, 1]$, subject to the conditions $y(0) = y(1) = 0$. Show that the corresponding E–L equation is $(y')^2 = \dfrac{d}{dx}(2yy')$ and that $y \equiv 0$ is a critical point. In fact, $y \equiv 0$ is the only critical point satisfying the boundary conditions. Further show that $y \equiv 0$ is neither a minimum nor a maximum for the functional J.

Exercise 6.28

The following example is known as the Dirichlet Principle. Let Ω be a given open bounded set in $\mathbb{R}^2$ with smooth boundary $\partial\Omega$. A smooth (at least C^2) function defined in Ω and satisfying the boundary condition $y(x) = g(x)$ for $x \in \partial\Omega$, where g is a given function, obviously defines a smooth surface in $\mathbb{R}^3$. The optimization problem involves in finding a smooth function y that minimizes the functional defined by $J(y) = \int_\Omega |\nabla y(x)|^2 \, dx$. If $\bar{y}$ is an optimal solution, show that $\bar{y}$ satisfies the Laplace equation $\Delta\bar{y}(x) = 0$ in Ω along with the boundary condition $\bar{y} = g$ on $\partial\Omega$. This equation is also satisfied by the electric potential in which a static two-dimensional electric field is distributed. The functional of the form $\dfrac{1}{2}\int_\Omega |\nabla y|^2 - \int_\Omega fv$ also represents the strain energy functional.

Exercise 6.29

Assume that the function $L : \mathbb{R}^n \to \mathbb{R}$ is continuous and satisfies the coercivity condition $\dfrac{L(q)}{|q|} \to \infty$ as $|q| \to \infty$. Consider the maximization problem

$$H(p) := \sup_{q \in \mathbb{R}^n} \{p \cdot q - L(q)\}.$$

Given $p \in \mathbb{R}^n$, show that there exists a $q = q(p) \in \mathbb{R}^n$ such that $H(p) = p \cdot q(p) - L(q(p))$. Find $H(p)$ when $L(q) = |q|^2/2$.

First-Order Partial Differential Equations | 199

Exercise 6.30

Let E be a closed subset of $\mathbb{R}^n$ and g be its indicator function: $g(x) = 0$ for $x \in E$ and $g(x) = \infty$ for $x \notin E$. Using the Hopf–Lax formula, show that the solution of the HJE

$$\begin{cases} u_t + |Du|^2 = 0 \text{ in } \mathbb{R}^n \times (0, \infty), \\ u(x, 0) = g(x) \text{ in } \mathbb{R}^n, \end{cases}$$

is given by $u(x,t) = (d(x))^2/(4t)$, where $d(x) = d(x, E)$, $x \in \mathbb{R}^n$, is the distance function from the set E.

Exercise 6.31

Let $\mathcal{M} = \{y \in C^1[-1, 1] : y(-1) = 0, y(1) = 1\}$. Show that $\inf_{y \in \mathcal{M}} J(y) = 0$, where the functional J is defined as $J(y) = \int_0^1 L(y, y') \, dx$, for the Lagrangian $L(y, y') = y^2(y' - 1)^2$. Further show that there is no $y \in \mathcal{M}$ such that $J(y) = 0$. Construct a Lipschitz continuous function $\tilde{y}$, satisfying the conditions $\tilde{y}(-1) = 0$, $\tilde{y}(1) = 1$, for which $J(\tilde{y}) = 0$.

Exercise 6.32

Consider the HJE

$$\begin{cases} u_t + (1/2)|Du|^2 = 0 \text{ in } \mathbb{R}^n \times (0, \infty), \\ u(x, 0) = g(x). \end{cases}$$

Find the solution for this IVP when $g(x) = |x|$ and $g(x) = -|x|$, using the Hopf–Lax formula.

Conservation Laws

Exercise 6.33

Consider the Burgers' equation $u_t + uu_x = 0$ for $x \in \mathbb{R}$, $t > 0$ with the initial condition $u(x, 0) = u_0(x)$, where u_0 is given by

$$u_0(x) = \begin{cases} 1 \text{ if } x \leq 0, \\ 1 - x \text{ if } 0 < x < 1, \\ 0 \text{ if } x \geq 1. \end{cases}$$

(a) Show that the characteristic curves do not meet till $t = 1$ and hence, find the solution $u(x, t)$ for all $x \in \mathbb{R}$ and $0 < t < 1$.

(b) Construct a curve of discontinuity $s(t)$ for $t \geqslant 1$ satisfying $s(1) = 1$. Construct a generalized solution which is smooth in the regions $x < s(t)$ and $x > s(t)$ for $t \geqslant 1$ and, satisfying the Rankine–Hugoniot condition across the curve $x = s(t)$.

Exercise 6.34

Construct a generalized solution for the IVP $u_t + uu_x = 0$ for $x \in \mathbb{R}$, $t > 0$ and $u(x, 0) = u_0(x)$, $x \in \mathbb{R}$, where is u_0 is given by

$$u_0(x) = \begin{cases} 0 & \text{if } x \leqslant 0, \\ 1 & \text{if } 0 \leqslant x \leqslant 1, \\ 0 & \text{if } x \geqslant 1. \end{cases}$$

Exercise 6.35

Construct the characteristics of the generalized Burgers' equation $u_t + (f(u))_x = 0$, $x \in \mathbb{R}$, $t > 0$, satisfying the initial condition $u(x, 0) = u_0(x)$ for $x \in \mathbb{R}$, where u_0 is a C^1 function and $f(u) = |u|^p/p$, $p \geqslant 2$. Find a C^1 solution in a region in the upper-half plane $t > 0$.

Exercise 6.36

Consider the Burgers' equation in the upper-half plane $t > 0$. Find the solution using the method of characteristics, in the appropriate regions in the upper-half plane, satisfying the following initial conditions:

(a) $u_0(x) = x^2$ and (b) $u_0(x) = -x|x|$, $x \in \mathbb{R}$.

Exercise 6.37

Consider the Burgers' equation[3] $u_t + uu_x = 0$, with the initial condition

$$u_0(x) = \begin{cases} 0 & \text{if } x < 0, \\ 1 & \text{if } x > 0. \end{cases}$$

Find the minimizer in the corresponding Lax–Oleinik formula and thus, write down the formula for the solution.

Exercise 6.38

Consider the Burgers' equation and find the solution using Lax–Oleinik formula with the initial values as in Exercises 6.33 and 6.34 and also with the initial value

$$u_0(x) = \begin{cases} 0 & \text{if } x < 0, \\ 1 & \text{if } x > 1, \end{cases}$$

and u_0 is non-decreasing in the interval $[0, 1]$ and $u_0 \in C^1(\mathbb{R})$.

---

[3] The solution of the inhomogeneous Burgers' equation $u_t + uu_x + x = 0$, with the initial condition $u(x, 0) = 1$ is interesting and leads to an example of *Lagrangian manifold* (see Ref. [18]).

Exercise 6.39

Find a C^1 solution in a region in the upper-half space of the two-dimensional Burgers' equation $u_t + u^2(u_x + u_y) = 0$ satisfying the initial condition $u(x, y, 0) = x + y$.

6.5 Solutions

Exercise 6.1

The given equation is a linear equation. The characteristic equations are given by $\dfrac{dx}{dt} = x + 2$ and $\dfrac{dy}{dt} = 2y$. If we use the variable x to describe the characteristics, then we have $\dfrac{dy}{dx} = \dfrac{2y}{x+2}$. Hence, $y = C(x+2)^2$, where C is a non-zero arbitrary constant. These are parabolas with common vertex at $(-2, 0)$. Now along the characteristic curve, we have

$$\frac{d}{dx}u(x, C(x+2)^2) = u_x + \frac{2C(x+2)^2 u_y}{x+2} = \frac{(x+2)u_x + 2yu_y}{x+2} = \frac{2u}{x+2}$$

provided $x \neq -2$. Note that $(0, -2)$ is a singularity of the PDE in the sense that the left-hand side of the equation vanishes. Now, fix $y_0 \neq 0$ and consider the characteristic curve through $(-1, y_0)$, which is given by $y = y_0(x+2)^2$ as $C = y_0$. Thus, u along this characteristic curve satisfies

$$\frac{du}{dx}(x, y_0(x+2)^2) = \frac{2u}{x+2}.$$

Integrating this, we get $u = \sqrt{|y_0|}(x+2)^2 = \sqrt{|y|}|x+2| = \sqrt{|y|}(x+2)$ for $x > -2$, using the given initial condition. This is valid for $y \neq 0$. Note that all the characteristic curves meet at the common point $x = -2$. We cannot solve the IVP when the initial curve is the x-axis. The initial conditions in parametric form are given by $x_0(s) = -1$, $y_0(s) = s$ and $u_0(s) = \sqrt{|s|}$, and the transversality condition is

$$a(x_0(s), y_0(s), u_0(s))\frac{dy_0}{ds} - b(x_0(s), y_0(s), u_0(s))\frac{dx_0}{ds} = 1 \neq 0$$

on the initial curve. But, if the initial curve is the x-axis, then the transversality condition is not satisfied at the point $(-2, 0)$, as the x-axis is a tangent to all the parabolas $y = C(x+2)^2$ (see Figure 6.1).

Note: The same procedure works if we replace the initial line $x = -1$ by any other line $x = \alpha$, $\alpha \neq -2$. However, if we consider a line $y = \alpha x + \beta$, $\alpha \neq 0$ as the initial curve, caution should be exercised as such a line may intersect a characteristic curve at two points. The same remark also applies to the line $y = \beta$ for $\beta > 0$.

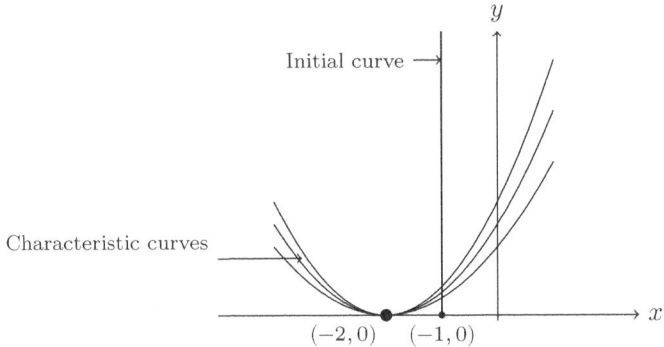

Figure 6.1 | Initial curve and characteristics.

Exercise 6.2

Note that the given equation is the Euler's equation satisfied by a homogeneous function of order 2. It is easy to see that the straight lines $y = cx$, c a constant, are the characteristic curves as the characteristic equation in implicit form is given by $\dfrac{dy}{dx} = \dfrac{y}{x}$. Along any of these curves, u satisfies

$$\frac{d}{dx}u(x,cx) = u_x(x,cx) + u_y(x,cx)\cdot c = u_x + \frac{yu_y}{x} = \frac{2u}{x},$$

whose solution is given by $u(x,cx) = kx^2$. Note that k depends on c, which itself varies from one characteristic to another. Thus, we have the general solution $u(x,y) = k\left(\dfrac{y}{x}\right)x^2$, where k is an arbitrary function.

(1) Now applying the initial condition $u = 1$ on the hyperbola $y = 1/x$, we get $1 = k\left(\dfrac{1}{x^2}\right)x^2$ or $k\left(\dfrac{1}{x^2}\right) = \dfrac{1}{x^2}$ and hence, the required solution is $u(x,y) = \dfrac{y}{x}x^2 = xy$.

(2) The initial condition in this case implies that

$$1 = u(\cos\theta, \sin\theta) = k\left(\frac{\sin\theta}{\cos\theta}\right)\cos^2\theta, \text{ for } \theta \in \mathbb{R}.$$

This immediately gives $k(\tan\theta) = \sec^2\theta$ and hence, $k(x) = 1 + x^2$, for $x \in \mathbb{R}$. Thus, the required solution is $u(x,y) = x^2 + y^2$.

There will be difficulties if, instead, we prescribe the initial condition on the x-axis as the characteristic curves do not intersect the x-axis, except at the origin. Now consider the initial curve $y = e^x$. Considering the function $f(x) = e^x - cx$, it is easy to see that the initial curve $y = e^x$ touches tangentially from below the characteristic curve $y = ex$, that is, when $c = e$. Indeed, the transversality condition is not satisfied on the initial curve at the point $(1, e)$. Further, this initial curve does not intersect any characteristic $y = cx$ when

$c < e$ and intersects exactly at two points when $c > e$. Hence, the problem cannot be solved by the method of characteristics. In fact, the IVP is not well-posed.

Exercise 6.3

The characteristic equation is $\dfrac{dy}{dx} = 1$ and hence, the characteristic curves are $y = x + C$. Using the initial curve $x = s, y = 2s$, we have $y = x + s$. These are curves parallel to the diagonal $y = x$. Along the characteristic curve u satisfies $\dfrac{du}{dx} = 1$, that is, $u(x, x + s) = x + k = x + k(s)$. Thus, $u(x,y) = x + k(y-x)$, where the function k to be determined. It is easy to see using the initial condition that $k(x) = \phi(x) - x$. Thus, the solution is $u(x,y) = 2x - y + \phi(y - x)$.

An alternative procedure is the following. The general solution of the homogeneous equation $u_x + u_y = 0$ is $u(x,y) = \psi(y - x)$ for any C^1 function ψ. This in turn gives the general solution of $u_x + u_y = 1$ as $u(x,y) = \psi(y - x) + x$ (or $\psi(y - x) + y$). Using the initial condition, we then obtain that $\phi(x) = u(x, 2x) = \psi(2x - x) + x$ or $\psi(x) = \phi(x) - x$. Thus, $u(x,y) = 2x - y + \phi(y - x)$, as before.

The same procedure works if we replace the initial curve by the line $y = \alpha x$ where α is any real number other than 1. In this case, we have $\phi(x) = u(x, \alpha x) = \psi(\alpha x - x) + x$, which gives $\psi(x) = \phi\left(\dfrac{x}{\alpha - 1}\right) - \dfrac{x}{\alpha - 1}$, as $\alpha \neq 1$. On the other hand, if we prescribe initial condition on the line $y = x$, which is a characteristic for the given equation, it becomes a characteristic Cauchy problem. In this scenario, we see that any function of the form $u(x,y) = x + \psi(y - x)$ is a solution of the equation satisfying $u(x,x) = x + \psi(0)$. Thus, the initial function cannot be arbitrary and we have non-uniqueness.

Exercise 6.4

If either $a = 0$ or $b = 0$, the given PDE reduces to an ODE. We may, therefore, assume that $ab \neq 0$. The characteristic equations are $\dfrac{dx}{dt} = a$ and $\dfrac{dy}{dt} = b$. Thus, a characteristic curve passing through a fixed point (x_0, y_0), that is, $x(0) = x_0, y(0) = y_0$, is given, in parametric form, by $x(t) = at + x_0, y(t) = bt + y_0$. Along the characteristic, u satisfies $\dfrac{du}{dt} = -cu$, which, upon solving, gives $u = ke^{-ct}$, where $k = k(x_0, y_0)$ is a constant that depends on the initial point (x_0, y_0). Thus, $u(x,y) = k(x - at, y - bt)e^{-ct}$ is the general solution where k is an arbitrary function. Eliminating t, we may write the general solution as $u(x,y) = f(bx - ay) \exp\left(-(c/2)(x/a + y/b)\right)$, where f is an arbitrary C^1 function.

Exercise 6.5

The initial conditions in parametric form can be written as

$$x_0(s) = s, \; y_0(s) = 0, \; u_0(s) = -(1 + s), \; \text{for } s \in (-1, \infty)$$

and, since $a = (x+1)^2$, $b = (y-1)^2$, the transversality condition

$$a(x_0(s), y_0(s))\frac{dy_0}{ds} - b(x_0(s), y_0(s))\frac{dx_0}{ds} = (s+1)^2 \cdot 0 - 1 \cdot 1 \neq 0$$

is satisfied. The characteristic equations are $\frac{dx}{dt} = (x+1)^2$ and $\frac{dy}{dt} = (y-1)^2$. Integrating these equations, using the initial conditions $x(0, s) = s$, $y(0, s) = 0$, we get $x = x(t, s) = -1 - (t - c_1)^{-1}$ and $y = y(t, s) = 1 - (t+1)^{-1}$, where $c_1 = (1+s)^{-1}$. Along the characteristic curve $x = x(t, s)$, $y = y(t, s)$, the unknown u satisfies

$$\frac{du}{dt} = (x+y)u = -((t-c_1)^{-1} + (t+1)^{-1})u.$$

This can be integrated using the initial condition $u(0, s) = u_0(s) = -(1+s)$ to obtain $u(t, s) = (t - c_1)^{-1}(t+1)^{-1}$. Solving for t, s in terms of x and y, the solution of the IVP is given by $u(x, y) = (x+1)(y-1)$.

Exercise 6.6

(1) It is similar to the Exercise 6.2 except for the initial curve. The characteristic curves are given by $y = cx$ and the general solution $u(x, y) = g\left(\frac{y}{x}\right)x^k$, where g is an arbitrary function.
To verify the transversality condition, note that the initial curve is parametrically represented as: $x_0(s) = s, y_0(s) = \alpha$. Since $a = x, b = y$, we have

$$a(x_0(s), y_0(s))y_0'(s) - b(x_0(s), y_0(s))x_0'(s) = s \cdot 0 - \alpha \cdot 1 = \alpha \neq 0.$$

Now, using the initial condition, we get $F(x) = u(x, \alpha) = g\left(\frac{\alpha}{x}\right)x^k$. By putting $\frac{\alpha}{x} = v$, we get the arbitrary function as $g(v) = F\left(\frac{\alpha}{v}\right)\left(\frac{v}{\alpha}\right)^k$. Thus, we get the solution to IVP

$$u(x, y) = F\left(\frac{\alpha x}{y}\right)\left(\frac{y}{\alpha x}\right)^k x^k = F\left(\frac{\alpha x}{y}\right)\left(\frac{y}{\alpha}\right)^k.$$

(2) The characteristics and the initial curves are exactly the same as in Exercise 6.1. The characteristic curves are given by $y = C(x+2)^2$. By the same procedure, we get the solution as $u = \sqrt{|y_0|}(x+2)^\alpha = \frac{\sqrt{|y|}}{|x+2|}(x+2)^\alpha$ for $x > -2$. Thus, u is defined for $x > -2$ and $y \neq 0$.

(3) On either of the co-ordinate axes, the given PDE is reduced and can easily be analysed. Thus, we may restrict the discussion to any of the four (open) quadrants. The characteristic curves in implicit form are the circles $x^2 + y^2 = c$ obtained by

solving $\dfrac{dy}{dx} = -\dfrac{x}{y}$. The initial curve is $x_0(s) = s, y_0(s) = 0$ and since $a = y, b = -x$, the transversality condition

$$a(x_0(s), y_0(s))y_0'(s) - b(x_0(s), y_0(s))x_0'(s) = 0 \cdot 0 - (-s) \cdot 1 \neq 0$$

is satisfied, as we are avoiding the origin. The equation for u in parametric form is $\dfrac{du}{dt} = 0$ with the initial condition $u(s, 0) = s^2$, which gives $u(t, s) = s^2$. Indeed $x(t, s) = s \cos t$, $y(t, s) = -s \sin t$ and hence, the solution is $u(x, y) = x^2 + y^2$. However, observe that the form of the solution is valid everywhere in $\mathbb{R}^2$.

We make the following remark. Even though each characteristic curve intersects the initial curve (x-axis) at two points, we do not face any problem in this case since u has same value on both intersecting points and u is constant along the characteristic curve. This is due to the fact that the initial value, $u_0(x) = x^2$, is an even function of x. The situation may change if we take a different initial function. For example, consider the case $u_0(x) = x$. The above procedure now yields the solution as

$$u(x, y) = \begin{cases} \sqrt{x^2 + y^2} & \text{if } x \geqslant 0, \\ -\sqrt{x^2 + y^2} & \text{if } x < 0. \end{cases}$$

Observe that for this solution, u_y does not exist at the origin $(0, 0)$. Thus, this solution is valid only in the punctured plane $\mathbb{R}^2 \setminus \{(0, 0)\}$ and the singularity at the origin is visible.

(4) The initial curve is $x_0(s) = 1, y_0(s) = s$ and the initial value is $u_0(s) = F(s)$. Since $a = x^2, b = -y^2$, the transversality condition

$$a(x_0(s), y_0(s))y_0'(s) - b(x_0(s), y_0(s))x_0'(s) = 1 \cdot 1 - (-s^2) \cdot 0 \neq 0$$

is satisfied. The characteristic curves are given by $\dfrac{dx}{dt} = x^2, \dfrac{dy}{dt} = -y^2$ which upon solving and using initial conditions, we get $x = x(t, s) = 1/(1-t), y = y(t, s) = s/(1-st)$. Inverting these equations, we arrive at $t = 1 - 1/x$ and $s = xy/(x + xy - y)$. Now the equation for u is $\dfrac{du}{dt} = 0$ and hence, u is constant along the characteristics. Using the initial values, we get $u(s; t) = F(s)$ and thus, the solution is $u(x, y) = F(xy/(x + xy - y))$, which is defined for all x, y such that $x + xy - y \neq 0$.

Exercise 6.7

(1) The characteristic curves are obtained by solving the equation $\dfrac{dy}{dx} = \dfrac{2xy}{x^2 - y^2 + 1}$. This ODE is of the form $Mdx + Ndy = 0$, where $M(x, y) = 2xy$ and $N(x, y) = -(x^2 - y^2 + 1)$. Since $M_y = 2x$ and $N_x = -2x$ are not equal, the ODE is not an exact

206 | Notes, Problems and Solutions in Differential Equations

equation. But, since $\dfrac{N_x - M_y}{M} = -\dfrac{2}{y}$ is a function of y alone, an integrating factor μ can be obtained by solving $\dfrac{\mu'(y)}{\mu(y)} = -\dfrac{2}{y}$. Hence, $\mu(y) = \dfrac{1}{y^2}$ and we have $\dfrac{\partial}{\partial y}(\mu M) = \dfrac{\partial}{\partial x}(\mu N)$. Let $\phi(x, y) = (x^2/y) + y + (1/y)$, $y \neq 0$. Then

$$d\phi = \phi_x \, dx + \phi_y \, dy = y^{-2}[2xy \, dx - (x^2 - y^2 + 1) \, dy] = 0,$$

using the given equation. Hence, the characteristic curves are the curves $\phi = C$ or $x^2 + y^2 + Cy + 1 = 0$ for appropriate constants C.

(2) The characteristic curves are given by the equation $\dfrac{dy}{dx} = -\dfrac{x^2 + y^2}{2xy}$, which is a homogeneous equation. Thus, introducing a new dependent variable $v = \dfrac{y}{x}$, the equation reduces to $\dfrac{dv}{dx} = \dfrac{-(1 + 3v^2)}{2xv}$, which is in the variable separable form. Solving this ODE and reverting back to the variables x and y, we see that characteristic curves are given by $x^3 + 3xy^2 = C$, where C is a constant.

Exercise 6.8

(1) The characteristic equation in the $x - t$ plane is given by (note that, the variable y is replaced by t) $\dfrac{dx}{dt} = x \cos t$. Therefore, the characteristic curves are $x = x(t) = Ae^{\sin t}$, which gives $A = xe^{-\sin t}$. Further, u satisfies $\dfrac{du}{dt} = 0$ along the characteristic curve $(t, x(t))$ and hence, u is constant along a characteristic. We thus have $u(x(t), t) = u(Ae^{\sin t}, t) = C(A)$, a constant depending on A. Putting $t = 0$ and using the initial condition, we get $C(A) = (1 + A^2)^{-1}$. Thus,

$$u(x, t) = C\left(xe^{-\sin t}\right) = \dfrac{e^{2 \sin t}}{x^2 + e^{2 \sin t}}.$$

(2) Solving the characteristic equation $\dfrac{dx}{dt} = x^2$, the characteristic curves $x = x(t) = (A - t)^{-1}$, which gives $A = (xt + 1)/x$. Again along the characteristic curve $u(x(t), t) = u((A - t)^{-1}, t) = C(A)$, a constant depends on A. Putting $t = 0$ and using the initial condition, we get $C(A) = \phi(A^{-1})$. Thus,

$$u(x, t) = C((xt + 1)/x) = \phi(x/(xt + 1)).$$

Note that the solution is defined only in the region $xt + 1 > 0$ of the upper-half plane.

(3) In order to determine the solution at a point (x_1, t_1), where $x_1 \in \mathbb{R}$ and $t_1 > 0$, we need to find the characteristic curve passing through it, using the characteristic equation $\dfrac{dx}{dt} = \dfrac{1}{1 + |x|}$ with the initial condition $x(t_1) = x_1$. Then find the point x_0 where this

curve meets the x-axis. The solution is then given by $u(x_1, t_1) = \phi(x_0)$. Because of the presence of the modulus sign in the characteristic equation, we need to work for the cases $x_1 > 0$ and $x_1 \leqslant 0$ separately. First note that any characteristic $x(t)$ is an increasing function of t. Consider the case $x_1 > 0$. As the solution $x(t)$ remains positive for t close to t_1, it satisfies the equation $\dfrac{dx}{dt} = \dfrac{1}{1+x}$. An integration then gives $x + x^2/2 = t + A$, where $A = x_1 + \dfrac{x_1^2}{2} - t_1$. This solution is valid for those t for which $x(t) \geqslant 0$. In particular, $x_0 = x(0)$ satisfies $x_0 + \dfrac{x_0^2}{2} = A$, from which we obtain $x_0 = -1 + (1 + 2x_1 + x_1^2 - 2t_1)^{\frac{1}{2}}$. Provided that $x_0 \geqslant 0$, which in turn requires the condition $x_1 \geqslant -1 + \sqrt{1 + 2t_1}$, the solution is given by

$$u(x_1, t_1) = \phi\left(-1 + (1 + 2x_1 + x_1^2 - 2t_1)^{\frac{1}{2}}\right).$$

On the other hand, we observe that $x(t_*) = 0$ for $t_* = -A$. If $t_* > 0$, then the characteristic $x(t)$ satisfies the inequality $x(t) < 0$ and satisfies the equation $\dfrac{dx}{dt} = \dfrac{1}{1-x}$, for $t < t_*$. An integration then shows that x satisfies $x - x^2/2 = t - t_*$ for $t < t_*$. In particular, $x_0 = x(0)$ satisfies $x_0 - x_0^2/2 = -t_*$, which in turn gives $x_0 = 1 - \sqrt{1 + 2t_*}$. The condition $t_* > 0$ implies that $x_1 \in [0, -1 + \sqrt{1 + 2t_1})$. Thus, the solution in this case is given by

$$u(x_1, t_1) = \phi\left(-1 + (1 + 2t_1 - 2x_1 - x_1^2)^{\frac{1}{2}}\right).$$

An exactly similar procedure can be adapted when $x_1 < 0$ and is simpler. The solution $x(t)$ of the characteristic solution satisfies the inequality $x(t) < x_1$ and thus satisfies the equation $\dfrac{dx}{dt} = \dfrac{1}{1-x}$, for $t < t_1$. Integrating this equation as before and after some simplifications, we obtain that $x_0 = x(0) = 1 - \sqrt{1 - 2B}$, where $B = x_1 - x_1^2/2 - t_1$, which is negative. Thus,

$$u(x_1, t_1) = \phi\left(1 - (1 + 2t_1 - 2x_1 - x_1^2)^{\frac{1}{2}}\right).$$

Since (x_1, t_1) is an arbitrary point, summing up the above discussion, we see that the solution is given by (remember that $t > 0$)

$$u(x, t) = \begin{cases} \phi\left(-1 + (1 - 2t + 2x + x^2)^{\frac{1}{2}}\right) & \text{if } x \geqslant -1 + \sqrt{1 + 2t}, \\ \phi\left(1 - (1 + 2t - 2x - x^2)^{\frac{1}{2}}\right) & \text{if } x \in \left[0, -1 + \sqrt{1 + 2t}\right), \\ \phi\left(1 - (1 + 2t - 2x + x^2)^{\frac{1}{2}}\right) & \text{if } x < 0. \end{cases}$$

It is not difficult to check that u is indeed a C^1 function in the upper-half plane and satisfies the initial condition.

(4) The characteristic curve satisfies the first-order linear ODE $\frac{dx}{dt} - x = t$, which can be solved to get $x = x(t) = Ae^t - t - 1$, where A is a constant. Now $u = u(t) = u(x(t), t)$ along the characteristics satisfies $\frac{du}{dt} = -t(x+1)u = -t(-t + Ae^t)u$. This can be integrated to get

$$u = C \exp\left(t^3/3 - A(te^t - e^t)\right),$$

where $C = C(A)$ is a constant that depends on the characteristics. Thus, we have

$$u(x,t) = C\left(e^{-t}(x+t+1)\right) \exp(t^3/3 - (x+t+1)(t-1)).$$

From the initial condition, we get $\phi(x) = u(x,0) = C(x+1)e^{x+1}$ and so $C(x) = \phi(x-1)e^{-x}$. Thus,

$$C(e^{-t}(x+t+1)) = \phi\left(e^{-t}(x+t+1) - 1\right) \exp\left(-e^{-t}(x+t+1)\right).$$

Hence, the required solution is

$$u(x,t) = \phi\left(e^{-t}(x+t+1) - 1\right) \exp[t^3/3 + (x+t+1)(1 - t - e^{-t})].$$

Exercise 6.9

(1) The initial space curve in the parametric form is given by $x_0(s) = s$, $y_0(s) = 0$, $u_0(s) = s$. Since $a = u, b = 1$, the transversality condition

$$a(x_0(s), y_0(s))y_0'(s) - b(x_0(s), y_0(s))x_0'(s) = s \cdot 0 - 1 \cdot 1 \neq 0$$

is satisfied. The characteristic curves are given by $x = x(t,s) = st + s$, $y = y(t,s) = t$, $u = u(t,s) = s$, which are obtained by solving the equations $\frac{dx}{dt} = u$, $\frac{dy}{dt} = 1$, $\frac{du}{dt} = 0$. We now obtain s, t in terms of x, y from the first two equations and substituting the same in the third equation, we get the solution as $u(x,y) = x/(y+1)$.

(2) The parametric form of the initial space curve is given by

$$x_0(s) = s, y_0(s) = s, u_0(s) = s/2,$$

where $0 < s \leqslant 1$. Since $a = u, b = 1$, the transversality condition

$$a(x_0(s), y_0(s))y_0'(s) - b(x_0(s), y_0(s))x_0'(s) = s/2 - 1 \neq 0$$

is satisfied for $0 < s \leqslant 1$. The characteristic curves are given by

$$x = x(t,s) = (t^2/2) + (st/2) + s, \; y = y(t,s) = t + s, \; u = u(t,s) = t + s/2,$$

which are obtained by solving the equations $\frac{dx}{dt} = u$, $\frac{dy}{dt} = 1$ and $\frac{du}{dt} = 1$. Now solving for s,t from the first two equations, in terms of x and y and, substituting them in the third equation, we get the required solution as $u(x,y) = (4y - 2x - y^2)/(2(2-y))$.

Exercise 6.10

(1) The PDE is exactly as in Exercise 6.9(1), but the initial value is discontinuous. This leads to an interesting phenomena, called a *rarefaction*. The initial space curve is given by $x_0(s) = s, y_0(s) = 0, u_0(s) = 0$ for $s < 0$ and $u_0(s) = 1$ for $s > 0$. Since $a = u, b = 1$, the transversality condition, for $s < 0$

$$a(x_0(s), y_0(s))y_0'(s) - b(x_0(s), y_0(s))x_0'(s) = 0 \cdot 0 - 1 \cdot 1 \neq 0$$

and for $s > 0$

$$a(x_0(s), y_0(s))y_0'(s) - b(x_0(s), y_0(s))x_0'(s) = 1 \cdot 0 - 1 \cdot 1 \neq 0$$

is satisfied. The trouble here is the discontinuity of the initial function u_0. The characteristic equations are given by $\dfrac{dx}{dt} = u, \dfrac{dy}{dt} = 1$ and $\dfrac{du}{dt} = 0$. Upon solving, we get

$$x = x(t, s) = u_0(s)t + s, \ y = y(t, s) = t, \ u = u(t, s) = s.$$

Thus, the characteristic curves are the lines parallel to the y-axis for $y < 0$ and are the lines parallel to the diagonal line $y = x$ for $y > 0$. This leaves the region $\{0 < x < t < 1\}$ in the upper-half plane, without any characteristics. Thus, we will not be able to obtain the solution by the classical method of the method of characteristics in this region. This region, empty of characteristics, is known as a *rarefaction* (see Figure 6.2[b]). Later, in the discussion of conservation laws, we learn how to obtain the solution in this region.

(2) In this case, the initial space curve is given by $x_0(s) = 1, y_0(s) = 0, u_0(s) = 1$ for $s < 0$ and $u_0(s) = 0$ for $s > 0$. Since $a = u, b = 1$, the transversality condition

$$a(x_0(s), y_0(s))y_0'(s) - b(x_0(s), y_0(s))x_0'(s) = u_0(s) \cdot 0 - 1 \cdot 1 \neq 0$$

is satisfied. Upon solving the characteristic equations as in (1), we get

$$x = x(t, s) = u_0(s)t + s, \ y = y(t, s) = t, \ u = u(t, s) = s.$$

In this case, the characteristic curves are the lines parallel to the y-axis for $y > 0$ and lines parallel to the diagonal $y = x$ for $y < 0$. Thus, a characteristic emanating from the x-axis for $y < 0$ intersects another characteristic emanating from the x-axis for $y > 0$ and results in a region, at points of which there are two values for the unknown u, transported from initial values through characteristics. This is a more serious situation than the case (1) above, leading to the concept of *discontinuities* or *shocks* in the solution (see Figure 6.2[c]).

(3) This is a comfortable situation. Here, the characteristic curves are the lines parallel to the y-axis for $y < 0$; the slopes of the lines increases continuously as the initial point moves from $y = 0$ to $y = 1$ and then lines parallel to the diagonal $y = x$. Hence, the problem can be solved (see Figure 6.2[a]).

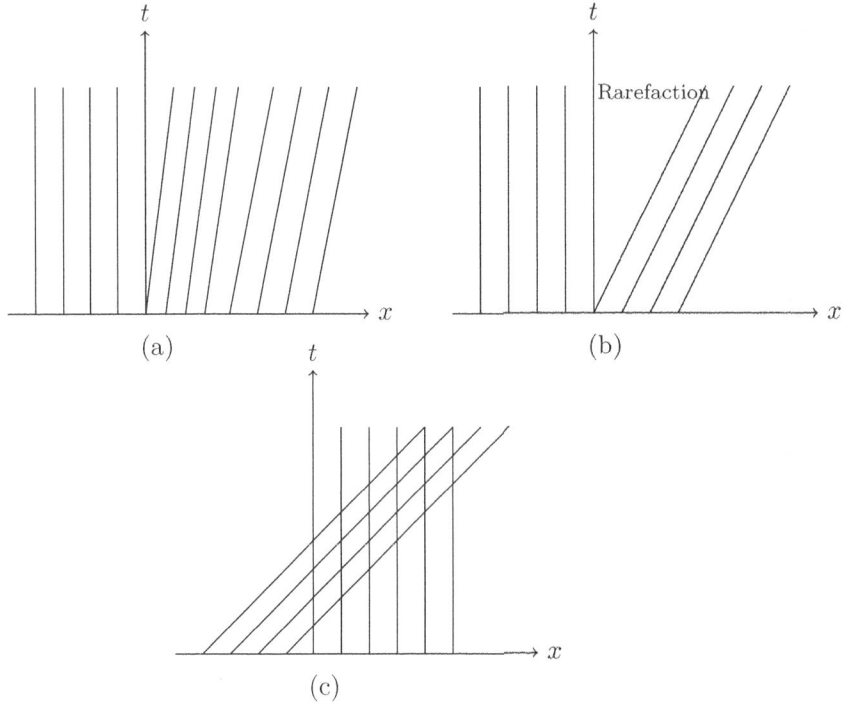

Figure 6.2 | Smooth, rarefaction and shock formation.

Exercise 6.11

The parametric form of the initial data is: $x_0(s) = s$, $y_0(s) = 0$ and $u_0(s) = a + bs^2$, and the characteristic equations are given by

$$\left.\begin{array}{l}\dfrac{dx}{dt} = F_p = 2p, \quad \dfrac{dy}{dt} = F_q = 2q, \quad \dfrac{du}{dt} = pF_p + qF_q = 2(p^2 + q^2) \\ \dfrac{dp}{dt} = -F_x - pF_z = 2x, \quad \dfrac{dq}{dt} = -F_y - qF_z = 2y.\end{array}\right\} \quad (6.4.1)$$

The initial values $p_0(s)$ and $q_0(s)$ be chosen to satisfy the equation $p_0(s)^2 + q_0(s)^2 = x_0(s)^2 + y_0(s)^2 = s^2$ and the strip condition $2bs = \dfrac{du_0}{ds} = p_0 \dfrac{dx_0}{ds} + q_0 \dfrac{dy_0}{ds} = p_0$. This implies that $p_0(s) = 2bs$ and $4b^2s^2 + q_0(s)^2 = s^2$. This forces the constant b to satisfy the condition $4b^2 \leq 1$ for a possible existence of a real solution. Apparently, there is no (real) solution if this condition is violated. We then obtain $q_0(s) = \pm \alpha s$, where $\alpha = \sqrt{1 - 4b^2}$. We first consider the positive sign. In equation (6.4.1), the equations satisfied by x and p together are linear and involve only x and p variables. They can be solved without much difficulty and we obtain

$$x(t, s) = s \cosh(2t) + 2bs \sinh(2t) \quad \text{and} \quad p(t, s) = s \sinh(2t) + 2bs \cosh(2t).$$

Similarly, $y(t,s) = \alpha s \sinh(2t)$ and $q(t,s) = \alpha s \cosh(2t)$. Next, we have $\dfrac{du}{dt} = 2(p^2 + q^2)$. Using the expressions for p and q, we obtain that $\dfrac{du}{dt} = s^2\left((1+2b)e^{4t} + (1-2b)e^{-4t}\right)$. An integration immediately shows that $u(t,s) = (s^2/4)\left[(1+2b)e^{4t} - (1-2b)e^{-4t}\right]$. The final task is to express the variables s and t in terms of x and y. Since the expressions for x and y are similar to those for p and q, it is not difficult to do this. Finally, the required solution is given by $u(x,y) = a + \alpha xy + b(x^2 - y^2)$.

If we replace α by $-\alpha$, that is, considering the negative sign, we obtain another solution. Thus, there are at least two solutions. In particular, if $b = \pm 1/2$, the two solutions are given by $u(x,y) = a \pm (1/2)(x^2 \pm y^2)$ and when $b = 0$, they are $u(x,y) = a \pm xy$.

Exercise 6.12

(1) We have $F(x,y,z,p,q) \equiv p^3 - q = 0$. The characteristic equations are given by

$$\begin{cases} \dfrac{dx}{dt} = F_p = 3p^2, \ \dfrac{dy}{dt} = F_q = -1, \ \dfrac{dz}{dt} = pF_p + qF_q = 3p^3 - q \\ \dfrac{dp}{dt} = -F_x - pF_z = 0, \ \dfrac{dq}{dt} = -F_y - qF_z = 0. \end{cases}$$

The initial space curve in parametric form is given by

$$x_0(s) = s, \ y_0(s) = 0, \ z_0(s) = 2s^{3/2}.$$

The initial values $p_0(s)$ and $q_0(s)$ be chosen to satisfy the equation

$$0 = F(s) = F(x_0(s), y_0(s), y_0(s), p_0(s), q_0(s)) = p_0(s)^3 - q_0(s)$$

and the strip condition is $z_0' = p_0 x_0' + q_0 y_0'$. It is straightforward to see that $p_0(s) = 3s^{1/2}$ and $q_0(s) = 27s^{3/2}$ satisfy the requirements. Thus, by solving the last two characteristic equations, we get $p(t,s) = 3s^{1/2}$ and $q(t,s) = 27s^{3/2}$. Now, the first three equations can be solved to obtain

$$x = x(t,s) = (1+27t)s, \ y = y(t,s) = -t, \ u = z = z(t,s) = 2s^{3/2}(1+27t).$$

Hence, $t = -y$ and $s = x/(1-27y)$. Replacing the variables s and t in terms of x and y in the expression for z, we get the solution as $u(x,y) = 2x^{3/2}(1-27y)^{-1/2}$, which is valid for $x > 0$ and $y < 1/27$.

(2) Put $v(x,y) = u(x,y) - (1/2)(x^2 + y^2)$. Then v satisfies $v_x = u_x - x$ and $v_y = u_y - y$ and, therefore,

$$v_x^2 + v_y^2 = u_x^2 + u_y^2 - 2(xu_x + yu_y) + x^2 + y^2 = x^2 + y^2.$$

But, this is the same equation as in Exercise 6.11. Further, v satisfies the initial condition $v(x,0) = u(x,0) - (1/2)x^2 = a + (b-1/2)x^2$. Hence, we obtain two solutions provided that $|b - 1/2| \leq 1/2$, that is, $0 \leq b \leq 1$. The two solutions are given by

$$v(x,y) = a \pm 2\sqrt{b(1-b)}\, xy + (b - 1/2)(x^2 - y^2)$$

or, in terms of u, we have

$$u(x,y) = v(x,y) + (1/2)(x^2 + y^2) = a + \left(\sqrt{b}\,x \pm \sqrt{1-b}\,y\right)^2.$$

In particular, when $b = 0$, the two solutions are a and $a + y^2$; when $b = 1$, they are $a + x^2$ and $a + x^2 + y^2$.

Exercise 6.13

Here, the PDE is $F(x, y, z, p, q) \equiv xp^2 + yq - z = 0$. The initial conditions in parametric form is $x_0(s) = s$, $y_0(s) = 1$, $z_0(s) = -s$. The initial values $p_0(s)$ and $q_0(s)$ satisfy the equation $x_0 p_0^2 + y_0 q_0 + z_0 = sp_0^2 + q_0 + s = 0$ and the strip condition $z_0' = p_0 x_0' + q_0 y_0'$. Thus, $p_0(s) = -1$ and hence, $q_0(s) = -2s$. The transversality condition evaluated on the initial curve is $F_p y_0'(s) - F_q x_0'(s) = -1 \neq 0$. The characteristic equations are given by

$$\frac{dx}{dt} = 2xp, \quad \frac{dy}{dt} = y, \quad \frac{dz}{dt} = pF_p + qF_q = 2xp^2 + q, \quad \frac{dp}{dt} = -p^2 + p, \quad \frac{dq}{dt} = 0.$$

The equations satisfied by y, p and q are decoupled and hence, can be immediately solved using the initial values. These solutions can then be used to solve the ODE for x to get

$$x(t;s) = s(2 - e^t), \quad y(t;s) = e^t, \quad p(t;s) = \frac{e^t}{e^t - 2}, \quad q(t;s) = -2s.$$

The given PDE now implies

$$z(t;s) = x(t;s)p^2(t;s) + y(t;s)q(t;s) = x\frac{y^2}{(y-2)^2} - 2y\frac{x}{(y-2)^2}$$

which can be simplified to get the required solution as $u(x, y) = xy/(y - 2)$.

Exercise 6.14

We have the characteristic equation given by $\dfrac{dx(t)}{dt} = u(x(t), t)$ and along the characteristic u satisfies $\dfrac{du}{dt}(x(t), t) = 0$. Thus, u is a constant along the characteristic, that is, $u(x(t), t) = u_0(x_0)$ if $x(t)$ is the characteristic curve emanating from x_0 on the x-axis. Thus, $\dfrac{dx(t)}{dt} = u_0(x_0)$. Hence, we have $x = x(t) = u_0(x_0)t + x_0 = x_0^3 t + x_0$.

The mapping $x_0 \to tx_0^3 + x_0$, $t \geq 0$, is bijective from $\mathbb{R}$ to $\mathbb{R}$. Thus, for each $x \in \mathbb{R}$, there is a unique x_0 such that $tx_0^3 + x_0 = x$. For $t > 0$, This is a cubic equation without the second-degree term, hence can be solved by Cardano's method to get $x_0 = 2^{-1/3} t^{-1/2} [q_1^{1/3} + q_2^{1/3}]$, where

$$q_1 = x\sqrt{t} + \sqrt{x^2 t + 4/27}, \quad q_2 = x\sqrt{t} - \sqrt{x^2 t + 4/27}.$$

The solution is given by $u(x, t) = x_0^3 = (x - x_0)/t$, $x \in \mathbb{R}$, $t > 0$.

Exercise 6.15

This is a generalized Burgers' equation. The initial curve and data, in parametric form, are given by $x_0(s) = 0$, $y_0(s) = s$, $u_0(s) = s$. Since $a = 1, b = u^2$, the transversality condition

$$a(x_0(s), y_0(s))y_0'(s) - b(x_0(s), y_0(s))x_0'(s) = 1 \cdot 1 - s^2 \cdot 0 \neq 0$$

is satisfied. The characteristic equations are given by

$$\frac{dx}{dt} = 1, \frac{dy}{dt} = u^2, \frac{du}{dt} = 0.$$

Upon solving, we get

$$x = x(t,s) = t, \ y = y(t,s) = s^2 t + s, \ u = u(t,s) = s.$$

The variables s and t can be easily written in terms of x and y and we get $t = x$ and $s = \dfrac{-1 \pm \sqrt{1+4xy}}{2x}$. Substituting these into the expression for u, we have

$$u(x,y) = \frac{-1 \pm \sqrt{1+4xy}}{2x},$$

for $x > 0$. Indeed, $u(0, y)$ is not defined and has to be understood in the limiting sense as $x \to 0$. The appropriate sign can be chosen using the initial condition $u(0, y) = y$. It is easy to verify, using L'Hospital rule or otherwise, that $u(0,y) = \lim\limits_{x \to 0} \dfrac{-1 + \sqrt{1+4xy}}{2x} = y$. It is not difficult to verify that u is a C^1 function in the region $\{x > 0, 1 + 4xy > 0\}$ and indeed the solution of the given equation.

Exercise 6.16

The initial data in parametric form is given by

$$x_0(s) = \cos s, \ y_0(s) = \sin s \text{ and } u_0(s) = 1, \ s \in [0, 2\pi).$$

The characteristic equations are given by

$$\frac{dx}{dt} = 1, \ \frac{dy}{dt} = 1 \text{ and } \frac{du}{dt} = 1.$$

Solving these equations using the above initial conditions, we obtain

$$x(t,s) = t + \cos s, \ y(t,s) = t + \sin s \text{ and } u(t,s) = 1 + t.$$

Since the expression for u contains only the t variable, we can obtain the same in terms of x and y using the first two expressions. We have $(x-t)^2 + (y-t)^2 = 1$ and, therefore,

t satisfies the quadratic equation $2t^2 - 2t(x+y) + (x^2 + y^2 - 1) = 0$. For t to be real, this forces the condition $|x - y| \leq \sqrt{2}$ on x, y. Therefore,

$$u(x, y) = 1 + t = 1 + (1/2)\left(x + y \pm \sqrt{2 - (x-y)^2}\right).$$

From this expression, we observe that both u_x and u_y become infinite as the point (x, y) approaches the lines $|x - y| = \sqrt{2}$. Thus, for u to be the required solution, we need to confine to the open strip $|x - y| < \sqrt{2}$. This can also be seen as follows. The characteristics are the straight lines parallel to the line $y = x$ in the plane. All these lines intersect the unit circle transversally, except the ones that are tangent to the unit circle at the points $(1/\sqrt{2}, -1/\sqrt{2})$ and $(-1/\sqrt{2}, 1/\sqrt{2})$. These points lie on the lines $|x - y| = \sqrt{2}$. Finally, the required solution is given by

$$u(x, y) = \begin{cases} 1 + (1/2)\left[x + y - \sqrt{2 - (x-y)^2}\right] & \text{if } x + y \geq 0, \\ 1 + (1/2)\left[x + y + \sqrt{2 - (x-y)^2}\right] & \text{if } x + y \leq 0, \end{cases}$$

in the region $\{|x - y| < \sqrt{2}\}$ in the plane, satisfying the condition $u = 1$ on the unit circle.

Exercise 6.17

Here, $F(x, y, z, p, q) = p^2 + q^2 - 1$. Thus, $F_x = F_y = F_z = 0$ and $F_p = 2p$, $F_q = 2q$. The characteristic curves are given by

$$\frac{dx}{dt} = 2p, \quad \frac{dy}{dt} = 2q, \quad \frac{dz}{dt} = pF_p + qF_q = 2, \quad \frac{dp}{dt} = 0, \quad \frac{dq}{dt} = 0.$$

The initial curve is $x_0(s) = s$, $y_0(s) = 1 - s$, $z_0(s) = 0$. Further, $p_0(s)$ and $q_0(s)$ satisfies the equation $p_0(s)^2 + q_0(s)^2 = 1$ and the strip condition $z_0'(s) = p_0(s)x_0'(s) + q_0(s)y_0'(s)$, which gives $p_0(s) = q_0(s)$. Solving, we get $p_0(s) = q_0(s) = \pm 1/\sqrt{2}$. Now solve the characteristic equations to get

$$p(t, s) = p_0(s) = \pm 1/\sqrt{2}, \quad q(t, s) = q_0(s) = \pm 1/\sqrt{2}.$$

Solve the first two equations to obtain

$$\frac{dx}{dt} = 2p_0(s) = \pm\sqrt{2}, \quad \frac{dy}{dt} = 2q_0(s) = \pm\sqrt{2}$$

and using the initial conditions, we get

$$x = x(t, s) = \pm\sqrt{2}t + s, \quad y = y(t, s) = \pm\sqrt{2}t + (1 - s).$$

Adding the two equations, we have $x + y = \pm 2\sqrt{2}t + 1$, which leads to $t = \pm(x + y - 1)/2\sqrt{2}$. Thus, the solution u is given by

$$u(x, y) = z(t, s) = 2t = \pm(1/\sqrt{2})(x + y - 1).$$

Exercise 6.18

The characteristic equations are given by

$$\frac{dx}{dt} = x^2, \quad \frac{dy}{dt} = y^2, \quad \frac{du}{dt} = (x+y)u.$$

In symbolic form, these can also be written as

$$\frac{dx}{x^2} = \frac{dy}{y^2} = \frac{du}{(x+y)u}.$$

A first integral can be obtained by the combination

$$\frac{dx - dy}{x^2 - y^2} = \frac{du}{(x+y)u}.$$

That is, $\dfrac{d(x-y)}{x-y} = \dfrac{du}{u}$. On integration, we have $\dfrac{x-y}{u} = C_1$, a constant. To get another integral, consider the following combination

$$\frac{y\,dx + x\,dy}{x^2 y + y^2 x} = \frac{du}{(x+y)u}.$$

This gives $\dfrac{d(xy)}{xy(x+y)} = \dfrac{du}{(x+y)u}$. Upon integrating, we get a second integral as $\dfrac{xy}{u} = C_2$. Hence, a general solution is given by

$$0 = F(C_1, C_2) = F\left(\frac{x-y}{u}, \frac{xy}{u}\right),$$

where F is an arbitrary function.

The second integral can also be written as $\dfrac{u}{xy} = \dfrac{1}{C_2}$. Hence another form of general solution is given by $1/C_2 = f(C_1)$ with f arbitrary. That is, $u = xy f\left(\dfrac{x-y}{u}\right)$. Again, we can write a third combination as

$$\frac{dx - dy}{x^2 - y^2} = \frac{y\,dx + x\,dy}{x^2 y + y^2 x} = \frac{d(xy)}{xy(x+y)}.$$

This gives $\dfrac{x-y}{xy} = C_3$ and hence, a general solution as above can be written as $u = xy\, g\left(\dfrac{x-y}{xy}\right)$.

Exercise 6.19

This exercise is a rephrase of Exercise 6.13. Here, we obtain the solution to the IVP from the complete integral by finding the envelope. The equation can be written as $F(x,y,z,p,q) \equiv xp^2 + yq - z = 0$. The characteristic equations are given by

$$\frac{dx}{dt} = 2xp, \ \frac{dy}{dt} = y, \ \frac{dz}{dt} = 2xp^2 + yq, \ \frac{dp}{dt} = -p^2 + p, \ \frac{dq}{dt} = 0.$$

In symbolic form, this is equivalent to

$$\frac{dx}{2xp} = \frac{dy}{y} = \frac{dz}{2xp^2 + yq} = \frac{dp}{-p^2 + p} = \frac{dq}{0}.$$

Thus, $q = a$ is a constant. From the PDE, we get $xp^2 + ay - z = 0$, which gives $p = \sqrt{\frac{z - ay}{x}}$. Formally, we have $dz = pdx + qdy$ and thus,

$$\frac{dx}{\sqrt{x}} = \frac{dz - ady}{p\sqrt{x}} = \frac{dz - ady}{\sqrt{z - ay}}.$$

On integration, we have $\sqrt{z - ay} = \sqrt{x} + b$, where b is another constant of integration. Squaring and rearranging, we get the complete integral as $z = x + ay + b^2 + 2b\sqrt{x}$. Indeed, we can easily verify that $u = z$ is a solution of the given PDE for any constants a, b. Now using the initial condition $y = 1, z = -x$, we get $-x = x + a + b^2 + 2b\sqrt{x}$. Differentiating the above expression with respect to x, we get $x = b^2/2$ and substituting in the above equation, we get the relation $a = -b^2/2$. Thus, the solution z satisfies $z = x - \frac{b^2}{2}y + b^2 + 2b\sqrt{x}$. We now find the envelope of the above one-parameter family of surfaces. In this direction, differentiate the above equation with respect to b to get $b = \frac{2\sqrt{x}}{y - 2}$. Substituting this value of b in the one-parameter family to the integral surface through the given curve as $u = z = \frac{xy}{y - 2}$, which is the same solution as obtained in Exercise 6.13. Note that this solution cannot be obtained from the complete integral for any choice of a, b. Such solutions are called as *singular solutions*.[4]

Exercise 6.20

The given equation is $F(x, y, z, p, q) \equiv 2xu + q^2 - x^2p - xyq = 0$ and, therefore, the characteristic equations in the symbolic form are given by

$$\frac{dx}{-x^2} = \frac{dy}{2q - xy} = \frac{dz}{-x^2p + 2q^2 - xyq} = \frac{dp}{-2z + qy} = \frac{dq}{-qx}.$$

Taking the first and fifth equations, we get $q = ax$ for some constant a. Substituting it in the PDE, we get $p = \frac{a^2x^2 + 2xz - ax^2y}{x^2}$. Substituting these values of p and q in the

---

[4] Given a complete integral, the envelope of the two parameter system is also a solution. Note that the envelope touches each integral surface along a curve tangentially and hence, it has the partial derivatives as that of the integral surface. Thus, the envelope satisfies the same PDE and this solution is called a *singular solution*. It is obtained by eliminating two arbitrary constants in the complete integral $f(x, y, z, a, b) = 0$ and its derivatives $f_a(x, y, z, a, b) = 0$ $f_b(x, y, z, a, b) = 0$.

First-Order Partial Differential Equations | 217

relation $dz = p\,dx + q\,dy$, and a bit of computation leads to the expression

$$d\left(\frac{z}{x^2}\right) = -a^2 d\left(\frac{1}{x}\right) + a\,d\left(\frac{y}{x}\right),$$

and upon an integration, we arrive at the solution $z = a^2 x + axy + bx^2$, where a, b are arbitrary constants.

A different solution can be obtained by looking for an envelope of the above family of solutions. Taking $b = \phi(a)$, we have a one-parameter family of solutions as $z + a^2 x - axy - \phi(a)x^2 = 0$. Differentiating with respect to a, we get $2a = y + \phi'(a)x$, where $\phi' = \dfrac{d\phi}{da}$. Substituting this in the above expression of the solution, we get

$$4z = -(y + \phi'(a)x)^2 x + 2(y + \phi'(a)x)xy + 4\phi(a)x^2$$
$$= x[y^2 + (4\phi(a) - (\phi'(a))^2)x^2].$$

Choosing ϕ such that $4\phi(a) = (\phi'(a))^2$, we obtain the solution $z = xy^2/4$, which is not in the class of solutions obtained above.

Exercise 6.21

The characteristic equations are given by $\dfrac{dx}{1} = \dfrac{dy}{-1} = \dfrac{dz}{x^2 + y^2}$. The first two equations give $y = -x + a$ or $x + y = a$, where a is a constant. Again combining the first two equations and the third equation, we get $\dfrac{x^2 dx - y^2 dy}{x^2 + y^2} = \dfrac{dz}{x^2 + y^2}$, which gives $z = \dfrac{1}{3}(x^3 - y^3) + b$. Indeed, it is a one-parameter family of solutions. To get a two-parameter family, we can add the term $a(x + y)$, since it is a solution to the homogeneous equation $u_x - u_y = 0$. We, thus, have complete integral $z = a(x + y) + (1/3)(x^3 - y^3) + b$.

Now to solve the IVP, using the initial values $y = x, z = x^2$, we get $x^2 = 2ax + b$. Differentiating with respect x, we get $x = a$, which can be substituted to get $b = -a^2$. Now substitute the value of b in the complete integral to get one-parameter family $z = a(x + y) + \dfrac{1}{3}(x^3 - y^3) - a^2$. To find the envelope, differentiate the above expression with respect to a, to get $a = \dfrac{x + y}{2}$, which in turn gives the solution to the IVP as

$$u = z = (1/4)(x + y)^2 + (1/3)(x^3 - y^3).$$

Exercise 6.22

Here, we have $F(t, x, z, r, p) = r + H(x, p)$, where $z = u, r = z_t, p = (p_1, \cdots, p_n), p_i = \dfrac{\partial z}{\partial x_i}$. Since there are n independent variables, the characteristic equations consist of $2n + 3$ equations. But we can decouple some of them to get the $2n$ required equations. Introduce the characteristic variables $t(\sigma), x_i(\sigma), z(\sigma)$ that satisfy the characteristic equations

$$\frac{dt}{d\sigma} = F_r = 1, \quad \frac{dx_i}{d\sigma} = F_{p_i} = D_{p_i} H(x, p), \quad \frac{dz}{d\sigma} = rF_r + p_i F_{p_i}$$

and
$$\frac{dr}{d\sigma} = -F_t - rF_z = 0, \quad \frac{dp_i}{d\sigma} = -F_{x_i} - p_i F_z = -F_{x_i} = -D_{x_i} H(x,p).$$

Using the first equation, we can eliminate the running variable σ in a characteristic curve and take $\sigma = t$ as the parameter along the curve. Thus, we get the set of $2n$ equations as

$$\frac{dx_i}{dt} = D_{p_i} H(x,p), \quad \frac{dp_i}{dt} = -D_{x_i} H(x,p).$$

This is a system of $2n$ first-order equations and is known as a Hamiltonian system. This terminology is derived from the classical mechanics. In Newtonian mechanics, H represents the total energy of the system and the above system is the Hamiltonian formalism corresponding to the Newtonian formalism of the motion of the particle which is a system of n second-order ODE (see Ref. [45] for details). Finally, z can be obtained by solving the ODE it satisfies.

Exercise 6.23

First note that the exit time τ is the *first* time the trajectory leaves the domain $[0,1] \times [-1,1]$ under consideration. The main and important observation is that it is enough to consider the straight line trajectories in the minimizing trajectories. To see this, define

$$w(t,x) = \min \left\{ \int_t^\tau L(t, x(s), x'(s)) \, ds : x(s) = x + v(s-t) \right\}$$
$$= \min_{v \in \mathbb{R}} \{(\tau - t) L(t, x, v)\}.$$

Since the class of straight lines that minimizes w is smaller than the class of Lipschitz continuous functions that defines u, we have $u(t,x) \leq w(t,x)$. To see the other way inequality, for given a Lipschitz x with $x(t) = x$, define a straight line $\tilde{x}(s) = x + v(s-t)$, where

$$v = \frac{1}{\tau - t} \int_t^\tau x'(s) \, ds = \frac{x(\tau) - x(t)}{\tau - t}$$

which is nothing but the secant connecting the end points of x and $x(\tau)$. Then, $\tilde{x}(t) = x$ and $\tilde{x}(\tau) = x(\tau)$, thus $\tilde{x}$ is an admissible trajectory for w. Further, it is trivial to see that

$$\int_t^\tau L(t, \tilde{x}(s), \tilde{x}'(s)) \, ds \leq \int_t^\tau L(t, x(s), x'(s)) \, ds.$$

Now minimizing over all Lipschitz trajectories, we see that $w(x,t) \leq u(x,t)$.

Optimal or minimizing solution v^*: Take any $(t,x) \in [0,1] \times (-1,1)$. If $|v| < 1$ (i.e., the slope of the line $x(s) = x + (s-t)v$), then clearly the exit time $\tau = 1$ (see Figure 6.3) and hence $(\tau - t)L(v) = (1-t)(1+v^2/4)$. Thus, in the region $|v| < 1$, the minimum occurs at $v = 0$ and $\min_{|v|<1}(\tau-t)L(v) = (\tau-t)L(0) = 1-t$. For $v \geqslant 1$, the line exits (see Figure 6.3) for $\tau \leqslant 1$ and thus $1 = x(\tau) = x + (\tau-t)v$, which gives the exit time $\tau = t + (1-x)/v$ and thus,

$$(\tau - t)L(v) = [(1-x)/v](1+v^2/4) = (1-x)((1/v) + (v/4)).$$

Now differentiate with respect to v to see that the minimum occurs at $v = 2$ and the minimal value $\min_{v \geqslant 1}(\tau-t)L(v) = (\tau-t)L(2) = 1-x$. Similar computations for $v \leqslant -1$ will lead to $\tau = t - (1+x)/v$, $v = -2$ and $\min_{v \leqslant -1}(\tau-t)L(v) = 1+x$. Thus, we have the result as presented in the problem statement. Note that u is not differentiable when $|x| = t$.

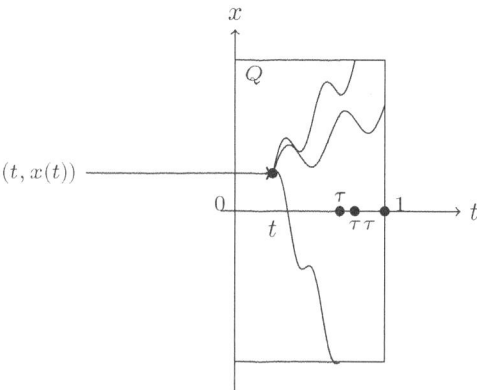

Figure 6.3 | Exit times.

When u is differentiable at any point (t, x), it is a direct computation to show that u satisfies the HJE with the boundary and initial conditions.

Note: In fact, HJE has infinitely many Lipschitz continuous solutions, which are as follows. For each $k \in \mathbb{N}$, we consider a function h_k defined on $[-1, 1]$. Geometrically, the graph of h_k consists of k isosceles triangles whose sides have slopes ± 1. Analytically, we first partition the interval $[-1, 1]$ into k subintervals $I_j = [k^{-1}(2j - k), k^{-1}(2j + 2 - k)]$ for $j = 0, 1, \ldots, k - 1$ and then define h_k by $h_k(x) = k^{-1} - |x - k^{-1}(2j + 1 - k)|$ for $x \in I_j$, $j = 0, 1, \ldots, k - 1$. The Lipschitz continuous function which is a solution of the HJE under consideration is given by $u_k(t, x) = \min\{h_k(x), 1 - t\}$. See, for instance, Ref. [45].

Exercise 6.24

(a) The Hamiltonian $H: \mathbb{R}^n \to \mathbb{R}$ is defined via the Legendre transformation as

$$H(p) = \sup_{q \in \mathbb{R}^n} \{p \cdot q - L(q)\}.$$

Since $L(q) = 1 + |q|^2/4$, we can write $p \cdot q - L(q) = |p|^2 - 1 - |p - q/2|^2$. Therefore, we find that the maximum is achieved at $q = 2p$ and thus, $H(p) = |p|^2 - 1$.

(b) If $u(x,t) = \min\left[\int_0^t L(w'(s))ds + g(w(0))\right]$, then by Hopf–Lax formula, we have

$$u(x,t) = \min_{y \in \mathbb{R}^n}\left[tL\left(\frac{x-y}{t}\right) + g(w(0))\right]$$

$$= \min_{y \in \mathbb{R}^n}\left[tL\left(\frac{x-y}{t}\right) + \frac{1}{2}|y|^2\right]$$

$$= \min_{y \in \mathbb{R}^n}\left[t\left(1 + \frac{|x-y|^2}{4t^2}\right) + \frac{1}{2}|y|^2\right].$$

Differentiating the term inside the minimum, it is straightforward to see that the minimum is achieved at the point $y = \dfrac{x}{2t+1}$ and we find that $u(x,t) = t + \dfrac{|x|^2}{2(2t+1)}$. Now a direct computation shows that u satisfies the HJE $u_t + H(Du) = 0$, that is, $u_t + |Du|^2 = 1$ and the initial condition $u(x,0) = \dfrac{|x|^2}{2}$.

Exercise 6.25

The problem can be converted to a minimization problem using the conservation of energy. Let the positive y-axis points downward and let A and B be placed at $(a, 0)$ and (x_f, y_f), respectively with $a < x_f$, $y_f > 0$ (see Figure 6.4[a]). Since the total energy initially is zero, we have at any point of time $\dfrac{mv^2}{2} - mgy = 0$, where $y = y(x)$ is a curve such that $y(a) = 0$, $y(x_f) = y_f$ and $v = y'$. Normalizing with suitable units so that we may assume $m = 1$, $g = 1/2$. Then, we get $v = \sqrt{y}$. Thus, the approximate distance travelled by the bead from a point $(x, y(x))$ to a close enough point $(x + \Delta x, y(x + \Delta x))$ along the curve $y = y(x)$ is $\sqrt{1 + (y'(x))^2}\Delta x$ and hence, the time taken during this travel is $\dfrac{\sqrt{1 + (y'(x))^2}\Delta x}{v} = \dfrac{\sqrt{1 + (y'(x))^2}\Delta x}{\sqrt{y}}$. Thus, the total time taken by the bead to travel from point A to the end point B along the curve $y = y(x)$ is

$$J(y) = \int_a^b L(y, y')dx,$$

where $L(y, y') = \dfrac{\sqrt{1 + (y')^2}}{\sqrt{y}}$ is the Lagrangian and $b = y_f$. Thus, we need to find a trajectory y so that $J(y)$ is minimum. If y minimizes J, then y satisfies the E–L equation

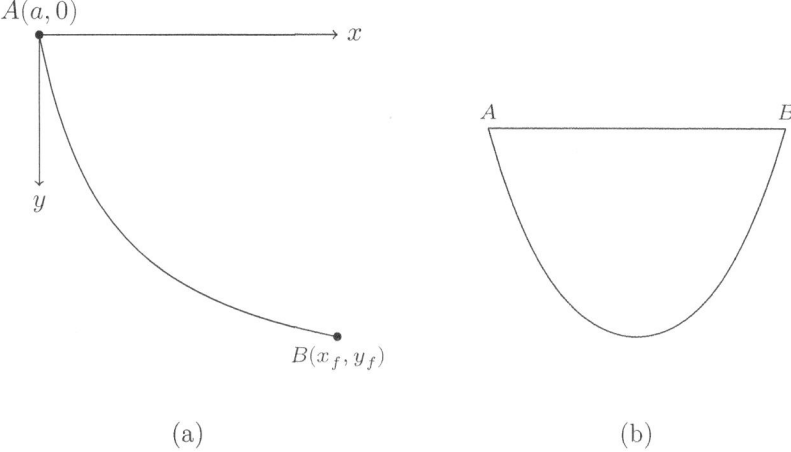

Figure 6.4 | (a) Brachistochrone and (b) catenary.

$L_y = \frac{d}{dx} L_{y'}$. We have

$$L_y = -\frac{\sqrt{1+(y')^2}}{2y\sqrt{y}}, \quad L_{y'} = \frac{y'}{\sqrt{y}\sqrt{1+(y')^2}}.$$

Therefore,

$$\frac{d}{dx} L_{y'} = L_{yy'} y' + L_{y'y'} y'' = \frac{y''}{\sqrt{y}[1+(y')^2]\sqrt{1+(y')^2}} - \frac{(y')^2}{2y\sqrt{y}\sqrt{1+(y')^2}}.$$

Hence, using the E–L equation, we get

$$\frac{y''}{\sqrt{y}[1+(y')^2]\sqrt{1+(y')^2}} - \frac{(y')^2}{2y\sqrt{y}\sqrt{1+(y')^2}} = -\frac{\sqrt{1+(y')^2}}{2y\sqrt{y}},$$

which simplifies to $-2yy'' = 1 + (y')^2$. This second-order equation can actually be integrated once to get the first-order equation $y'[1+(y')^2] = $ constant. A solution in parametric form can be represented as

$$x(t) = d + c(t - \sin t), \quad y(t) = c(1 - \cos t),$$

where c and d are arbitrary constants. These optimal curves are *cycloids*, that is, the locus of a fixed point on a circle (e.g., the wheel of a cycle; hence the name cycloid) when it rolls without slipping on the horizontal axis. It is interesting to remark that the first solution by Johann Bernoulli was based on Snell's law of light refraction (see Ref. [52]).

Exercise 6.26

How do we describe this as an optimization problem and what is the object to be minimized? Let $y : [a, b] \to (0, \infty)$ with $y(a) = A, y(b) = B$ as the end points of the possible shape of the chain denoted by $y = y(x)$. The only force acting on the chain is the potential energy (due to gravitation) and the chain will take the shape of minimal potential energy. The length of the chain corresponding to a small interval $[x, x + \Delta x]$ is $\sqrt{1 + (y'(x))^2}\Delta x$. Thus, if we assume that the chain has unit mass, then the mass of this part of the chain is its length numerically. Thus, the potential energy (with normalization of $g = 1$) is $mgh = y(x)\sqrt{1 + (y'(x))^2}\Delta x$ and hence, the total potential energy functional is given by

$$J(y) = \int_a^b y(x)\sqrt{1 + (y'(x))^2}dx.$$

Thus, we have $L(y, y') = y\sqrt{1 + (y')^2}$. By a direct computation, we get $L_y = \sqrt{1 + y'^2}$, $L_{y'} = \dfrac{yy'}{\sqrt{1 + (y')^2}}$ and a further computation leads to

$$\frac{d}{dx}L_{y'} = L_{yy'}y' + L_{y'y'}y'' = \frac{(y')^2(1 + (y')^2) + yy''}{(1 + (y')^2)\sqrt{1 + (y')^2}}.$$

Using the E–L equation, we get the equation for y as $yy'' = 1 + (y')^2$. The general solution of the above equation is given by $y(x) = c\cosh(x/c)$, where c is an arbitrary constant, assuming the chain does not touch the ground. The catenary curve was first obtained by Johann Bernoulli in 1670. The name catenary is derived from the Latin word *catena*, meaning *chain*. ∎

Exercise 6.27

Here $L(y, y') = y(y')^2$. The E–L equations are given by $L_y = \dfrac{d}{dx}L_{y'}$. This gives $(y')^2 = \dfrac{d}{dx}(2yy')$ and clearly $y = 0$ is a solution. To prove the uniqueness of the solution, integrate the E–L equations and use the boundary condition to get $\int_0^1 (y')^2\, dx = 0$. Thus, $(y')^2 = 0$ and hence, y is constant. From the boundary condition, we get $y \equiv 0$.

By choosing $y(x) > 0$ in $(0, 1)$, we get $J(y) > 0$ and if we choose $y(x) < 0$ in $(0, 1)$, we get $J(y) < 0$. Thus, $y = 0$ is neither a minimum nor a maximum. The message is that the critical points need not be extremal. ∎

Exercise 6.28

Choose $v \in C^2(\overline{\Omega})$ with $v(x) = 0$ on $\partial\Omega$. The class of admissible functions are $y \in C^2(\overline{\Omega})$ with $y(x) = g(x)$ on $\partial\Omega$. Then the function $y + \tau v$ is also an admissible function for any

$\tau \in \mathbb{R}$. For y, v fixed, define $f(\tau) = J(y + \tau v)$. We compute $f'(0)$. We have

$$f(\tau) = J(y + \tau v) = \int_\Omega \nabla(y + \tau v)(x) \cdot \nabla(y + \tau v)(x) \, dx$$
$$= J(v) + 2\tau \int_\Omega \nabla y(x) \cdot \nabla v(x) \, dx + \tau^2 \int_\Omega \nabla v(x) \cdot \nabla v(x) \, dx.$$

Thus,
$$f'(0) = \lim_{\tau \to 0} \frac{J(y + \tau v) - J(y)}{\tau} = 2 \int_\Omega \nabla y(x) \cdot \nabla v(x) \, dx.$$

Now, if $\bar{y}$ is minimal for the functional J, that is, $J(\bar{y}) \leq J(y)$ for all admissible functions, it is easy to see that $f'(0) \geq 0$ (respectively $f'(0) \leq 0$) by choosing $\tau > 0$, $\tau \to 0$ (respectively $\tau < 0$, $\tau \to 0$). Thus, we have $f'(0) = 0$. In other words, we get

$$\int_\Omega \nabla \bar{y}(x) \cdot \nabla v(x) \, dx = 0.$$

The above identity is true for all the chosen test functions v. An integration by parts (or Green's theorem) will lead to

$$\int_\Omega \Delta \bar{y}(x) \cdot v(x) \, dx = 0.$$

Since v arbitrary, we see that $\Delta \bar{y} = 0$ with $\bar{y} = g$ on $\partial \Omega$.

Exercise 6.29

Fix $p \in \mathbb{R}^n$. We claim that $p \cdot q - L(q) \to -\infty$ as $|q| \to \infty$. Intuitively, this is due to the fact that the first term has linear growth, whereas by coercivity, L has super-linear growth.

We need to show that given any $C > 0$, there is an $r > 0$ such that $p \cdot q - L(q) < -C$ for all $|q| > r$. Now by the assumption on L, there is an $r_1 > 0$ such that $L(q) > (C + |p|)|q|$ for all $|q| > r_1$. Put $r = \max(r_1, 1)$. Then, for $|q| > r$, we have

$$p \cdot q - L(q) \leq |p||q| - L(q) < |q|(|p| - (C + |p|)) < -C,$$

as required.

Next, by the definition of H, we have $H(p) \geq -L(0)$, by choosing $q = 0$. Using the above claim, we see that there is an $M > 0$ such that $p \cdot q - L(q) < -L(0)$ for all $|q| > M$. Thus, it suffices to take the supremum in the closed ball of radium M, in the definition of H. More precisely

$$H(p) := \sup_{q \in \mathbb{R}^n} \{p \cdot q - L(q)\} = \sup_{q \in \overline{B_M(0)}} \{p \cdot q - L(q)\}.$$

Since now, the supremum is taken over a compact set, the supremum is achieved in $\overline{B_M(0)}$. That is, for a given $p \in \mathbb{R}^n$, there exists a $q = q(p) \in \mathbb{R}^n$ such that $H(p) = p \cdot q(p) - L(q(p))$. Since $q = q(p)$ is a point where the maximum occurs, we have $\nabla_q(p \cdot q - L(q)) = 0$ at $q = q(p)$. This gives $p = q(p)$ when $L(q) = \frac{|q|^2}{2}$. Thus,
$$H(p) = p \cdot p - L(p) = \frac{|p|^2}{2} = L(p).$$

Note: If $L(q) = |q|^2$, then we get $H(p) = \frac{|q|^2}{4}$. Also, we can start with H and then compute L in an exactly similar way.

Exercise 6.30

Here, $H(p) = |p|^2$ and hence, by Exercise 6.29, we have $L(q) = \frac{|q|^2}{4}$. Thus, by Hopf–Lax formula, the solution is given by

$$u(x,t) = \min_{y \in \mathbb{R}^n} \left[tL\left(\frac{x-y}{t}\right) + g(y) \right]$$
$$= \min_{y \in E} \left[tL\left(\frac{x-y}{t}\right) \right],$$

since $g = \infty$ if $y \notin E$. Thus $u(x,t) = \frac{(d(x))^2}{4t}$.

Exercise 6.31

Clearly $J(y) \geqslant 0$ for all $y \in \mathcal{M}$. Also, observe that if $y \in \mathcal{M}$ and $0 < y'(\xi) < 1$ for some $\xi \in (-1, 1)$, then $J(y) > 0$, as then in a small interval near ξ, we have $y \neq 0$ and $y' \neq 1$, implying that $L(y, y') > 0$ in this interval. We now show that $J(y)$ can be made arbitrarily small by choosing an appropriate $y \in \mathcal{M}$.

Let $\varepsilon > 0$. Consider the function y_ε defined on $[-1, 1]$ by

$$y_\varepsilon(x) = \begin{cases} 0 & \text{if } x \leqslant 0, \\ \dfrac{2x^2}{\varepsilon} - \dfrac{x^3}{\varepsilon^2} & \text{if } 0 < x < \varepsilon, \\ x & \text{if } \varepsilon \leqslant x \leqslant 1. \end{cases}$$

It is not difficult to verify that $y_\varepsilon \in \mathcal{M}$. Also, a bit lengthy computation shows that $J(y_\varepsilon) = c\varepsilon^3$, for some positive constant c. This shows that the infimum of $J(y)$ is zero, as required. Further, if $y \in \mathcal{M}$, then there is a $\xi \in (-1, 1)$ such that $y'(\xi) = 1/2$. As noted above, we see that $J(y) > 0$. Thus, there is no $y \in \mathcal{M}$ for which $J(y) = 0$.

Heuristically, we can obtain a Lipschitz continuous function by taking the limit of y_ε as $\varepsilon \to 0$. Define the function $\tilde{y}$ on $[-1, 1]$ by

$$\tilde{y}(x) = \begin{cases} 0 & \text{if } x \leqslant 0, \\ x & \text{if } 0 < x \leqslant 1. \end{cases}$$

It is straightforward to verify that $\tilde{y}$ is Lipschitz continuous and satisfies the boundary conditions $\tilde{y}(-1) = 0$ and $\tilde{y}(1) = 1$. Further, an easy computation shows that $J(\tilde{y}) = 0$.

Exercise 6.32

First, observe that $|x|$ is not semi-concave, but $-|x|$ is semi-concave. Here, the Hamiltonian $H(p) = \frac{1}{2}|p|^2$. Hence (see Exercise 6.29), $L(q) = \frac{1}{2}|q|^2$.

(a) Take $g(x) = |x|$. By Hopf–Lax formula, we have

$$u(x,t) = \min_{y \in \mathbb{R}^n} \left\{ \frac{|x-y|^2}{2t} + |y| \right\}.$$

Now consider the term

$$\frac{|x-y|^2}{2t} + |y| = \frac{|x|^2}{2t} + \frac{|y|^2}{2t} - \frac{x \cdot y}{t} + |y|.$$

Since we are minimizing with respect to y and x is fixed, the minimum depends only on the last three terms in the above expression. Indeed, if $|x| \leqslant t$, then by Cauchy–Schwarz inequality, we have $\frac{x \cdot y}{t} \leqslant |y|$ and hence, $\frac{|y|^2}{2t} - \frac{x \cdot y}{t} + |y| \geqslant 0$. For this, the minimum occurs at $y = 0$ and thus, $u(x,t) = \frac{|x|^2}{2t}$ for $|x| \leqslant t$.

For $|x| \geqslant t$, differentiate to get $D_y \left(\frac{|x-y|^2}{2t} + |y| \right) = \frac{y-x}{t} + \frac{y}{|y|} = 0$, for $y \neq 0$. This gives $y = (x/|x|)(|x| - t)$. This is a minimum point as the second derivative is positive. Thus, $u(x,t) = |x| - \frac{t}{2}$ for $|x| \geqslant t$. Thus, we have the solution

$$u(x,t) = \begin{cases} \dfrac{|x|^2}{2t} & \text{if } |x| \leqslant t, \\ |x| - \dfrac{t}{2} & \text{if } |x| \geqslant t. \end{cases}$$

(b) Now consider $g(x) = -|x|$. By Hopf–Lax formula, we have

$$u(x,t) = \min_{y \in \mathbb{R}^n} \left\{ \frac{|x-y|^2}{2t} - |y| \right\}.$$

We can find the minimum point by a direct computation to get $y = (x/|x|)(|x| + t)$ and, therefore, the solution is given by $u(x,t) = -(|x| + t/2)$.

Exercise 6.33

Here, $f(u) = u^2/2$. The characteristics through a point (x,t) in the upper-half plane are given by

$$x = \begin{cases} x_0 + t & \text{for } x_0 \leq 0, \\ x_0 + (1 - x_0)t & \text{for } 0 < x_0 < 1, \\ x_0 & \text{for } x_0 \geq 1. \end{cases} \quad (6.4.2)$$

Thus, the characteristic $x = t$, passing through $(0,0)$, meets a characteristic $x = x_0 + (1 - x_0)t$, $0 < x_0 < 1$, only when $t = 1$. Thus, no two characteristics meet until $t = 1$. This proves (a). Further, for each point (x,t), $0 < t < 1$, there is a unique backward characteristic passing through it and the solution there can be defined. Thus, the solution is given by

$$u(x,t) = \begin{cases} 1 & \text{for } x \leq t, \\ \dfrac{1-x}{1-t} & \text{for } t < x < 1, \\ 0 & \text{for } x \geq 1. \end{cases} \quad (6.4.3)$$

Note that u_0 is not differentiable at $x = 0$ and $x = 1$. This is reflected in the solution across the lines $x = t$ and $x = 1$.

For $t \geq 1$, the method of characteristics cannot be used, as there is more than one characteristics passing through any point in this regime. A shock $x = s(t)$, $t \geq 1$ may be constructed in this situation. The shock speed, that is the slope of this curve, should satisfy the Rankine–Hugoniot condition. Thus, $s(t) = (1+t)/2$, $t \geq 1$, which is the required line of discontinuity. Now define $u(x,t)$ for $t \geq 1$ by

$$u(x,t) = \begin{cases} 1 & \text{if } x < s(t), \\ 0 & \text{if } x > s(t). \end{cases} \quad (6.4.4)$$

It is easily checked that u given by equation (6.4.4) satisfies the Burgers' equation and the jump condition, that is, the Rankine–Hugoniot condition. Thus, the complete solution of IVP is given by equations (6.4.3) and (6.4.4). This answers the question in (b).

Exercise 6.34

We have $f(u) = u^2/2$. The characteristics are given by

$$x = \begin{cases} x_0 & \text{for } x_0 < 0, \\ x_0 + t & \text{for } 0 \leq x_0 \leq 1, \\ x_0 & \text{for } x_0 > 1. \end{cases} \quad (6.4.5)$$

Observe that no point (x,t) such that $0 < x < t$ is connected by a backward characteristic. This region is covered by a rarefaction wave. Thus, we have

$$u(x,t) = \begin{cases} 0 & \text{for } x < 0, \\ x/t & \text{for } 0 \leq x < t. \end{cases} \quad (6.4.6)$$

Now the characteristic $x = t$ through the origin meets the characteristic $x = 1$ when $t = 1$. Thus, a shock develops at $x = 1$. Using the jump condition, the speed of this shock is given by $\sigma = \dfrac{f(u_r) - f(u_\ell)}{u_r - u_\ell} = \dfrac{0 - 1/2}{0 - 1} = 1/2$. Thus, the shock is given by $x = 1 + t/2$. The characteristic $x = t$ meets this shock when $t = 2$. Thus, for $t < 2$, we have

$$u(x,t) = \begin{cases} 0 & \text{for } x < 0, \\ x/t & \text{for } 0 \leqslant x < t, \\ 1 & \text{for } t \leqslant x < 1 + t/2, \\ 0 & \text{for } x > 1 + t/2. \end{cases} \quad (6.4.7)$$

For $t \geqslant 2$, we need to extend the shock $x = 1 + t/2$, keeping in mind the jump condition to be satisfied. Let $x = s(t)$, $t \geqslant 2$ denote this shock, requiring that $s(2) = 2$. The jump condition gives

$$\frac{ds}{dt} = \frac{f(u_r) - f(u_\ell)}{u_r - u_\ell}.$$

Here, we have $u_r = 0$ and $u_\ell = s(t)/t$, coming from the solution x/t, which is valid for $0 \leqslant x < t$. Therefore, $\dfrac{ds}{dt} = \dfrac{s(t)}{2t}$. This gives, using the condition $s(2) = 2$, $s(t) = \sqrt{2t}$. Thus, for $t \geqslant 2$, the solution is given by

$$u(x,t) = \begin{cases} x/t & \text{for } 0 \leqslant x < s(t), \\ 0 & \text{for } x > s(t). \end{cases} \quad (6.4.8)$$

This together with equation (6.4.7) describes the complete solution.

This example describes in a way various possibilities that can occur in seeking a correct solution of a conservation law. Observe that the straight lines $x/t = a$, $0 < a < 1$, through the origin and meeting the shock $x = s(t)$, $t \geqslant 2$ are *not* characteristics. A schematic diagram exhibiting the various regions in the upper-half $x - t$ plane with regard to this question is shown in Figure 6.5.

Exercise 6.35

Here, we have $f(u) = |u|^p/p$, $p \geqslant 2$. When $p = 2$, this is the Burgers' equation. It is easily checked that $f'(u) = \operatorname{sgn}(u)|u|^{p-1}$ and $f''(u) = (p-1)|u|^{p-2}$. Thus, f' is strictly increasing and f is convex. Thus, the characteristics are given by

$$x = x_0 + f'(u(x_0))t.$$

Though it is not possible to solve the above equation for x_0 explicitly, we can assert its unique existence for a region in the upper-half plane. The solution is then given by $u(x,t) = u_0(x_0)$.

Since f' is strictly increasing, its inverse is also easy to compute. We have

$$g(u) = (f')^{-1}(u) = \operatorname{sgn}(u)|u|^{\frac{1}{p-1}}.$$

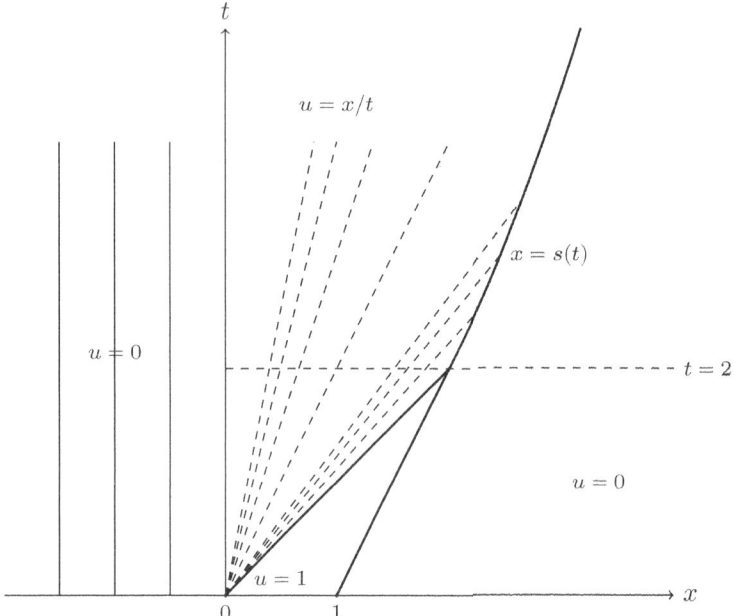

Figure 6.5 | Schematic description of $x - t$ plane of Exercise 6.34.

Though f is not uniformly convex, we can nevertheless determine its Legendre transform: $L(u) = ug(u) - f(g(u)) = (p-1)|u|^{\frac{p}{p-1}}/p$. Thus, we can also use Lax–Oleinik formula to obtain the solution; of course, in general, the minimizer may not explicitly be found.

Exercise 6.36

(a) Here, $u_0(x) = x^2$. The characteristics are given by

$$x = x_0 + x_0^2 t.$$

For $t > 0$, we can solve this equation for x_0 and we find that $x_0 = \left(-1 \pm \sqrt{1 + 4xt}\right)/2t$ and the solution is given

$$u(x,t) = \left(1 + 2xt - \sqrt{1 + 4xt}\right)/(2t^2).$$

The choice of the sign in choosing x_0 ensures that $u(x,t) \to u_0(x)$ as $t \to 0+$. The term under the radical sign must be non-negative. Hence, we have $1 + 4xt > 0$. The boundary of this region, $1 + 4xt = 0$, is one part of the hyperbola and the derivatives $u_x(x,t)$ and $u_t(x,t)$ become unbounded as (x,t) approaches this boundary. It is easily checked that u is indeed a C^1 solution in this region of the upper-half plane satisfying the initial condition.

(b) The procedure is the same as in (a) above with change in the initial condition: $u_0(x) = -x|x|$, $x \in \mathbb{R}$. The characteristics are given by $x = x_0 - x_0|x_0|t$. For $x = 0$, we take $x_0 = 0$, as the other solution $x_0 = \pm 1/t$ easily gets rejected. For $x > 0$, we observe that either $0 < x_0 < 1/t$ or $-1/t < x_0 < 0$. A simple observation shows that the negative solution is not permissible and the positive solution is given by $x_0 = \left(1 - \sqrt{1-4xt}\right)/2t$, provided that $1 - 4xt > 0$. The solution is then given by $u(x,t) = u_0(x_0) = -x_0^2$. In a similar way, we obtain the solution for $x < 0$. Summing up, the solution at a point (x, t), $t > 0$ is given by $u(0, t) = 0$ and for $x \neq 0$, by

$$u(x,t) = -\mathrm{sgn}(x)\left(1 - 2|x|t - \sqrt{1 - 4|x|t}\right)/(2t^2), \tag{6.4.9}$$

provided that $1 - 4|x|t > 0$. Here, $\mathrm{sgn}(x)$ denotes the sign of x. The region in the upper-half plane where the above solution is valid is bounded by the parts of the hyperbolas $1 - 4|x|t = 0$, $x \neq 0$.

Exercise 6.37

Here, we have $f(u) = u^2/2$. Therefore, $L(u) = u^2/2$ and $f'(u) = g(u) = u$. Also, $w_0(x) = 0$ if $x < 0$ and $w_0(x) = x$ if $x \geq 0$. Thus, the function that is to be minimized is given by (the point (x, t) is in the upper-half plane and is fixed, but arbitrary)

$$\begin{cases} \dfrac{(x-y)^2}{2t} & \text{if } y < 0, \\ \dfrac{(x-y)^2}{2t} + y & \text{if } y \geq 0. \end{cases}$$

Though this is a simple calculus problem, care need to be exercised in identifying the correct minimizer, as the function to be minimized has different expressions in different regions. A simple way to do this is to find all possible minimizers in each of the regions and then compare the minimum values of the function therein. This remark applies to similar exercises as follows.

Thus, the minimizer $y = y(x, t)$ can be obtained as

$$y = \begin{cases} x & \text{if } x < 0, \\ 0 & \text{if } 0 \leq x < t, \\ x - t & \text{if } 0 < t \leq x. \end{cases}$$

Plugging this expression into the Lax–Oleinik formula, we obtain the solution to IVP:

$$u(x,t) = \begin{cases} 0 & \text{if } x < 0, \\ x/t & \text{if } 0 \leq x < t, \\ 1 & \text{if } 0 < t \leq x, \end{cases}$$

which represents a rarefaction wave.

Exercise 6.38

First, consider the case of the initial condition as in Exercise 6.33. The corresponding w_0 is then given by

$$w_0(x) = \begin{cases} x & \text{if } x < 0, \\ x(1-x/2) & \text{if } 0 \leq x \leq 1, \\ 1/2 & \text{if } x > 1. \end{cases}$$

The corresponding function that is to be minimized is given by

$$\begin{cases} \dfrac{(x-y)^2}{2t} + y & \text{if } y < 0, \\ \dfrac{(x-y)^2}{2t} + y(1-y/2) & \text{if } 0 \leq y \leq 1, \\ \dfrac{(x-y)^2}{2t} + 1/2 & \text{if } y > 1. \end{cases}$$

The cases $0 < t < 1$ and $t \geq 1$ need to be analysed separately. Let $0 < t < 1$. It follows that the minimizer $y = y(x,t)$ is given by

$$y = \begin{cases} x - t & \text{if } x < t < 1, \\ \dfrac{x-t}{1-t} & \text{if } t \leq x < 1, \\ x & \text{if } x \geq 1 \text{ and } t < 1. \end{cases}$$

Thus, for $0 < t < 1$, the solution is given by

$$u(x,t) = \begin{cases} 1 & \text{if } x < t < 1, \\ \dfrac{1-x}{1-t} & \text{if } t \leq x < 1, \\ 0 & \text{if } x \geq 1 \text{ and } t < 1. \end{cases}$$

For $t \geq 1$, we still have $y = x - t$ if $x < t$, at which the minimum value of the function is $x - t/2$. For $x > t \geq 1$, the middle part of the function to be minimized becomes redundant and the minimum value of the function in the region $x > 1$ is $1/2$, achieved at $y = x$. Thus, if $x - t/2 < 1/2$, that is, $x < (1+t)/2$, then the minimizer is $y = x - t$. If $x > (1+t)/2$, then the minimizer is $y = x$. If we put together all these pieces into the Lax–Oleinik formula, we arrive at the formula (remember $t \geq 1$):

$$u(x,t) = \begin{cases} 1 & \text{if } x < (1+t)/2, \\ 0 & \text{if } x > (1+t)/2. \end{cases}$$

Of course, this is the same as the solution already derived in Exercise 6.33.

Next, consider the initial value given in Exercise 6.34. Now w_0 is given by

$$w_0(x) = \begin{cases} 0 & \text{if } x < 0, \\ x & \text{if } 0 \leqslant x < 1, \\ 1 & \text{if } x \geqslant 1. \end{cases}$$

Thus, the function that to be minimized is given by

$$\begin{cases} \dfrac{(x-y)^2}{2t} & \text{if } y < 0, \\ \dfrac{(x-y)^2}{2t} + y & \text{if } 0 \leqslant y < 1, \\ \dfrac{(x-y)^2}{2t} + 1 & \text{if } y > 1. \end{cases} \quad (6.4.10)$$

In the first region, the minimizer is $y = x$, if $x < 0$, with minimum value 0; if $x \geqslant 0$, then the minimizer is $y = 0$, with minimum value $x^2/(2t)$.

In the second region, the minimizer is $y = x - t$, if $t \leqslant x < t + 1$, with minimum value $x - t/2$; if either $x < t$ or $x \geqslant t + 1$, then the minimizer is either $y = 0$ or $y = 1$, with corresponding minimum value.

In the third region, the minimizer is $y = x$, if $x \geqslant 1$, with minimum value 1; otherwise, the minimum value is larger than 1.

With these observations and proceeding as in the previous case, we see that the solution is given by

$$u(x,t) = \begin{cases} 0 & \text{if } x < 0, \\ x/t & \text{if } 0 \leqslant x < t, \\ 1 & \text{if } t \leqslant x < 1 + t/2, \\ 0 & \text{if } x > 1 + t/2. \end{cases} \quad (6.4.11)$$

We observe that there is region of rarefaction wave when $0 \leqslant x < t$ and the line $x = 1 + t/2$ represents a shock. It is also easy to verify that the jump condition is satisfied across the line of discontinuity $x = 1 + t/2$. However, the lines $x/t = a$, $1/2 < a < 1$ in the rarefaction region, meet this shock $x = 1 + t/2$ for some $t > 2$ and the jump condition is violated, as can be easily seen. Thus, the solution given by equation (6.4.11) is valid for only $t \leqslant 2$. In order to find the solution for $t \geqslant 2$, we proceed as follows (see Figure 6.5).

Let $t \geqslant 2$. In the region $0 \leqslant x < t$, the minimizer of the function in equation (6.4.10) is $y = 0$ with minimum value $x^2/(2t)$. Thus, if $x^2/(2t) < 1$, the minimizer is $y = 0$, and the solution is $u(x,t) = x/t$. If, on the other hand, $x^2/(2t) > 1$, the solution is $u(x,t) = 0$. Thus, consider the curve $s(t) = \sqrt{2t}$, $t \geqslant 2$ and define

$$u(x,t) = \begin{cases} x/t & \text{if } 0 \leqslant x < s(t), \\ 0 & \text{if } x > s(t). \end{cases} \quad (6.4.12)$$

It is easily checked that u defined by equation (6.4.12) is indeed the solution of the Burgers' equation, satisfying the jump condition across the shock $x = s(t)$, in the region $t \geqslant 2$.

Finally, consider the case of the initial data $u_0 \in C^1(\mathbb{R})$ given by

$$u_0(x) = \begin{cases} 0 & \text{if } x < 0, \\ 1 & \text{if } x > 1, \end{cases}$$

and in the interval $[0, 1]$, u_0 is non-decreasing. Thus, the method of characteristics asserts that there is a C^1 solution in the entire upper-half plane. We now see that the same conclusion also follows from the Lax–Oleinik formula.

The function that is to be minimized is given by

$$\begin{cases} \dfrac{(x-y)^2}{2t} & \text{if } y < 0, \\ \dfrac{(x-y)^2}{2t} + w_0(y) & \text{if } 0 \leqslant y \leqslant 1, \\ \dfrac{(x-y)^2}{2t} + y + k - 1 & \text{if } y > 1, \end{cases}$$

where $w_0(y) = \displaystyle\int_0^y u_0(\eta)\,d\eta$ and $k = \displaystyle\int_0^1 u_0(\eta)\,d\eta$. Observe that the function $y \mapsto u_0(y) + y/t$ is strictly increasing in the interval $[0, 1]$, with minimum value 0 and maximum value $1 + 1/t$. Thus, the equation $u_0(y) + y/t = x/t$ has a unique root y for each $x \in [0, 1+t]$. Denote this root by $\tilde{y} = \tilde{y}(x, t)$. Then, the minimizer is

$$y = \begin{cases} x & \text{if } x < 0, \\ \tilde{y} & \text{if } 0 \leqslant x \leqslant 1 + t, \\ x - t & \text{if } x > 1 + t, \end{cases}$$

and the solution is given by

$$u(x, t) = \begin{cases} 0 & \text{if } x < 0, \\ \dfrac{x - \tilde{y}}{t} = u_0(\tilde{y}) & \text{if } 0 \leqslant x \leqslant 1 + t, \\ 1 & \text{if } x > 1 + t. \end{cases}$$

It is straightforward to check that u is C^1 in the upper-half plane and $u(x, t) \to u_0(x)$ as $t \to 0$ and all $x \in \mathbb{R}$.

Exercise 6.39

The characteristics passing through a point (x, y, t), $t > 0$ are obtained by solving the equations

$$\frac{dx}{dt} = u^2, \quad \frac{dy}{dt} = u^2, \quad \frac{du}{dt} = 0.$$

As u is a constant along any characteristics, they are given by

$$x(t) = x_0 + tu_0^2(x_0, y_0), \; y(t) = y_0 + tu_0^2(x_0, y_0), \; u(x,y) = u_0(x_0, y_0),$$

where (x_0, y_0) is the point where the characteristics meet the plane $t = 0$. As $u_0(x,y) = x + y$, if we are able to express $x_0 + y_0$ in terms of x, y and t, we are done. We have $x = x_0 + t(x_0 + y_0)^2$ and $y = y_0 + t(x_0 + y_0)^2$. Therefore, we obtain the quadratic equation $2t(x_0 + y_0)^2 + (x_0 + y_0) - (x + y) = 0$ for $x_0 + y_0$. Solving this quadratic equation and substituting it in the expression for u, we find that the required solution is given by

$$u(x,y) = (4t)^{-1}\left(\sqrt{1 + 8t(x+y)} - 1\right),$$

in the region $\{t > 0, \; 1 + 8t(x+y) > 0\}$ of the upper-half space.

7 Classification of Partial Differential Equations

7.1 Introduction

In many branches of mathematics, we do classify the objects of study. For example, in linear algebra, we classify the square matrices into symmetric matrices, orthogonal matrices and others. Similarly, we may group a particular set of partial differential equations (PDE) whose solutions have common qualitative and/or quantitative properties. Among the important equations of mathematical physics, the three main equations, namely the wave equation, Laplace equation and the heat equation, are classified as follows:.

Wave Equation: This equation is a prototype of *hyperbolic equations*. In one dimension, this equation models many real-world problems, such as small transversal vibrations of a string, longitudinal vibrations of a rod, electrical oscillations in a wire, torsional oscillations of shafts and oscillations in gases.

In two dimensions, the wave equation models vibrations of a membrane, and in three dimensions, it models the propagation of sound waves, light waves and electromagnetic waves in a medium.

Heat Equation: It models the heat diffusion in a medium and is a prototype of *parabolic equations*. It also describes the diffusion of a chemical substance in a medium. As such, it is also called *diffusion equation*.

Laplace Equation: A steady-state process, that is, a physical process that does not change with time, is generally modelled by a Laplace or Poisson equation and is classified as an *elliptic equation*. A physical process that is in equilibrium state is also described by an elliptic equation.

Another reason for the classification of PDE is more mathematical in nature by the consideration of a comparison with ordinary differential equation (ODE). Typically, an ODE is associated with a Cauchy problem or an IVP. Consider such a problem for a second-order ODE:

$$u'' + bu' + cu = g, \ u(x_0) = u_0, \ u'(x_0) = u_1.$$

The existence and uniqueness results are proved under suitable conditions on the coefficients a, b, c and *arbitrary* u_0, u_1. Consider now the case of a second-order PDE:

$$\sum_{i,j=1}^{n} a_{ij}(x)\frac{\partial^2 u}{\partial x_i \partial x_j} + \sum_{i=1}^{n} a_i(x)\frac{\partial u}{\partial x_i} = f(x), \ x \in \Omega. \tag{7.1.1}$$

The corresponding Cauchy problem consists of prescribing the Cauchy or initial data on an $n-1$ dimensional surface Γ of class C^k, $k \geq 2$, lying in Ω, where Γ is represented by the equation $F(x) = 0$ with the additional condition that $|\nabla F(x)| \neq 0$ for all $x \in \Gamma$, that is,

$$u(x) = u_0(x), \ \frac{\partial u}{\partial \nu}(x) = u_1(x), \ x \in \Gamma.$$

In general, $u_0(x)$ and $u_1(x)$ are not arbitrary, and this depends on the nature of Γ in relation to the given PDE. This leads to the notion of *characteristic* and *non-characteristic surfaces* and then the classification of PDE.

A point $\bar{x} \in \Gamma$ is called a *characteristic point* for equation (7.1.1) if

$$(\nabla F(\bar{x}))^T A(\bar{x}) \nabla F(\bar{x}) = 0,$$

where $A(x) = [a_{ij}(x)]$. If all points of Γ are characteristic points, then Γ is called a *characteristic surface*.

Having said so much about classification, there are many important equations that lie outside such a process of classification. Some examples are the Schrödinger equation of quantum mechanics: $i\hbar \frac{\partial \psi}{\partial t} = -\frac{\hbar^2}{2m_0} \Delta \psi + V\psi$, where $\hbar$ is the Planck's constant and V is a potential. The Korteweg–de Vries (KdV) equation is $u_t + 6uu_x + u_{xxx} = 0$, and the Airy equation is $u_t + u_{xxx} = 0$.

In the modern theory of the subject, the question of classification has undergone many important changes. Especially, the linear PDEs are now classified into *hypoelliptic* and *non-hypoelliptic* equations. Unfortunately, it is not possible to describe these classes in simple terms; the nature of the *fundamental solution* is required for such a classification. Roughly speaking, the hypoelliptic class includes elliptic and parabolic equations and the non-hypoelliptic class includes hyperbolic equations.

7.2 Classification

We mainly consider the second-order linear PDE and briefly mention about higher-order equations. Let us begin with a simple situation of a second-order equation with constant coefficients, linear principal part.

$$\sum_{i,j=1}^{n} a_{ij} \frac{\partial^2 u}{\partial x_i \partial x_j} + \text{l.o.t.} = 0.$$

Here, $a_{ij} \neq 0$ for at least one pair i, j and l.o.t. represents lower-order terms. Since we are interested in only the transformation of the second-order terms, it is immaterial at this stage whether l.o.t. are linear or non-linear, but the second-order terms are linear.

The principal part can be written in matrix notation as

$$\left(\nabla_x^{\text{T}} A \nabla_x\right) u,$$

where the matrix $A = [a_{ij}]$ is real, symmetric, and $\nabla_x = \left(\frac{\partial}{\partial x_1} \cdots \frac{\partial}{\partial x_n}\right)^{\text{T}}$ denotes the gradient vector and the superscript T denotes the transpose of a matrix or vector; the subscript x is used to emphasize that the derivatives are taken with respect to the x variables.

There is an orthogonal matrix (also called a rotation matrix) R such that $R^{\text{T}} A R = D$, where D is the diagonal matrix

$$D = \text{diag}\left[\lambda_1 \; \cdots \; \lambda_p \; -\lambda_{p+1} \; \cdots \; -\lambda_{p+q} \; 0 \; \cdots \; 0\right].$$

Here, $\lambda_1, \cdots, \lambda_p$ are the positive eigenvalues of the matrix A; $-\lambda_{p+1}, \cdots, -\lambda_{p+q}$ are its negative eigenvalues and the remaining r eigenvalues of A are zero. Thus, $p + q + r = n$, either $p \geq 1$ or $q \geq 1$, as A is a non-zero symmetric matrix.[1] Thus, the quadratic form associated with A, namely $\xi^{\text{T}} A \xi = \sum_{i,j=1}^{n} a_{ij} \xi_i \xi_j$, takes the form

$$\sum_{i=1}^{p} \lambda_i \tilde{\xi}_i^2 - \sum_{i=p+1}^{p+q} \lambda_i \tilde{\xi}_i^2,$$

---

[1] If A is a real symmetric matrix whose only eigenvalue is 0, then A is the zero matrix.

after changing the variable ξ to $\tilde{\xi} = R^T\xi$. Now, put $\eta_i = \sqrt{\lambda_i}\tilde{\xi}_i$ for $i = 1, 2, \ldots, p+q$ and $\eta_i = \tilde{\xi}_i$ for $i > p+q$. The quadratic form associated with A thus reduces to a *canonical form*:

$$\sum_{i=1}^{p} \eta_i^2 - \sum_{i=p+1}^{p+q} \eta_i^2.$$

Unfortunately, there are more than one way of obtaining a canonical form of a given quadratic form. For example, the quadratic form $q(x_1, x_2) = x_1^2 + x_1 x_2 + x_2^2$ in $\mathbb{R}^2$ can be reduced to the following canonical forms:

- $q = (x_1 + x_2/2)^2 + (\sqrt{3}x_2/2)^2$.
- $q = (\sqrt{3}x_1/2)^2 + ((x_1/2) + x_2)^2$.
- $q = \left(\sqrt{3}(x_1+x_2)/2\right)^2 + ((x_1 - x_2)/2)^2$.

However, the good and important point about the reduction of a quadratic form to a canonical form is that the non-negative integers p and q are *invariant*. This is the *Sylvester's law of inertia*.

Similarly, if we change the variables x to $\tilde{x} = R^T x$, then $\nabla_x u = R\nabla_{\tilde{x}} u$. Thus, the principal part of equation (7.1.1) becomes

$$\sum_{i=1}^{p} \lambda_i \frac{\partial^2 u}{\partial \tilde{x}_i^2} - \sum_{i=p+1}^{p+q} \lambda_i \frac{\partial^2 u}{\partial \tilde{x}_i^2}.$$

Now, change $\tilde{x}$ to y, where $y_i = \tilde{x}_i/\sqrt{\lambda_i}$ for $i = 1, 2, \ldots, p+q$ and $y_i = \tilde{x}_i$ for $i > p+q$. Then, the principal part of equation (7.1.1) becomes

$$\sum_{i=1}^{p} \frac{\partial^2 u}{\partial y_i^2} - \sum_{i=p+1}^{p+q} \frac{\partial^2 u}{\partial y_i^2}. \tag{7.2.1}$$

This is referred to as a *canonical form* of the principal part of equation (7.1.1).

We now consider the following three cases that occur most in many applications:

- $p = n$, $q = r = 0$.
- $p = n-1$, $q = 1$, $r = 0$.
- $p = n-1$, $q = 0$, $r = 1$.

In $\mathbb{R}^3$, the level sets of the quadratic $\lambda_1 x_1^2 + \lambda_2 x_2^2 + \lambda_3 x_3^2$, where λ_i are real numbers, is called an *ellipsoid* if all the λ_i are positive; a *hyperboloid* if one λ_i is positive and the other two are negative or vice versa. The same terminology is adapted for PDE also. The PDE (7.1.1) is called an *elliptic* equation if $p = n$ (or $q = n$) and *hyperbolic* equation if $p = n - 1$, $q = 1$ (or $p = 1$, $q = n - 1$).

In $\mathbb{R}^3$, the level set of the function $x_1^2 + x_2^2 \pm x_3$ is called a *paraboloid*. Observe the inclusion of the *linear* x_3 term. Thus, PDE (7.1.1) is called a *parabolic* equation if $p = n - 1$, $r = 1$, *and the reduced equation* (7.2.1) *contains a first-order derivative term in a specified variable*. If the latter condition is not satisfied, then the equation should not be classified as parabolic. We shall see an example later.

Now moving to the case of variable coefficients in equation (7.1.1), we wish to reduce, if possible, the principal part to its canonical form, as in equation (7.2.1). If equation (7.1.1) is posed in a domain Ω of $\mathbb{R}^n$, we can *freeze* the coefficients a_{ij} at a point $x_0 \in \Omega$. We then obtain an orthogonal matrix $R(x_0)$, and the eigenvalues also depend on x_0. However, such a procedure poses problems with smoothness of the eigenvalues, and it is also difficult to track their signs; smoothness of $R(x)$ as x varies in Ω is also required. For example, consider the *Tricomi equation*

$$y u_{xx} + u_{yy} = 0.$$

In the region $y > 0$, this is elliptic, and in the region $y < 0$, it is hyperbolic. In conclusion, it will not be possible to work with such a procedure.

We now try to look for a different strategy and seek a change of variables of the form $y_j = \phi_j(x)$, $x \in \Omega$. To begin with, we require this transformation to be non-singular, that is, the Jacobian $\det \left[\dfrac{\partial \phi_i}{\partial x_j} \right]$ should not vanish in Ω. The principal part

$$\sum_{i,j=1}^n a_{ij}(x) \frac{\partial^2 u}{\partial x_i \partial x_j}$$

gets transformed to

$$\sum_{i,j=1}^n \tilde{a}_{ij} \frac{\partial^2 u}{\partial y_i \partial y_j},$$

where

$$\tilde{a}_{ij} = \sum_{k,m=1}^n a_{km} \frac{\partial \phi_i}{\partial x_k} \frac{\partial \phi_j}{\partial x_m}.$$

We now count the number of unknowns and equations. For the said reduction, we require $\tilde{a}_{ij} = 0$ for all $i \neq j$. Using symmetry, therefore, there are $n(n-1)/2$ equations. Thus, if $n > 3$, then $n(n-1)/2 > n$, and we get into an overdetermined system. If $n = 3$, then $n(n-1)/2 = n$. However, then, we also require that $\tilde{a}_{ii} = \pm 1$ or 0, leading again to an overdetermined system. Thus, we are left with only the case of $n = 2$.

7.3 Second-Order Linear Equations in Two Variables

Consider a second-order linear equation in two variables x and y:

$$Lu = a(x,y)u_{xx} + 2b(x,y)u_{xy} + c(x,y)u_{yy} \\ + d(x,y)u_x + e(x,y)u_y + f(x,y)u = g(x,y). \quad (7.3.1)$$

The principal part is

$$L_0 u = a(x,y)u_{xx} + 2b(x,y)u_{xy} + c(x,y)u_{yy}. \quad (7.3.2)$$

Look for non-singular change of variables $\tilde{x} = \phi(x,y)$ and $\tilde{y} = \psi(x,y)$ with $\phi_x \psi_y - \phi_y \psi_x \neq 0$.

The present case is simple and just one quantity, namely the discriminant, $ac - b^2$ of the quadratic determines the *type* of the equation. Equation (7.3.1) is called

- *elliptic* if $ac - b^2 > 0$.
- *hyperbolic* if $ac - b^2 < 0$.
- *parabolic* if $ac - b^2 = 0$.

As remarked earlier, an equation would be called a parabolic equation *only* if it contains some specific first-order term in the reduced form. The principal part L_0 transforms to $\tilde{L}_0 = \tilde{a}\frac{\partial^2}{\partial \tilde{x}^2} + 2\tilde{b}\frac{\partial^2}{\partial \tilde{x} \partial \tilde{y}} + \tilde{c}\frac{\partial^2}{\partial \tilde{y}^2}$, where

$$\left. \begin{array}{l} \bullet \ \tilde{a} = a\phi_x^2 + 2b\phi_x \phi_y + c\phi_y^2. \\ \bullet \ \tilde{b} = a\phi_x \psi_x + b(\phi_x \psi_y + \phi_y \psi_x) + c\phi_y \psi_y. \\ \bullet \ \tilde{c} = a\psi_x^2 + 2b\psi_x \psi_y + c\psi_y^2. \end{array} \right\} \quad (7.3.3)$$

Furthermore, we have $\tilde{a}\tilde{c} - \tilde{b}^2 = (ac - b^2)(\phi_x \psi_y - \phi_y \psi_x)^2$.

7.3.1 Reduction to Canonical Form

Since ϕ_x and ϕ_y (similarly for the function ψ), we see from equation (7.3.3) that ϕ_x/ϕ_y or ϕ_y/ϕ_x is a root of the quadratic equation $a\zeta^2 + 2b\zeta + c = 0$. Depending on the sign of the discriminant, we now consider the following cases.

Hyperbolic Case: This is the case when $ac - b^2 < 0$. Let ζ_1 and ζ_2 be the distinct roots of the quadratic equation $a\zeta^2 + 2b\zeta + c = 0$. Consider the first-order equations $\phi_x - \zeta_1 \phi_y = 0$ and $\psi_x - \zeta_2 \psi_y = 0$. The transformation $(x, y) \mapsto (\phi, \psi)$ is non-singular that follows from $\zeta_1 \neq \zeta_2$. It now follows that $\tilde{a} = \tilde{c} = 0$ (see equation (7.3.3)) and L_0 reduces to $\dfrac{\partial^2}{\partial \tilde{x} \partial \tilde{y}}$. The curves $\phi = $ constant and $\psi = $ constant are called the *characteristic curves* of the hyperbolic operator L. Finally, put $\xi = (\tilde{x} + \tilde{y})/2$ and $\eta = (\tilde{x} - \tilde{y})/2$. Then, L_0 reduces to $\dfrac{\partial^2}{\partial \xi^2} - \dfrac{\partial^2}{\partial \eta^2}$, the wave operator.

Parabolic Case: This is the case $ac - b^2 = 0$. Assume $a \neq 0$. Now, the quadratic equation $a\zeta^2 + 2b\zeta + c = 0$ has only one (double) real root $\zeta = -b/a$. There is only one family of characteristic curves $\phi = $ constant, in this case, and ϕ satisfies the first-order equation $a\phi_x + b\phi_y = 0$ or, equivalently, $b\phi_x + c\phi_y = 0$. Choose the new co-ordinates as $\tilde{x} = \phi(x, y)$ and $\tilde{y} = \psi(x, y)$, where ψ is any function that makes the transformation non-singular. This makes $\tilde{a} = \tilde{b} = 0$ (see equation (7.3.3)). Choosing $\psi(x, y) = x$ gives $\tilde{c} = a \neq 0$, and the principal part reduces to $\dfrac{\partial^2}{\partial \tilde{y}^2}$.

Caution: If the full reduction, including the first-order terms, contains the term $\dfrac{\partial}{\partial \tilde{x}}$, then the operator L is said to be parabolic; otherwise, it is classified as *weakly hyperbolic*. The operator $\dfrac{\partial^2}{\partial x^2} + 2 \dfrac{\partial^2}{\partial x \partial y} + \dfrac{\partial^2}{\partial y^2}$ is weakly hyperbolic, *not* parabolic.

Elliptic Case: Here, $ac - b^2 > 0$. In this case, the quadratic equation $a\zeta^2 + 2b\zeta + c = 0$ has complex conjugate roots $\zeta, \bar{\zeta}$. The first-order equation $\phi_x - \zeta \phi_y$ has complex coefficients now. Even to assert the existence of a solution to this first-order equation, we need to make additional assumptions of analyticity on the coefficients and invoke *Cauchy–Kovalevsky theorem*. This itself is a hard theorem.

However, in practice, when the coefficients are (complex) constants or simple functions such as polynomials, it is quite possible to write down a solution. We see that $\psi = \bar{\phi}$ satisfies $\psi_x - \bar{\zeta} \psi_y$. Now, write $\phi = X + iY$, where $i = \sqrt{-1}$. Equating $\tilde{b}$ to 0^2 (see equation (7.3.3)) and performing some computations, we see that the principal part L_0 reduces to $\dfrac{\partial^2}{\partial X^2} + \dfrac{\partial^2}{\partial Y^2}$, which is Laplace operator in two dimensions.

---

[2] That $\tilde{b} = 0$ follows from the Cauchy–Riemann equations, if we assume ϕ is analytic.

7.4 Higher-Order Equations

We shall now briefly discuss the classification of higher-order equations. Consider an mth-order linear partial differential operator

$$L = L(x, D) = \sum_{|\alpha| \le m} a_\alpha(x) D^\alpha,\ x \in \Omega \subset \mathbb{R}^n,$$

with $\sum_{|\alpha|=m} |a_\alpha(x)| > 0$. Here, Ω is an open set in $\mathbb{R}^n$. The *principal part* of L is the operator

$$L_m = L_m(x, D) = \sum_{|\alpha|=m} a_\alpha(x) D^\alpha.$$

The *principal symbol* or *characteristic form* is the homogeneous polynomial

$$Q_m(x, \xi) = \sum_{|\alpha|=m} a_\alpha(x) \xi^\alpha,$$

and the *full symbol* is the polynomial

$$Q(x, \xi) = \sum_{|\alpha| \le m} a_\alpha(x) \xi^\alpha.$$

The *characteristic set* or the *characteristic variety* of L is defined by

$$\mathrm{char}(L) = \{(x, \xi) \in \Omega \times \mathbb{R}^n \setminus \{0\} : Q_m(x, \xi) = 0\}.$$

The operator L is said to be *elliptic* if $\mathrm{char}(L) = \emptyset$. The operator L is said to be an *operator with simple characteristics* if $\mathrm{char}(L) \neq \emptyset$, and if $(x_0, \xi_0) \in \mathrm{char}(L)$, then $\dfrac{\partial Q_m}{\partial \xi_j}(x_0, \xi_0) \neq 0$ for some j. More complete results can be found in the literature regarding both the elliptic operators and operators with simple characteristics.

The general references are [45], [16], [35] and [23].

7.5 Exercises

Exercise 7.1

Determine the types of the following equations and reduce them to canonical form.

(a) $e^{2x} u_{xx} + 2e^{x+y} u_{xy} + e^{2y} u_{yy} = 0$.

(b) $u_{xx} + 2u_{xy} + 4u_{xz} + 5u_{zz} + u_x + 2u_y = 0$ (in $\mathbb{R}^3$).
(c) $u_{xx} - 2(\sin x)u_{xy} - (\cos^2 x)u_{yy} - (\cos x)u_y = 0$.
(d) $y^2 u_{xx} + x^2 u_{yy} = 0$ $(xy \neq 0)$.
(e) $x^2 u_{xx} + 2xy u_{xy} + y^2 u_{yy} = 0$ $(x > 0)$.
(f) $\dfrac{\partial}{\partial x}\left((1-x)^2 \dfrac{\partial u}{\partial x}\right) = \dfrac{1}{a^2}(1-x)^2 \dfrac{\partial^2 u}{\partial t^2}$ $(a \neq 0)$.
(g) $(\sin^4(2x))u_{xx} + 4(\sin^4(2x))u = u_{tt} = 0$.
(h) $u_{xx} - 2u_{xy} - 3u_{yy} + u_y = 0$.

Exercise 7.2

Solve the characteristic Cauchy problem:

$$u_{xy} = f(x, y),\ (x, y) \in \mathbb{R}^2,$$

$$u(x, 0) = u_0(x),\ u_y(x, 0) = u_1(x),\ x \in \mathbb{R},$$

where f, u_0 and u_1 are given functions.

7.6 Solutions

Exercise 7.1

(a) The discriminant is $ac - b^2 = e^{2x+2y} - e^{2(x+y)} = 0$, and thus, the quadratic equation $e^{2x}\zeta^2 + 2e^{x+y}\zeta + e^{2y} = 0$ has only one real root: $\zeta = -\dfrac{e^{x+y}}{e^{2x}} = -e^{y-x}$. The characteristic family is $\phi = $ constant, where ϕ satisfies the first-order equation $\phi_x - \zeta\phi_y = 0$ or $e^x \phi_x + e^y \phi_y = 0$. It is straightforward to check that $\phi(x, y) = (1/2)(e^{-x} - e^{-y})$ is a solution. Let $\tilde{x} = (1/2)(e^{-x} - e^{-y})$, and it is convenient to choose the second (independent) variable as $\tilde{y} = (1/2)(e^{-x} + e^{-y})$. Observe that $\tilde{y} > |\tilde{x}|$ and $x = -\log(\tilde{y} + \tilde{x})$, $y = -\log(\tilde{y} - \tilde{x})$. The given equation therefore reduces (see equation (7.3.3)) to

$$u_{\tilde{y}\tilde{y}} + \dfrac{\tilde{y}}{\tilde{y}^2 - \tilde{x}^2} u_{\tilde{x}} + \dfrac{\tilde{x}(3\tilde{y}^2 + \tilde{x}^2)}{(\tilde{y}^2 - \tilde{x}^2)^2} u_{\tilde{y}} = 0.$$

Since the final reduction contains a first-order term in $\tilde{x}$ variable, we conclude that the given equation is parabolic.

(b) Since the equation has only constant coefficients, we can use the matrix notation to write the given equation as

$$(\nabla^T A \nabla)u + c^T \cdot \nabla u = 0,$$

where A is the symmetric 3×3 matrix, $A = \begin{bmatrix} 1 & 1 & 2 \\ 1 & 0 & 0 \\ 2 & 0 & 5 \end{bmatrix}$; ∇ denotes the gradient (column) vector with respect to x, y and z variables; c^{T} is the row vector $\begin{bmatrix} 1 & 2 & 0 \end{bmatrix}$ with superscript T denoting the transpose of a matrix or a vector and $\cdot$ denotes the scalar or dot product in $\mathbb{R}^3$. The eigenvalues of A are 1, $\lambda_1 = (5 + 3\sqrt{5})/2$ and $\lambda_2 = (5 - 3\sqrt{5})/2$. The corresponding orthonormal eigenvectors are

$$r_1 = (2/3) \begin{bmatrix} 1 \\ 1 \\ -1/2 \end{bmatrix}, \; r_2 = \mu_1 \begin{bmatrix} 1 \\ \lambda_1^{-1} \\ -2\lambda_2^{-1} \end{bmatrix}, \; r_3 = \mu_2 \begin{bmatrix} 1 \\ \lambda_2^{-1} \\ -2\lambda_1^{-1} \end{bmatrix},$$

where μ_1 and μ_2 are normalizing constants. Since two of the eigenvalues are positive and one is negative, the given equation is hyperbolic.

To obtain the canonical form, let R be the orthogonal matrix $R = \begin{bmatrix} r_1 & r_2 & r_3 \end{bmatrix}$, and consider the change of variables

$$\begin{bmatrix} \xi \\ \eta \\ \zeta \end{bmatrix} = R^{\mathrm{T}} \begin{bmatrix} x \\ y \\ z \end{bmatrix}.$$

If we denote by $\widetilde{\nabla}$ the gradient vector with respect to the (ξ, η, ζ) variables, then we have $\nabla = R\widetilde{\nabla}$ and the given equation reduces to

$$0 = (\nabla^{\mathrm{T}} A \nabla)u + c^{\mathrm{T}} \cdot \nabla u = (\widetilde{\nabla}^{\mathrm{T}} R^{\mathrm{T}} A R \widetilde{\nabla})u + (c^{\mathrm{T}} R) \cdot \widetilde{\nabla} u.$$

That is, $u_{\xi\xi} + \lambda_1 u_{\eta\eta} + \lambda_2 u_{\zeta\zeta} + d_1 u_\xi + d_2 u_\eta + d_3 u_\zeta = 0$, where d_1, d_2 and d_3 are the elements of the row vector $d^{\mathrm{T}} = c^{\mathrm{T}} R$. Finally, let $\tilde{x} = \xi$, $\tilde{y} = \lambda_1^{-1/2} \eta$ and $\tilde{z} = |\lambda_2|^{-1/2} \zeta$. In these new variables, we obtain the given equation in a canonical form:

$$u_{\tilde{z}\tilde{z}} - u_{\tilde{x}\tilde{x}} - u_{\tilde{y}\tilde{y}} - d_1 u_{\tilde{x}} - (d_2/\sqrt{\lambda_1}) u_{\tilde{y}} + (d_3/\sqrt{|\lambda_2|}) u_{\tilde{z}} = 0.$$

(c) Here, the discriminant $ac - b^2 = -\cos^2 x - \sin^2 x = -1 < 0$, and therefore, the equation is hyperbolic. For the reduction to a canonical form, first, observe that the roots of the quadratic equation $\zeta^2 - 2(\sin x)\zeta - \cos^2 x = 0$ are $\sin x \pm 1$. Thus, the characteristic families are given by $y - \cos x \pm x = $ constant. Introduce the new variables $\xi = y - \cos x + x$ and $\eta = y - \cos x - x$. The given equation now reduces to

$u_{\xi\eta} = 0$. This equation is easily solved, and we see that the general solution is given by $u = F(\xi) + G(\eta)$. Going back to the x and y variables, we see that

$$u(x,y) = F(y - \cos x + x) + G(y - \cos x - x),$$

where F and G are arbitrary C^2 functions, is the general solution of the given equation.

(d) On the co-ordinate axis $x = 0$ or $y = 0$, the given equation reduces to an ODE. Thus, we restrict the discussion to the first quadrant $x > 0$, $y > 0$, the analysis being similar in other quadrants. Here, $ac - b^2 = x^2 y^2 > 0$, and hence the equation is elliptic. The roots of the quadratic equation $x^2 \zeta^2 + y^2 = 0$ are complex: $\zeta = \pm i(x/y)$. The characteristic equation is given by $\phi_x - i(x/y)\phi_y = 0$. One solution of this equation is $\phi(x,y) = y^2 + ix^2$. Thus, we introduce the new variables $\tilde{x} = y^2$ and $\tilde{y} = x^2$. The given equation then reduces to

$$u_{\tilde{x}\tilde{x}} + u_{\tilde{y}\tilde{y}} + \frac{1}{2\tilde{x}} u_{\tilde{x}} + \frac{1}{2\tilde{y}} u_{\tilde{y}} = 0,$$

which is the required canonical form.

(e) The given equation reduces to an ODE on either of the co-ordinate axes. Thus, we restrict the analysis to the first quadrant $x > 0$, $y > 0$. The discriminant is given by $ac - b^2 = x^2 y^2 - (xy)^2 = 0$, and there is only one real characteristic, which is given by $\phi_x + \frac{y}{x}\phi_y = 0$. One solution of this first-order equation is $\phi(x,y) = \log(y/x)$ (the reader should analyse the cases of the second and fourth quadrants, where $xy < 0$). Now, consider the new variables $\xi = \log(y/x)$, $\eta = x$. The choice of the variable η is arbitrary, and non-singularity of the transformation is the only requirement. The reader should explore other possibilities. We have $x = \eta$ and $y = \eta e^\xi$. In the new co-ordinates, the given equation transforms into $u_{\eta\eta} = 0$. Since the final reduction *does not* contain a first-order term in the ξ variable, the given equation is weakly hyperbolic and *not* parabolic.

It is also easy to solve the given equation. The general solution is given by $u(\xi, \eta) = F(\xi) + G(\xi)\eta$, where F and G are functions of ξ only. In x and y variables, the general solution can therefore be written as

$$u(x,y) = F(\log y - \log x) + xG(\log y - \log x).$$

(f) Rewrite the given equation as $u_{tt} - a^2 u_{xx} - \frac{2}{1-x} u_x = 0$, assuming $x \neq 1$. This can be easily reduced to a canonical form by changing the variable t to at. For $x = 1$, both the sides of the equation become 0, assuming that the derivatives of u are bounded in a neighbourhood of $x = 1$.

(g) The discriminant is given by $ac - b^2 = -\sin^4(2x)$, which is negative for $x \in (0, \pi)$. Thus, the given equation is hyperbolic. We restrict the attention to this interval for the reduction to a canonical form.

The roots of the quadratic equation are $\pm\sin^2(2x)$, and the characteristics satisfy the first-order equation $\phi_t \pm \sin^2(2x)\phi_x = 0$. We take $\phi(x,t) = t \pm \frac{1}{2}\cot(2x)$ as solutions. This leads us to the new variables as $\xi = t + (1/2)\cot(2x)$ and $\tau = t - (1/2)\cot(2x)$. This implies that $x = (1/2)\cot^{-1}(\xi - \tau)$ and $t = (\xi + \tau)/2$. A somewhat lengthy computation then gives the canonical form:

$$u_{\xi\tau} = \sin^5(2x)\cos(2x)(u_\xi - u_\tau) + \sin^4(2x)\, u.$$

The trigonometric functions in the coefficients are easily expressed in terms of ξ and τ using the relation $\xi - \tau = \cot(2x)$ and the familiar trigonometric identities.

(h) Here, $ac - b^2 = -3 - 1 = -4 < 0$, and hence the equation is hyperbolic. The roots of the quadratic equation $\zeta^2 - 2\zeta - 3 = 0$ are $3, -1$. Thus, we choose the characteristic variables $\tilde{x} = 3x + y$ and $\tilde{y} = x - y$ as the new variables. This gives the canonical form of the given equation as $u_{\tilde{x}\tilde{y}} + \frac{1}{16}(u_{\tilde{x}} - u_{\tilde{y}}) = 0$.

Exercise 7.2

Note that the line $y = 0$ (also $x = 0$) is a characteristic of the given equation. Integrating the given equation, we find that

$$u(x,y) = F(x) + G(y) + \int_0^x \int_0^y f(\xi, \eta)\, d\eta d\xi,$$

where F and G are arbitrary functions. Using the data, we obtain that

$$u_0(x) = F(x) + G(0) \text{ and } u_1(x) = G'(0) + \int_0^x f(\xi, 0)\, d\xi.$$

In particular, we have $G(0) = 0$, $u_1(0) = G'(0)$ and $u_1'(x) = f(x, 0)$ for $x \in \mathbb{R}$. This shows that the functions f and u_1 cannot be arbitrary but satisfy a relation between them. We thus find the solution u as

$$u(x,y) = u_0(x) + G(y) + \int_0^x \int_0^y f(\xi, \eta)\, d\eta d\xi,$$

where G is an arbitrary C^2 function satisfying the relations $G(0) = 0$ and $G'(0) = u_1(0)$ and f and u_1 satisfy the relation $u_1'(x) = f(x, 0)$ for $x \in \mathbb{R}$. Thus, there is also non-uniqueness of the solution.

8
Laplace and Poisson Equations

8.1 Introduction

We briefly recall some of the definitions, formulas and results from the book Ref.[45] which are relevant for solving the exercises of this chapter. A reader can refer the above-cited book or any other book with similar content for more detailed proofs and discussion. The most general form of a second-order linear partial differential equation (PDE) in n variables is given by

$$Lu \equiv \sum_{i,j=1}^{n} a_{ij}(x)D_{ij}u + \sum_{i=1}^{n} b_i(x)D_i u + c(x)u + d(x) = 0, \qquad (8.1.1)$$

where $x \in \Omega$, an open set in $\mathbb{R}^n$, $a_{ij} = a_{ji}$. The operator L is said to be *uniformly elliptic* if there exists an $\alpha > 0$ such that $\sum_{i,j=1}^{n} a_{ij}(x)\xi_i \xi_j \geqslant \alpha \sum_{i=1}^{n} |\xi|^2$, for all $x \in \Omega$ and $\xi \in \mathbb{R}^n$. Recall that $\chi_L(x, \xi) = \sum_{i,j=1}^{n} a_{ij}(x)\xi_i \xi_j$ is the *characteristic form*, also called the *principal symbol* associated with the operator L. An important elliptic operator is $L = \Delta = \sum_{i=1}^{n} \frac{\partial^2}{\partial x_i^2}$, the *Laplace operator*. This operator has many interesting properties — mean value property (MVP) and minimum and maximum principles. There is also a notion of a *fundamental solution*. This is not specific to the Laplace operator Δ, but every constant coefficient differential operator possesses a fundamental solution. A restricted definition is the following. A locally integrable function E is called a *fundamental solution* of L if $\langle E, L'\psi \rangle = \psi(0)$ for all smooth functions ψ with compact support, where L' is the adjoint operator of L. Symbolically, this is written as $LE = \delta$, the *Dirac delta function*. Since the operator Δ is self-adjoint, we have $\int E \Delta \psi = \psi(0)$. Interestingly, the fundamental solution is not a solution of the Laplace equation $\Delta E = 0$ in a strict sense but very useful in the construction/representation of solutions to the Laplace and Poisson equations. Note that $\Delta E = 0$ in $\mathbb{R}^n \setminus \{0\}$, and hence E indeed has a singularity at the origin. A fundamental solution is not unique.

A fundamental solution of $-\Delta$ is given by

$$\phi(x) = \begin{cases} -\dfrac{1}{2\pi} \log |x| & \text{if } n = 2, \\ \dfrac{1}{n(n-2)\omega_n |x|^{n-2}} & \text{if } n \geqslant 3, \end{cases} \tag{8.1.2}$$

where ω_n is the volume of the unit ball in $\mathbb{R}^n$. Throughout this chapter, we use ϕ to denote the fundamental solution of $-\Delta$. As it is customary (though it is not the case always) in the analysis of PDE, we use $-\Delta$. Furthermore, note that $-\Delta$ is a self-adjoint positive operator. The fundamental solution ϕ and its first-order derivatives $D\phi$ are locally integrable, whereas the second-order derivatives are not locally integrable. Thus, we need to be careful at the singularity when we deal with the second-order derivatives in the analysis. This leads to difficulty in solving the *Poisson equation* $-\Delta u = f$ in a smooth class, when f is just a continuous function. In fact, there is non-existence.

Some definitions are in order. A function $u \in C^2(\Omega)$ is said to be *harmonic* in Ω, if it satisfies the *Laplace equation* $\Delta u = 0$ in Ω; u is said to be *sub-harmonic* (similarly, *super-harmonic*) in Ω if it satisfies the inequality $\Delta u \geqslant 0$ (similarly, $\Delta u \leqslant 0$) in Ω. These notions are fundamental in the study of existence and uniqueness results for the Laplace equation via *Perron's method*. We have the following result.

Theorem 8.1. *Let $u \in C^2(\Omega)$ be sub-harmonic. Then, for any ball $B = B_R(y) \Subset \Omega$, we have*

$$u(y) \leqslant \frac{1}{n\omega_n R^{n-1}} \int_{\partial B} u(x)\, d\sigma(x) = \frac{1}{|\partial B|} \int_{\partial B} u\, d\sigma(x) \tag{8.1.3}$$

$$u(y) \leqslant \frac{1}{\omega_n R^n} \int_B u(x)\, dx = \frac{1}{|B|} \int_B u(x)\, dx. \tag{8.1.4}$$

The above inequalities are reversed if u is super-harmonic. Finally, if u is harmonic, then we have

$$u(y) = \frac{1}{|\partial B|} \int_{\partial B} u\, d\sigma = \frac{1}{|B|} \int_B u(x)\, dx. \tag{8.1.5}$$

Equation (8.1.5) is referred to as the *mean value formula,* and a harmonic function u is said to satisfy the MVP. A partial converse of Theorem 8.1 also holds:

Theorem 8.2 (Converse of MVP). Let $u \in C^2(\Omega)$ satisfy

$$u(x) = \frac{1}{|\partial B|} \int_{\partial B} u \, d\sigma = \frac{1}{|B|} \int_B u \, dx, \qquad (8.1.6)$$

for each ball $B = B_r(x) \Subset \Omega$, and then u is harmonic.

Some remarks are in order. The MVP can be defined for continuous functions, without C^2 assumption. In fact, we can prove that a continuous function u is harmonic if it satisfies MVP. Thus, MVP is a defining property of harmonic functions.

Maximum and Minimum Principles: We know that for a one-variable function, if the derivative vanishes at an interior point, then the sign of the second derivative determines whether the point is a local maximum or local minimum. We look for similar results in higher dimensions, and the differential operators play the role of the second derivatives. Thus, in general, maximum and minimum principles are trademarks of the second-order differential operators, though such results are available for higher-order operators.

Theorem 8.3 (Strong Maximum Principle). Assume Ω is a bounded, open and connected set in $\mathbb{R}^n$, and u is a bounded sub-harmonic function in Ω. If there is a $y \in \Omega$ such that $u(y) = \sup_\Omega u$, then u is a constant. That is, a bounded sub-harmonic function cannot assume an interior maximum unless it is a constant.

If u is super-harmonic, then by applying the above theorem to $-u$, we see that a non-constant super-harmonic function cannot assume an interior minimum and a non-constant harmonic function cannot assume both interior minimum and maximum.

The above theorem can be stated in a slightly weaker form, which is referred to as *weak maximum and weak minimum principles.*

Theorem 8.4. Assume Ω is a bounded region in $\mathbb{R}^n$. For $u \in C^2(\Omega) \cap C(\overline{\Omega})$, the following statements hold:

(i) if u is sub-harmonic in Ω, then $\sup_\Omega u = \max_{\overline{\Omega}} u = \max_{\partial \Omega} u$.

(ii) if u is super-harmonic in Ω, then $\inf_\Omega u = \min_{\overline{\Omega}} u = \min_{\partial \Omega} u$.

(iii) if u is harmonic in Ω, then $\min_{\partial \Omega} u \leqslant u(x) \leqslant \max_{\partial \Omega} u$, for all $x \in \Omega$.

The maximum and minimum principles also hold for more general second-order elliptic equations. Consider a second-order elliptic partial differential operator L of the form

$$L = \sum_{i,j=1}^{n} a_{ij} \frac{\partial^2}{\partial x_i \partial x_j} + \sum_{i=1}^{n} b_i \frac{\partial}{\partial x_i} + c, \qquad (8.1.7)$$

where $a_{ij} = a_{ji}$, b_i and c are real-valued functions defined on Ω and are assumed to be smooth.

Theorem 8.5. Suppose the operator L is elliptic in Ω and $u \in C^2(\Omega)$ takes its supremum at some point $x_0 \in \Omega$. Then, $Lu(x_0) \leqslant c(x_0) u(x_0)$.

We now state two interesting results, namely *Harnack's inequality* and *Hopf's lemma*. Point-wise comparison of a non-negative harmonic function away from the boundary is the theme of the Harnack's inequality. As an application, we get *Hopf's lemma* for the ball which gives an estimate for the normal derivative of the solution at the boundary.

Theorem 8.6 (Harnack's Inequality). Suppose Ω be a region in $\mathbb{R}^n$ and $V \Subset \Omega$. If u is a non-negative harmonic function in Ω, then there exists a constant $C > 0$ depending only on V and n such that

$$\sup_V u \leqslant C \inf_V u. \qquad (8.1.8)$$

In particular,

$$\frac{1}{C} u(y) \leqslant u(x) \leqslant C u(y), \quad \text{for all } x, y \in V.$$

Lemma 8.7 (Hopf's Lemma). Let $u \in C^2(B_R(x_0)) \cap C\left(\overline{B_R(x_0)}\right)$ be harmonic in $B_R(x_0)$ and $x_* \in \partial B_R(x_0)$ be a minimum point of u in $\overline{B_R(x_0)}$. Then,

$$-\frac{\partial u}{\partial \nu}(x_*) \geqslant 2^{1-n} \left(\frac{u(x_0) - u(x_*)}{R} \right) > 0,$$

provided that the one-sided normal derivative

$$\frac{\partial u}{\partial \nu}(x_*) = \lim_{t \to 0+} \frac{u(x_* - t\nu) - u(x_*)}{t}$$

exists.

Green's Function and Representation Formula: We are interested in the solvability of the boundary value problems, in particular, the Dirichlet problem

$$\begin{cases} -\Delta u = f \text{ in } \Omega, \\ u = g \text{ on } \partial\Omega. \end{cases} \tag{8.1.9}$$

If $\Omega = \mathbb{R}^n$, and f is a C^2 function with compact support, then it can be proved that the convolution of f with the fundamental solution ϕ, that is, $u = f * \phi$, satisfies the *Poisson equation* $-\Delta u = f$ in $\mathbb{R}^n$. In other words, the solution is represented as

$$u(x) = \int_{\mathbb{R}^n} \phi(x-y) f(y) \, dy.$$

Symbolically, $(-\Delta)^{-1}$ has an integral representation with the kernel $\phi(x-y)$. Recall the basic notion that the integral is the inverse of the differential operator $\dfrac{d}{dx}$. Now to deal with the boundary value problem, we need a kernel that should also take care of the boundary values. More precisely, we need to have a kernel $G(x, y)$ that has the same singularity as that of the fundamental solution (to recover the equation) but having some prescribed boundary values. This is the idea behind the Green's function. Suppose $G(x, \cdot)$ solves the *measure-valued* PDE

$$-\Delta_y G = \delta_x \text{ in } \Omega$$
$$G = 0 \text{ in } \partial\Omega,$$

where δ_x is the Dirac delta function concentrated at x. It means that

$$-\int_\Omega G(x,y) \Delta\psi(y) \, dy = \psi(x),$$

for all smooth ψ with compact support in Ω. In general, the solvability of G is as difficult as that of the original problem.

Let $u \in C^2(\overline{\Omega})$. An application of the Green' formula yields the identity:

$$u(x) = \int_\Omega \Delta u(y) \phi(y-x) \, dy - \int_{\partial\Omega} u(y) \frac{\partial \phi}{\partial \nu}(y-x) \, d\sigma(y)$$
$$+ \int_{\partial\Omega} \phi(y-x) \frac{\partial u}{\partial \nu}(y) \, d\sigma(y), \tag{8.1.10}$$

for $x \in \Omega$. However, for the unknown term $\dfrac{\partial u}{\partial \nu}$, this identity would have provided us with a formula for the solution of the Dirichlet problem (8.1.9). To get rid of this term, we add a *corrector function*, denoted by $\phi^x(y)$, to the fundamental solution ϕ as follows:

Let $x \in \Omega$ be fixed, and let the function $\phi^x(y)$ solve the problem

$$\Delta_y \phi^x = 0 \text{ in } \Omega \text{ and } \phi^x(y) = \phi(y-x) \text{ for } y \in \partial\Omega.$$

If we now define $G(x,y) = \phi(x-y) - \phi^x(y)$ for $x, y \in \Omega$, $x \neq y$, we obtain the *Green's representation formula* for the solution of the Dirichlet problem (8.1.9):

$$u(x) = \int_\Omega f(y)\, G(x,y)\, dy - \int_{\partial\Omega} g(y) \frac{\partial G}{\partial \nu}(y-x)\, d\sigma(y), \qquad (8.1.11)$$

for $x \in \Omega$. If the boundary $\partial\Omega$ is sufficiently smooth, the existence and uniqueness of the Green's function G can be proved by solving the problem for the corrector function. Furthermore, it can be shown that G is symmetric and has smooth normal derivative $\dfrac{\partial G}{\partial \nu}$.

However, there are certain domains where we can find the corrector function quite easily by using the geometry of the domain and reflection. In particular, the Green's function can be found explicitly in the case of a ball and upper half-space using the symmetry of the domain. This plays a powerful tool in the analysis of potential theory (Laplace and Poisson equations), more generally in the second-order elliptic equations together with the formal Green's representation formula as given below.

Green's Representation Formula: Suppose u solves the boundary value problem (8.1.9). Then, u satisfies

$$u(x) = \int_\Omega f(y) G(x,y)\, dy - \int_{\partial\Omega} g(y) \frac{\partial G}{\partial \nu}(x,y)\, d\sigma(y), \ x \in \Omega. \qquad (8.1.12)$$

Green's Function for the Upper Half-Space: Consider the upper half-space

$$\mathbb{R}_+^n = \{x = (x_1, \cdots x_n) \in \mathbb{R}^n \ : \ x_n > 0\}.$$

In this case, the Green's function is given by

$$G(x,y) = \phi(y-x) - \phi(y-\tilde{x}),$$

where $\tilde{x} = (x_1, \cdots, x_{n-1}, -x_n)$ is the reflection of x through the $x_n = 0$ space. Define the kernel

$$K(x,y) = \frac{2x_n}{n\omega_n |x-y|^n}, \quad x \in \mathbb{R}^n_+, \ y \in \mathbb{R}^n_+.$$

This is known as *Poisson kernel* for $-\Delta$ in $\mathbb{R}^n_+$. If $u \in C^2(\overline{\Omega})$ solves

$$\Delta u = 0 \ \text{ in } \mathbb{R}^n_+, \quad u = g \text{ on } \partial \mathbb{R}^n_+, \tag{8.1.13}$$

then u has the representation formula (known as *Poisson formula* for the upper half-space):

$$u(x) = \frac{2x_n}{n\omega_n} \int_{\partial \mathbb{R}^n_+} \frac{g(y)}{|x-y|^n} d\sigma(y) = \int_{\partial \mathbb{R}^n_+} K(x,y) g(y) \, d\sigma(y), \tag{8.1.14}$$

for $x \in \mathbb{R}^n_+$. Conversely, if g is a bounded continuous function defined on $\mathbb{R}^{n-1}$, then the above formula gives the smooth solution to the boundary value problem, where the boundary values are interpreted as limiting values.

Green's Function for a Ball: Here, the Green's function is given by

$$G(x,y) = \phi(y-x) - \phi(|x|(y-\tilde{x})),$$

where $\tilde{x} = \dfrac{r^2 x}{|x|^2}$ is the reflection point of x through the boundary $\partial B_r(0)$. The *Poisson kernel* for $-\Delta$ in the ball $B_r(0)$ can be computed as

$$K(x,y) = \frac{r^2 - |x|^2}{n\omega_n r} \frac{1}{|x-y|^n}.$$

Furthermore, if u solves the Dirichlet problem

$$\Delta u = 0 \text{ in } B_r(0), \quad u = g \text{ on } \partial B_r(0),$$

then we get the *Poisson formula* for the ball

$$u(x) = \frac{r^2 - |x|^2}{n\omega_n r} \int_{\partial B_r(0)} \frac{g(y)}{|x-y|^n} d\sigma(y) = \int_{\partial B_r(0)} K(x,y) g(y) \, d\sigma(y), \tag{8.1.15}$$

for $x \in B_r(0)$. We extend the same idea to construct the Green's function for some other similar domains, in the exercises. Before ending this brief introduction, we make a couple of more definitions that are useful in the exercises. We have already seen that MVP does not require any smoothness, other than continuity, and we can define harmonicity for continuous functions. Let Ω be an open set in $\mathbb{R}^n$. A function $u \in C(\Omega)$ is said to be *sub-harmonic* if

$$u(x) \leqslant \frac{1}{|B_r(x)|} \int_{B_r(x)} u(y)\,dy,$$

for all $x \in \Omega$ and $r > 0$ such that $B_r(x) \Subset \Omega$. A function $u \in C(\Omega)$ is said to be *super-harmonic* if $-u$ is sub-harmonic. A function $u \in C(\Omega)$ is said to be harmonic if u is both sub and super-harmonic; equivalently, if u satisfies the MVP. We have the following maximum principle under this weak assumption of regularity.

Proposition 8.8. Suppose Ω is a bounded open and connected subset of $\mathbb{R}^n$. If $u \in C(\overline{\Omega})$ is sub-harmonic, then the weak maximum principle holds for u, that is,

$$\max_{\overline{\Omega}} u = \max_{\partial \Omega} u.$$

Similarly, if u is super-harmonic, then the weak minimum principle holds for u. Finally, if u satisfies the MVP, then

$$\min_{\partial \Omega} u \leqslant u(x) \leqslant \max_{\partial \Omega} u,$$

for all $x \in \overline{\Omega}$.

Hölder Continuous Functions: Let $x_0 \in \Omega$, where Ω is a bounded domain in $\mathbb{R}^n$ and $0 < \alpha < 1$. A function $f : \Omega \to \mathbb{R}$ is said to be *Hölder continuous* of order α at x_0 if there exists a constant $C > 0$ such that[1]

$$|f(x) - f(x_0)| \leqslant C|x - x_0|^\alpha, \tag{8.1.16}$$

for all $x \in \Omega$. If $\alpha = 1$ in equation (8.1.16), then the function f is said to be *Lipschitz continuous*. A standard example is $f(x) = |x|^\alpha$, $0 < \alpha < 1$, which is Hölder continuous of order α.

Uniform and Local Hölder Continuity: A function $f : \Omega \to \mathbb{R}$ is said to be *uniformly Hölder continuous* of order α in Ω if there exists a constant $C > 0$ such that

$$|f(x) - f(y)| \leqslant C|x - y|^\alpha,$$

for all $x, y \in \Omega$. The function f is locally Hölder continuous if f is uniformly Hölder continuous in every compact subset of Ω. When $\alpha = 1$, the function f is said to be *uniformly Lipschitz continuous*. It is also clear that a uniformly Hölder continuous

---

[1] If $\alpha > 1$ and satisfies equation (8.1.16), it is an interesting fact that f is a constant function.

function is also uniformly continuous. We denote by $C^{0,\alpha}(\overline{\Omega})$ the space of all uniformly Hölder continuous functions of order α in $\overline{\Omega}$ and define

$$\|f\|_{0,\alpha} = \|f\|_0 + \sup_{x,y\in\overline{\Omega},\, x\neq y} \frac{|f(x)-f(y)|}{|x-y|^\alpha},$$

for $f \in C^{0,\alpha}(\overline{\Omega})$, where $\|f\|_0 = \sup_{x\in\overline{\Omega}}|f(x)|$ is the sup-norm. It is not difficult to verify that $C^{0,\alpha}(\overline{\Omega})$ is a Banach space equipped with the above norm.

The general references are [45], [25], [19], [40], [55], [42], [23], [34].

8.2 Exercises

Exercise 8.1

Let $u, v \in C^2(\overline{\Omega})$, where Ω is a bounded, open set in $\mathbb{R}^n$ with smooth boundary $\partial\Omega$. Using the divergence theorem, prove the following identities:

(a) $\displaystyle\int_\Omega \Delta u \, dx = \int_{\partial\Omega} \frac{\partial u}{\partial \nu} \, d\sigma(x).$

(b) $\displaystyle\int_\Omega v \Delta u \, dx = -\int_\Omega \nabla u \cdot \nabla v \, dx + \int_{\partial\Omega} \frac{\partial u}{\partial \nu} v \, d\sigma(x).$

(c) $\displaystyle\int_\Omega (v\Delta u - u\Delta v) \, dx = \int_{\partial\Omega} \left(\frac{\partial u}{\partial \nu} v - u \frac{\partial v}{\partial \nu}\right) d\sigma(x).$

Here, $\dfrac{\partial u}{\partial \nu} = \nabla u \cdot \nu$ is the normal derivative, and $\nabla = \left(\dfrac{\partial}{\partial x_1}, \cdots, \dfrac{\partial}{\partial x_n}\right)$ is the *grad* operator. The above results are known as *Green's identities*.

Exercise 8.2

For the case $n = 2$, write down the Laplace operator Δ in polar coordinates.

Exercise 8.3

This exercise describes the spherical symmetry of the Laplace operator. Let R be a rotation matrix, that is, $RR^T = R^T R = I$, and u be harmonic in $\mathbb{R}^n$. Define v by $v(x) = u(Rx)$. Show that v is also harmonic in $\mathbb{R}^n$.

Exercise 8.4

Let $v(r) = u(|x|)$, where $r = |x|$. Show that $\Delta u \equiv v''(r) + \dfrac{n-1}{r} v'(r)$. Solve the ordinary differential equation (ODE) $v''(r) + \dfrac{n-1}{r} v'(r) = 0$ to obtain the fundamental solution ϕ.

Exercise 8.5

Let ϕ be the fundamental solution of $-\Delta$. Show that there exists a constant $C > 0$ such that
$|D\phi(x)| \leq \dfrac{C}{|x|^{n-1}}$, $|D^2\phi(x)| \leq \dfrac{C}{|x|^n}$ and $x \neq 0$. Furthermore, analyse the local integrability of the functions ϕ, $D\phi$ and $D^2\phi$.

Exercise 8.6

Let $f \in C_c^2(\mathbb{R}^n)$ and ϕ be the fundamental solution of $-\Delta$. Define $I_\varepsilon = \displaystyle\int_{B_\varepsilon(0)} \phi(y)(\Delta f)(x-y)\,dy$, $\varepsilon > 0$. Show that there exists a constant $C > 0$ such that

$$|I_\varepsilon| \leq \begin{cases} C\varepsilon^2 |\log \varepsilon| & \text{if } n = 2, \\ C\varepsilon^2 & \text{if } n \geq 3. \end{cases}$$

Also, compute $\dfrac{\partial \phi}{\partial \nu}$ on $\partial B_\varepsilon(0)$.

Exercise 8.7

Let Ω be a domain in $\mathbb{R}^2$ symmetric about the x-axis, and let $\Omega^+ = \{(x,y) : y > 0\}$ be the upper part of Ω. Assume $u \in C(\overline{\Omega^+})$ is harmonic in Ω^+ with $u = 0$ on $\partial\Omega^+ \cap \{y = 0\}$. Define for $(x,y) \in \Omega$,

$$v(x,y) = \begin{cases} u(x,y) & \text{if } y \geq 0, \\ -u(x,-y) & \text{if } y < 0. \end{cases}$$

Show that v is harmonic. A similar result holds for domains in $\mathbb{R}^n$.

Exercise 8.8

Let $u \in C^2(\Omega) \cap C^0(\overline{\Omega})$ be a solution of

$$Lu = \Delta u + \sum_{k=1}^n a_k(x)\dfrac{\partial u}{\partial x_k} + c(x)u = 0 \text{ in } \Omega,$$

with $c(x) < 0$ in Ω and $u = 0$ on $\partial\Omega$, and c and a_k's are smooth functions. Show that $u \equiv 0$.

Exercise 8.9

Let u be a solution of the Dirichlet problem: $-\Delta u = \lambda u$ in Ω, $u = 0$ on $\partial\Omega$, where λ is a scalar and Ω is a bounded open set in $\mathbb{R}^n$. If $\lambda \leq 0$, prove that $u \equiv 0$.

Exercise 8.10

Let $u \in C^2(\overline{B_1(0)})$ solve $-\Delta u = f$ in $B_1(0)$, $u = 0$ on $\partial B_1(0)$, where $f \in C(\overline{B_1(0)})$. Show that there exists $C > 0$ such that

$$\max_{x \in B_1(0)} |u(x)| \leq C \max_{x \in \overline{B_1(0)}} |f(x)|.$$

More generally, if u solves $-\Delta u = f$ in $B_1(0)$, $u = g$ on $\partial B_1(0)$, then

$$\max_{B_1(0)} |u| \leq C \left(\max_{\partial B_1(0)} |g| + \max_{B_1(0)} |f| \right).$$

Exercise 8.11

Let u be a non-negative harmonic function in $\mathbb{R}^n$.

(a) By using the Poisson's formula (8.1.15) for the ball and MVP, show that

$$\frac{R^{n-2}(R^2 - |x|^2)}{(R+|x|)^n} u(0) \leq u(x) \leq \frac{R^{n-2}(R^2 - |x|^2)}{(R-|x|)^n} u(0),$$

for any $R > 0$ and $|x| < R$.

(b) By letting $R \to \infty$ in (a) above, conclude that u is a constant.

This gives a stronger form of the Liouville's theorem: If u is harmonic in $\mathbb{R}^n$ and bounded below (or above), then u is a constant function.

Exercise 8.12

Find the Green's function for the Dirichlet problem of Δ for the following domains:

(a) $\Omega = \mathbb{R}^{n,k+} = \{x = (x_1, x_2, \ldots, x_n) \in \mathbb{R}^n : x_j > 0 \text{ for } 1 \leq j \leq k\}$, where $1 \leq k \leq n$. When $n = 3$ and $k = 2$, this domain is called a *dihedral angle*, and when $n = k = 3$, it is the first octant in $\mathbb{R}^3$.

(b) The upper-half ball: $\Omega = \{x \in B_R(0) : x_1 > 0\}$.

(c) $\Omega = \{x \in B_R(0) \subset \mathbb{R}^n : x_j > 0 \text{ for } 1 \leq j \leq k\}$, where $1 \leq k \leq n$.

(d) The strip $\Omega = \{x = (x_1, x_2, \ldots, x_n) \in \mathbb{R}^n : 0 < x_1 < 1\}$ in $\mathbb{R}^n$ ($n \geq 3$).

Exercise 8.13

Let $A = [a_{ij}]$ be a real $n \times n$ matrix with zero trace and $a \in \mathbb{R}^n$. What can be said about a harmonic function in $\mathbb{R}^n$ that satisfies the inequality $u(x) \leq \sum_{i,j=1}^{n} a_{ij} x_i x_j + a \cdot x$ for all $x \in \mathbb{R}^n$? Justify your answer.

Exercise 8.14

Let u be a harmonic function in $\mathbb{R}^n$. Describe the range of u.

Exercise 8.15

If u is a harmonic function in $\mathbb{R}^n$ satisfying $|u(x)| \leq C(1 + |x|^s)$, for some non-negative real s and all $x \in \mathbb{R}^n$, show that u is a polynomial of degree at most $[s]$, where $[s]$ denotes the integer part of s.

Exercise 8.16

Let Ω be an open, bounded set in $\mathbb{R}^n$. Suppose $u \in C^2(\Omega) \cap C^0(\overline{\Omega})$ satisfies $\Delta u = -1$ in Ω, $u = 0$ on $\partial\Omega$. Show that $u(x) \geq \dfrac{1}{2n}(d(x, \partial\Omega))^2$ for each $x \in \Omega$.

Exercise 8.17

If $x \in \mathbb{R}^n$, write $x = (x', x_n)$, $x' \in \mathbb{R}^{n-1}$. Let u be the unique solution of $\Delta u = 0$ in $B_1(0)$ and $u = \phi$ on $\partial B_1(0)$, where $\phi \in C(\partial B_1(0))$. If ϕ satisfies the relation $\phi(x', x_n) = -\phi(x', -x_n)$, show that u also satisfies the relation $u(x', x_n) = -u(x', -x_n)$.

Exercise 8.18

Let u be harmonic in $B_1^+ = \{x \in B_1(0) : x_n > 0\}$ and $u = 0$ on $x_n = 0$. Extend u to a harmonic function in $B_1(0)$.

Exercise 8.19

Let Ω be an open set in $\mathbb{R}^n$ and $u \in C(\Omega)$. Show that u is sub-harmonic if and only if for every bounded, open, connected $\Omega' \subset \Omega$ and every harmonic function v in Ω' with $v = u$ on $\partial\Omega'$, the inequality $u \leq v$ holds.

Exercise 8.20

Suppose u is a non-negative harmonic function in an open connected set Ω in $\mathbb{R}^n$. Show that either $u \equiv 0$ in Ω or $u > 0$ in Ω.

Exercise 8.21

Suppose Ω is a bounded, open set in $\mathbb{R}^n$ and $u \in C^2(\Omega) \cap C^1(\overline{\Omega})$ is a solution of the Neumann problem: $\Delta u = 0$ in Ω, $\dfrac{\partial u}{\partial \nu} = 0$ on $\partial\Omega$. Show that u is identically a constant.

Exercise 8.22

Suppose Ω is a bounded, open, connected subset of $\mathbb{R}^n$. If $u \in C(\overline{\Omega}) \cap C^2(\Omega)$ is a solution of $\Delta u = u^3 + f$ with $f < 0$ in Ω and $u \geq 0$ on $\partial\Omega$, show that $u > 0$ in Ω.

Exercise 8.23

Suppose Ω is an open subset of $\mathbb{R}^n$ and $u \in C^1(\overline{\Omega})$. If $\int_{\partial B} \frac{\partial u}{\partial \nu} d\sigma = 0$ for every ball $B \Subset \Omega$, show that u is harmonic in Ω.

Exercise 8.24

Let $u \in C^2(B_1(0)) \cap C^1(\overline{B_1(0)})$ be a solution of the mixed problem

$$\Delta u = -1 \text{ in } B_1(0),$$
$$u = 0 \text{ on } \partial B_1(0) \cap \{x_n > 0\},$$
$$\frac{\partial u}{\partial \nu} = -u \text{ on } \partial B_1(0) \cap \{x_n < 0\}.$$

Show that $u \geq 0$ in $\overline{B_1(0)}$ and $u > 0$ on $\partial B_1(0) \cap \{x_n < 0\}$.

Exercise 8.25

Give an example of a C^2 function u in $\mathbb{R}^n$ such that $u > 0$ and $\Delta u - \lambda u \geq 0$ in $\mathbb{R}^n$, for some $\lambda > 0$. Can such a function be bounded? Justify your solution.

Exercise 8.26

Let Ω be a bounded or unbounded domain in $\mathbb{R}^n$ and $u \in C(\overline{\Omega}) \cap C^2(\Omega)$ be bounded and satisfy $\Delta u - \lambda u \geq 0$ for some $\lambda > 0$. Prove that $u \leq \sup_{\partial \Omega} u_+$ in $\overline{\Omega}$. Here, $u_+ = \max\{u, 0\}$. By convention, the sup is zero if $\partial \Omega$ is empty.

Exercise 8.27

Let $\Omega = \mathbb{R}^n_+$, $u \in C(\overline{\Omega}) \cap C^2(\Omega)$ be bounded and $\Delta u - \lambda u \geq 0$ in Ω for some $\lambda > 0$. If $x_0 \in \partial \Omega$ is such that $u(x_0) = \sup_{\partial \Omega} u_+ > 0$ and $\frac{\partial u}{\partial \nu}(x_0)$ exist, show that $\frac{\partial u}{\partial \nu}(x_0) \leq -\sqrt{\lambda} u(x_0) < 0$. Here $u_+ = \max\{u, 0\}$.

Exercise 8.28

Let u be harmonic in $B_R(0) \subset \mathbb{R}^n$, $n \geq 3$. For $x \in \mathbb{R}^n$, $|x| > R$, define

$$U(x) = \left(\frac{R}{|x|}\right)^{n-2} u\left(\frac{R^2 x}{|x|^2}\right).$$

The function U is called the *Kelvin transform* of u. Show that U is harmonic in the region $|x| > R$ and that $U(x) \to 0$ as $|x| \to \infty$. Furthermore, write down u in terms of U.

Exercise 8.29

Consider the ball $B_r(x_0)$ in $\mathbb{R}^3$. If $y \notin \overline{B_r(x_0)}$, compute $\displaystyle\int_{B_r(x_0)} \frac{dx}{|x-y|}$.

Exercise 8.30

Let Ω be an open subset of $\mathbb{R}^n$. Call a function $u \in C(\Omega)$ *subharmonic* (respectively, *superharmonic*) in Ω, if the mean value inequality

$$u(x) \leqslant \text{ (respectively, } \geqslant \text{)} \frac{1}{|B_r(x)|} \int_{B_r(x)} u(y)\,dy$$

holds for all $x \in \Omega$ and $r > 0$ such that $B_r(x) \Subset \Omega$. See Exercise 8.19.

If Ω is a bounded, open and connected subset of $\mathbb{R}^n$ and $u \in C(\overline{\Omega})$ is subharmonic (respectively, superharmonic) in Ω, show that u satisfies the weak maximum principle (respectively, weak minimum principle):

$$\max_{\overline{\Omega}} u = \max_{\partial\Omega} u \quad \text{(respectively, } \min_{\overline{\Omega}} u = \min_{\partial\Omega} u\text{)}.$$

Exercise 8.31

Suppose Ω is an open subset of $\mathbb{R}^n$ and $u \in C(\Omega)$ satisfies MVP:

$$u(x) = \frac{1}{|B_r(x)|} \int_{B_r(x)} u(y)\,dy,$$

for all $x \in \Omega$, and $r > 0$ such that $B_r(x) \Subset \Omega$. Show that u is a harmonic function in Ω, that is, $\Delta u(x) = 0$ for all $x \in \Omega$.

Remark: This exercise shows that MVP for a continuous function is equivalent to harmonicity. In particular, a continuous function satisfying MVP is actually a C^∞ function, in fact, an analytic function.

Exercise 8.32

Let u be a C^2 function in a domain Ω in $\mathbb{R}^2$. Let $(x,y) \in \Omega$. Let $A(x-h, y-k)$, $B(x+h, y-k)$, $C(x+h, y+k)$ and $D(x-h, y+k)$ be four points. Suppose $u(x,y) = \frac{1}{4}[u(A) + u(B) + u(C) + u(D)]$ for all $h, k > 0$ such that the rectangular region formed by A, B, C and D is in Ω. Then, show that u is harmonic in Ω. The converse need not be true.

Exercise 8.33

Let ϕ be the fundamental solution of $-\Delta$ and $f \in C_c^\infty(\mathbb{R}^n)$. Consider the function u defined as the convolution

$$u(x) = (\phi * f)(x) = \int_{\mathbb{R}^n} \phi(x-y) f(y)\, dy, \quad x \in \mathbb{R}^n.$$

(1) If $n = 2$ and $f \not\equiv 0$, show that u is an unbounded function.
(2) If $n \geqslant 3$, show that u is a bounded function for any f.

Exercise 8.34

Let Ω be a bounded, open set in $\mathbb{R}^n$ with smooth boundary $\partial\Omega$ and $\rho_1, \rho_2 \in C(\partial\Omega)$. Define the functions u_1 and u_2 by

$$u_1(x) = \int_{\partial\Omega} \phi(x-\xi)\rho_1(\xi)\, d\sigma(\xi) \quad \text{and} \quad u_2(x) = \int_{\partial\Omega} \rho_2(\xi) \frac{\partial \phi}{\partial \nu_\xi}(x-\xi)\, d\sigma(\xi),$$

for $x \in \mathbb{R}^n \setminus \partial\Omega$. Here, ν_ξ denotes the outward unit normal at $\xi \in \partial\Omega$. The functions u_1 and u_2 are referred to as *single-layer potential* and *double-layer potential* with *densities* ρ_1 and ρ_2, respectively. Show that $u_1, u_2 \in C^\infty(\mathbb{R}^n \setminus \partial\Omega)$ and they are harmonic in $\mathbb{R}^n \setminus \partial\Omega$.

Exercise 8.35

[Local Version of Potential Theory] Let Ω be a bounded, open set in $\mathbb{R}^n$, $n \geqslant 3$ with smooth boundary $\partial\Omega$. Let $x_0 \in \Omega$, and suppose that the function $f : \Omega \to \mathbb{R}$ is locally Hölder continuous of order $\alpha \in (0,1)$ around x_0, that is, f satisfies the estimate $|f(x) - f(x_0)| \leqslant A|x - x_0|^\alpha$, for all x in a neighbourhood of x_0 and for some positive constant A. Consider the Newtonian potential v defined by

$$v(x) = \int_\Omega \phi(x-y) f(y)\, dy, \quad x \in \Omega,$$

where $\phi(x) = (n(n-2)\omega_n)^{-1} |x|^{2-n}$ for $x \in \mathbb{R}^n$, $x \neq 0$, the fundamental solution of $-\Delta$. Show that v has continuous derivatives up to order 2 and satisfies the relation $-\Delta v(x_0) = f(x_0)$.

8.3 Solutions

Exercise 8.1

Recall the divergence theorem (integration by parts): For $u \in C^1(\overline{\Omega})$, we have

$$\int_\Omega \frac{\partial u}{\partial x_i} = \int_{\partial \Omega} u \nu_i,$$

for $i = 1, 2, \ldots, n$. Apply the above formula to $\frac{\partial u}{\partial x_i}$ in place of u and sum over $i = 1$ through n to get (a). Now, applying the divergence theorem to the product uv, we get

$$\int_\Omega \left(\frac{\partial u}{\partial x_i} v + \frac{\partial v}{\partial x_i} u \right) = \int_{\partial \Omega} uv \nu_i.$$

Now, apply this to $\frac{\partial u}{\partial x_i}$ in place of u and again sum over $i = 1$ through n to get the identity (b). If we interchange u and v in (b) and subtract the resulting identities one from the other, we get (c). ∎

Exercise 8.2

Write the polar coordinates, $x = r \cos \theta$, $y = r \sin \theta$, and view $u(x, y) = u(r, \theta)$. A straightforward computation of u_{xx} and u_{yy} in terms of the derivatives of r and θ will lead to $u_{xx} + u_{yy} = u_{rr} + \frac{1}{r} u_r + \frac{1}{r^2} u_{\theta\theta}$. ∎

Exercise 8.3

Let $y = Rx$, and so $y_j = R_{jk} x_k$ (repeated indexes need to be summed over $1, 2, \ldots, n$ and R_{jk} denote the entities of the matrix R). For $i = 1, 2, \ldots, n$, we have $v_{x_i}(x) = u_{y_j}(y) R_{jk} \delta_{ik} = u_{y_j}(y) R_{ji}$ and $v_{x_i x_i}(x) = u_{y_\ell y_j}(y) \, R_{\ell k} \delta_{ik} \, R_{ji} = u_{y_l y_j}(y) R_{\ell i} R_{ji}$. By assumption on R, we have $R_{\ell i} R_{ji} = (RR^T)_{j\ell} = \delta_{j\ell}$, the Kronecker delta. Thus, summing over i, we get $\Delta v(x) = \Delta u(Rx) = 0$. ∎

Exercise 8.4

We have $\frac{\partial r}{\partial x_i} = \frac{x_i}{r}$, and thus, $u_{x_i} = v_r \frac{x_i}{r}$ and $u_{x_i x_i} = \frac{x_i^2}{r^2} v_{rr} + \left(\frac{1}{r} - \frac{x_i^2}{r^2} \right) v_r$, for $i = 1, 2, \ldots, n$. Summing over i results in the stated ODE. Next, put $w(r) = v_r$, and then w

satisfies the first-order ODE $w_r(r) = \dfrac{n-1}{r} w(r)$. Solving the ODE, we get $v_r = w(r) = cr^{1-n}$, and performing one more integration, we get

$$v(r) = \begin{cases} c \log r + c_1, & \text{if } n = 2 \\ \dfrac{c}{r^{n-2}} + c_1, & \text{if } n \geqslant 3, \end{cases}$$

where c and c_1 are arbitrary constants. Using the specific value for c and taking $c_1 = 0$, we arrive at the fundamental solution for the minus Laplacian: For $x \neq 0$

$$\phi(x) = \begin{cases} -\dfrac{1}{2\pi} \log r & \text{if } n = 2, \\ \dfrac{1}{n(n-2)\omega_n r^{n-2}} & \text{if } n \geqslant 3, \end{cases}$$

where $r = |x|$ and ω_n is the surface area of the unit sphere in $\mathbb{R}^n$. Indeed, to prove the relation $-\Delta \phi = \delta_0$, that is, $-\displaystyle\int \phi \Delta \psi = \psi(0)$ for all smooth functions with compact support, requires further computing and analysis.

Exercise 8.5

Let $n \geqslant 3$. We have, for $x \neq 0$, $\dfrac{\partial}{\partial x_i} |x|^{2-n} = (2-n)|x|^{-n} x_i$ and

$$\dfrac{\partial^2}{\partial x_i \partial x_j} |x|^{2-n} = (2-n)|x|^{-n} \delta_{ij} + n(n-2)|x|^{-n-2} x_i x_j,$$

for $i, j = 1, 2, \ldots, n$. For $n = 2$, we similarly have

$$\dfrac{\partial}{\partial x_i} \log r = r^{-2} x_i \quad \text{and} \quad \dfrac{\partial^2}{\partial x_i \partial x_j} \log r = r^{-2} \delta_{ij} - 2 r^{-4} x_i x_j,$$

for $i, j = 1, 2$ and $r > 0$. Thus, we arrive at the stated estimates for $|D\phi|$ and $|D^2 \phi|$. For the statements regarding integrability, we observe that the singularities of ϕ, $|D\phi|$ and $|D^2 \phi|$ at the origin are like the singularity of $\dfrac{1}{|x|^{\alpha}}$. Also, $\log |x|$ is locally integrable in $\mathbb{R}^2$ around the origin, since we have

$$\int_{B_1(0)} \log |x| = 2\pi \int_0^1 r \log r \, dr = -\pi/2.$$

In general, for $n \geqslant 2$, we have $\displaystyle\int_{B_1(0)} |x|^{-\alpha} dx = |S_1(0)| \int_0^1 r^{-\alpha} r^{n-1} \, dr$. This one-dimensional integral is finite if and only if $\alpha - n + 1 < 1$ or equivalently $\alpha < n$. Using this, we conclude that ϕ and $D\phi$ are locally integrable. Next, we consider the local integrability of the functions $\frac{\partial^2}{\partial x_i \partial x_j}|x|^{2-n}$, when $n \geq 3$. For $i = j$, we have $\frac{\partial^2}{\partial x_i^2}|x|^{2-n} = (n-2)|x|^{-n}(n\theta_i^2 - 1)$, where θ_i are angular variables. Therefore,

$$\int_{B_1(0)} \left|\frac{\partial^2}{\partial x_i^2}|x|^{2-n}\right| dx = \int_{S_1(0)} |n\theta_i^2 - 1|\, d\sigma(\theta) \int_0^1 r^{-n} r^{n-1}\, dr = \infty.$$

We conclude that $\frac{\partial^2}{\partial x_i^2}|x|^{2-n}$ is not locally integrable. If $i \neq j$, we have $\frac{\partial^2}{\partial x_i \partial x_j}|x|^{2-n} = n(n-2)|x|^{-n}\theta_i\,\theta_j$. Similar computation as above shows that $\frac{\partial^2}{\partial x_i \partial x_j}|x|^{2-n}$ is also not locally integrable.

For $n = 2$, we have $\left|\frac{\partial^2}{\partial x_1 \partial x_2}\log|x|\right| = |x|^{-2}|\sin 2\theta|$ and $\left|\frac{\partial^2}{\partial x_i^2}\log|x|\right| = |x|^{-2}|\cos 2\theta|$, for $i = 1, 2$, where θ is the angular variable. By performing similar computations, we conclude that $\frac{\partial^2 \phi}{\partial x_i \partial x_j}$ is also not locally integrable, for all $i, j = 1, 2, \ldots, n$ and $n \geq 2$.

Exercise 8.6

Since $f \in C_c^2(\mathbb{R}^n)$, we have $|I_\varepsilon| \leq C \int_{B_\varepsilon(0)} |\phi(y)|\, dy$. Now, for $n = 2$, we have

$$\int_{B_\varepsilon(0)} |\phi(y)|\, dy \leq C \int_0^\varepsilon |r \log r|\, dr.$$

For ε small, it is easy to see that $|r \log r|$ is an increasing function in $[0, \varepsilon]$ and hence the estimate. For $n \geq 3$, we get

$$\int_{B_\varepsilon(0)} |\phi(y)|\, dy \leq C \int_0^\varepsilon r^{2-n} r^{n-1}\, dr \leq C\varepsilon^2.$$

On $\partial B_\varepsilon(0)$, the outward normal is given by $\nu = \frac{x}{\varepsilon}$. Hence, from the computation of $D\phi$, for $x \in \partial B_\varepsilon(0)$, we get

$$\frac{\partial \phi}{\partial \nu} = D\phi(x) \cdot \nu = \begin{cases} -\dfrac{1}{2\pi\varepsilon} & \text{if } n = 2, \\ -\dfrac{1}{n\omega_n \varepsilon^{n-1}} & \text{if } n \geq 3. \end{cases}$$

Note that the right-hand side here is nothing but $-|\partial B_\varepsilon(0)|^{-1}$.

Exercise 8.7

Since $v(x,y) = u(x,y)$ is harmonic in the upper part and, by definition v is also harmonic in the lower part. In fact, we have

$$\Delta v(x,y) = \begin{cases} \Delta u(x,y) & \text{if } y > 0, \\ -\Delta u(x,-y) & \text{if } y < 0. \end{cases}$$

Thus, it remains to show that $\Delta v(x,0) = 0$. By definition, v is continuous. We use MVP to prove the harmonicity. In this direction, we have

$$\frac{1}{|B|} \int_B v(x,y)\, dx\, dy = \frac{1}{|B|} \left[\int_{B_+} u(x,y)\, dx\, dy - \int_{B_-} u(x,-y)\, dx\, dy \right] = 0,$$

where $B = B_r(x,0)$ is the ball of radius r centred at $(x,0)$ and contained in Ω and B_+ and B_- are, respectively, the upper and lower parts of B. Again, by continuity of u, we get $v(x,0) = 0$. Thus, MVP is satisfied for all balls with centre at the interface.

Hence, it only remains to check the MVP on a ball that intersects the interface with centre not on the interface. This is done by making the following observation (slightly general form of MVP) in the proof of converse of MVP: If a continuous function w satisfies the MVP at all points and for all balls in some neighbourhood of the points, then w is harmonic. Thus, if the centre lies outside the interface, we can always find a neighbourhood that does not intersect the interface and the MVP is satisfied. See Exercise 8.31.

Exercise 8.8

If possible, assume u attains a (local) maximum (or a minimum) different from 0, at some point $x_0 \in \Omega$. Replacing u by $-u$, if necessary, we may assume that $u(x_0) > 0$. Then, $\frac{\partial u}{\partial x_k}(x_0) = 0$ and $\Delta u(x_0) \leq 0$. Therefore,

$$0 = Lu(x_0) = \Delta u(x_0) + c(x_0)u(x_0) < 0,$$

a contradiction. Thus, $u \equiv 0$ in Ω.

Exercise 8.9

Multiplying the PDE by u and integrating by parts, we get

$$\int_\Omega |\nabla u|^2 = \lambda \int_\Omega |u|^2.$$

Thus, if $\lambda \leqslant 0$, then we get $\int_\Omega |\nabla u|^2 = 0$ and hence u is a constant. Since $u = 0$ on $\partial\Omega$, we have $u \equiv 0$ in Ω.

Remark: In other words, the boundary value problem has no solution for non-positive constants λ. These are eigenvalue problems that come under the class of spectral problems of compact operators. Using compact operator theory and weak formulation, it is established that there is a sequence of constants $\lambda_j \to \infty$ for which the above boundary value problem has solutions. Furthermore, for each j, the set of solutions forms a finite dimensional subspace of $L^2(\Omega)$. This subspace is called eigenspace corresponding to the eigenvalue λ_j, and the elements in the subspace are called eigenfunctions. Moreover, for $j = 1$, the eigenspace is one dimensional. In other words, the solution is unique up to multiplicative constants.

Exercise 8.10

Let $M = \max\limits_{B_1(0)} |f|$. Let v and $\tilde{u}$, respectively, solve $-\Delta v = 1$ and $-\Delta \tilde{u} = M$ in $B_1(0)$ with the boundary conditions $v = 0$ and $\tilde{u} = 0$ on $\partial B_1(0)$. Put $C = \max\limits_{B_1(0)} |v|$. By uniqueness, we have $\tilde{u} = Mv$ and hence $|\tilde{u}| \leqslant CM$. By linearity, we have

$$\Delta(u - \tilde{u}) = M - f \geqslant 0 \text{ in } B_1(0), \ u - \tilde{u} = 0 \text{ on } \partial B_1(0).$$

Hence, by the maximum principle, we get $u - \tilde{u} \leqslant 0$ which implies $u \leqslant CM$. Again, by linearity, we get

$$\Delta(u + \tilde{u}) = -(M + f) \leqslant 0 \text{ in } B_1(0), \ u + \tilde{u} = 0 \text{ on } \partial B_1(0).$$

Thus, by the minimum principle, we get $u + \tilde{u} \geqslant 0$, which implies $-u \leqslant CM$. Hence the first inequality.

For the second part, split the solution u as $u = z + w$, where z satisfies $-\Delta z = f$ in $B_1(0)$ and $z = 0$ on $\partial B_1(0)$. Apply the first part to z to get

$$\max_{x \in B_1(0)} |z(x)| \leqslant C \max_{x \in B_1(0)} |f(x)|.$$

On the other hand, w is harmonic with boundary values g, and thus, by the maximum principle, we have

$$\max_{x \in B_1(0)} |w(x)| \leqslant C \max_{x \in \partial B_1(0)} |g(x)|.$$

Hence, the general case follows.

Exercise 8.11

(a) Let $R > 0$ and $|x| < R$. Applying the Poisson's formula for the ball $B_R(0)$, we have

$$u(x) = \frac{R^2 - |x|^2}{n\omega_n R} \int_{\partial B_R(0)} \frac{u(y)}{|x-y|^n}\, d\sigma(y).$$

Using the hypothesis that $u \geq 0$ and the inequality $R - |x| \leq |x - y| \leq R + |x|$ for all $y \in \partial B_R(0)$, we obtain

$$\frac{R^{n-2}(R^2 - |x|^2)}{(R+|x|)^n} \frac{1}{n\omega_n R^{n-1}} \int_{\partial B_R(0)} u(y)\, d\sigma(y) \leq u(x)$$

$$\leq \frac{R^{n-2}(R^2 - |x|^2)}{(R-|x|)^n} \frac{1}{n\omega_n R^{n-1}} \int_{\partial B_R(0)} \frac{u(y)}{|x-y|^n}\, d\sigma(y).$$

Applying now MVP to the two integrals above, the inequalities in (a) follow.

(b) The inequality in (a) can be rewritten as

$$\frac{1 - |x|^2/R^2}{(1 + |x|/R)^n} u(0) \leq u(x) \leq \frac{1 - |x|^2/R^2}{(1 - |x|/R)^n} u(0).$$

Fixing x and letting $R \to \infty$, we see that $u(x) = u(0)$. Since x is arbitrary, this proves that u is a constant function.

In case u is bounded above, say $u \leq m$ for some constant m, apply (a) to the non-negative harmonic function $m - u$ and let $R \to \infty$; in case u is bounded below, say $u \geq m$, apply (a) to the non-negative harmonic function $u - m$ and let $R \to \infty$. In either case, we conclude that u is a constant function.

An Alternative Solution: We show an alternative way of proving the result which uses only MVP. Suppose u is a non-negative harmonic function in $\mathbb{R}^n$, and let $x, y \in \mathbb{R}^n$ and $r > 0$ be arbitrary. Using MVP, we have

$$u(x) = \frac{1}{\omega_n r^n} \int_{B_r(x)} u(z)\, dz$$

$$\leq \frac{1}{\omega_n r^n} \int_{B_R(y)} u(z)\, dz \quad (R = r + |x-y|)$$

$$= \frac{\omega_n R^n}{\omega_n r^n} \frac{1}{\omega_n R^n} \int_{B_R(y)} u(z)\, dz = (R/r)^n u(y).$$

Letting $r \to \infty$, we conclude that $u(x) \leq u(y)$. Similarly, $u(y) \leq u(x)$, and therefore, $u(x) = u(y)$. Since x and y are arbitrary, we conclude that u is a constant function.

A consequence of the Liouville's theorem is that the image or range of a non-constant harmonic function in $\mathbb{R}^n$ is $\mathbb{R}$.

Exercise 8.12

(a) To get an idea, first, let $k = 2$; the case $k = 1$ is described in Section 8.1. For $x \in \mathbb{R}^n$, define $\tilde{x}(1) = (-x_1, x_2, \ldots, x_n)$, $\tilde{x}(2) = (x_1, -x_2, \ldots, x_n)$ and $\tilde{x}(1,2) = (-x_1, -x_2, \ldots, x_n)$. These are, respectively, the reflections of x through the hyperplane $x_1 = 0$, $x_2 = 0$ and both $x_1 = 0$, $x_2 = 0$. For fixed $x \in \Omega$, consider the corrector function defined by $\phi^x(y) = \phi(y - \tilde{x}(1)) + \phi(y - \tilde{x}(2))$ for $y \in \overline{\Omega}$. It is evident that ϕ^x is harmonic in Ω. However, it does not satisfy the required boundary condition, namely $\phi^x(y) = \phi(y - x)$ for $y \in \partial\Omega$. To achieve this, we add another term and consider the corrector function as

$$\phi^x(y) = \phi(y - \tilde{x}(1)) + \phi(y - \tilde{x}(2)) - \phi(y - \tilde{x}(1,2)).$$

It is now easy to check that this is the required corrector function and the Green's function is given by $G(x, y) = \phi(x - y) - \phi^x(y)$.

The general case can now be treated easily. For a multi-index $\alpha = (\alpha_1, \alpha_2, \ldots, \alpha_k)$ with each α_j being either 0 or 1, define $x(\alpha) \in \mathbb{R}^n$ for $x \in \mathbb{R}^n$ by

$$(x(\alpha))_j = \begin{cases} x_j & \text{if } j > k, \\ (-1)^{\alpha_j} x_j & \text{if } j \leq k. \end{cases}$$

The corrector function is defined by $\phi^x(y) = -\sum_{\alpha \neq 0}(-1)^{|\alpha|}\phi(y - x(\alpha))$, where $|\alpha| = \alpha_1 + \cdots + \alpha_k$. The Green's function in this case can also be written as $G(x,y) = \sum_{\alpha}(-1)^{|\alpha|}\phi(y - x(\alpha))$, for $x, y \in \Omega$, $x \neq y$.

(b) The boundary of $\partial\Omega$ has two parts: One curved boundary that is the surface of the sphere in $\{x_1 > 0\}$ and another that is a flat boundary in the hyperspace $\{x_1 = 0\}$. Any point $x \in \Omega$ generates three reflection points: $\zeta_1 = R^2 x/|x|^2$, which is the reflection through the sphere $\{|x| = R\}$; $\zeta_2 = (-x_1, x_2, \ldots, x_n)$, which is the reflection through the hyperspace $\{x_1 = 0\}$ and $\zeta_3 = R^2 \zeta_2/|\zeta_2|^2 = R^2 \zeta_2/|x|^2$, which is the reflection of ζ_2 through the sphere $\{|x| = R\}$. See Figure 8.1(a).

Fix $x \in \Omega$ and define the function ϕ^x by

Laplace and Poisson Equations | 269

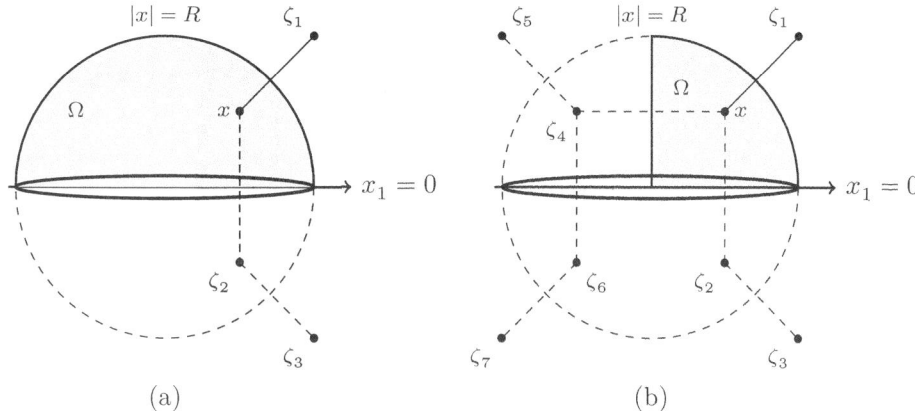

Figure 8.1 | Reflection points through (a) upper semi-ball and (b) quarter-ball.

$$\phi^x(y) = \phi\left(|x|(y-\zeta_1)/R\right) + \phi(y-\zeta_2) - \phi\left(|x|(y-\zeta_3)/R\right), y \in \overline{\Omega}.$$

As the points ζ_1, ζ_2 and ζ_3 lie outside Ω, the function ϕ^x is harmonic in Ω. We next find the values of ϕ^x on the boundary $\partial\Omega$. If $y \in \partial\Omega$ is on the curved boundary, then $|y| = R$ and $y_1 > 0$. A simple calculation shows that

$$\frac{|x|}{R}|y-\zeta_1| = |y-x| \text{ and } \frac{|x|}{R}|y-\zeta_3| = |y-\zeta_2|.$$

Thus, $\phi^x(y) = \phi(y-x)$ in this case. On the other hand, if $y \in \partial\Omega$ is on the flat boundary, then $|y| < R$ and $y_1 = 0$. In this case, we have, by symmetry, $|y-\zeta_1| = |y-\zeta_3|$. Thus, again, $\phi^x(y) = \phi(y-x)$. Therefore, ϕ^x is a required corrector function and the Green's function is given by

$$G(x,y) = \phi(x-y) - \phi^x(y)$$
$$= \phi(x-y) - \phi\left(|x|(y-\zeta_1)/R\right) - \phi(y-\zeta_2) + \phi\left(|x|(y-\zeta_3)/R\right),$$

for $x, y \in \Omega$, $x \neq y$.

(c) The arguments are similar to those in (b) above, but the computations are lengthy as there are more reflection points; see Figure 8.1(b). We use the notations as in (a) above. For any $x \in \Omega$, let $x(\alpha)$ be as in (a). Then, all $x(\alpha) \in B_R(0)$ and $x(0) = x$, $|x(\alpha)| = |x|$. Let $\zeta(\alpha)$ be the reflection of $x(\alpha)$ through the sphere $\{|x| = R\}$: $\zeta(\alpha) = R^2 x(\alpha)/|x(\alpha)|^2 = R^2 x(\alpha)/|x|^2$.

Now, fixing an $x \in \Omega$, define the function ϕ^x by

$$\phi^x(y) = -\sum_{\alpha \neq 0}(-1)^{|\alpha|}\phi(y - x(\alpha)) + \sum_{\alpha}(-1)^{|\alpha|}\phi(y - \zeta(\alpha)),$$

for $y \in \overline{\Omega}$. As all the points $x(\alpha)$, $\alpha \neq 0$ and all $\zeta(\alpha)$ lie outside Ω, the function ϕ^x is harmonic in Ω. Taking a point $y \in \partial\Omega$ and analysing each term in the definition of ϕ^x carefully, we see that $\phi^x(y) = \phi(y - x)$. Thus, ϕ^x is the required corrector function and the Green's function is given by $G(x, y) = \phi(x - y) - \phi^x(y)$. We can write

$$G(x, y) = \sum_{\alpha}(-1)^{|\alpha|}[\phi(y - x(\alpha)) - \phi(y - \zeta(\alpha))],$$

for $x, y \in \Omega$, $x \neq y$.

(d) Though the domain in question is simple, we see that we need an infinite number of reflection points to compute the Green's function. For $x \in \Omega$, define $\zeta_0 = x - 2x_1 e_1$ and $\zeta_1 = \zeta_0 + 2e_1$, where $e_1 = (1, 0, \ldots, 0)$. Note that ζ_0 and ζ_1 lie outside Ω and are reflection points of x through $\{x_1 = 0\}$ and $\{x_1 = 1\}$. Thus, the function $\phi(y - \zeta_0) + \phi(y - \zeta_1)$ for $y \in \Omega$ is harmonic there. Also, $\phi(y - \zeta_0) = \phi(y - x)$ on $\{x_1 = 0\}$ and $\phi(y - \zeta_1) = \phi(y - x)$ on $\{x_1 = 1\}$. However, this cannot be the corrector function, as $\phi(y - \zeta_0) \neq 0$ on $\{x_1 = 1\}$ and $\phi(y - \zeta_1) \neq 0$ on $\{x_1 = 0\}$. We thus need to add terms corresponding to the reflection points of ζ_0 and ζ_1. Apparently, this process cannot be terminated at a finite stage.[2] Now, define $\eta_k = x + 2k e_1$ and $\zeta_k = \zeta_0 + 2k e_1$ for $k \in \mathbb{Z}$. Note that $\eta_0 = x$; η_1 is the reflection point of ζ_0 through $\{x_1 = 1\}$, η_{-2} is the reflection point of ζ_1 through $\{x_1 = 0\}$ and so on. The Green's function is given by

$$G(x, y) = \sum_{k=-\infty}^{\infty}[\phi(y - \eta_k)) - \phi(y - \zeta_k)],$$

for $x, y \in \Omega$, $x \neq y$. Since $n \geqslant 3$, the convergence of this infinite series can be established.

Note that in (b) and (c) above, the Green's function is a combination of Green's function for the upper half-space and that of a ball.

Exercise 8.13

Let v denote the function defined on the right-hand side of the inequality. Clearly, the linear second term is harmonic. The first term can be written as $\sum_{i=1}^{n} a_{ii} x_i^2 + \sum_{\substack{i,j=1 \\ i \neq j}}^{n} a_{ij} x_i x_j$. In the expression above, the second term is harmonic, whereas the Laplacian of the first term is nothing but twice the trace of A. In essence, v is harmonic. Applying Exercise 8.11 to the non-negative harmonic function $v - u$, we conclude that $u(x) = \sum_{i,j=1}^{n} a_{ij} x_i x_j + a \cdot x + c$, where c is a constant.

---

[2] This procedure can be used to construct the Green's function for an annulus. The reflection points are now more difficult to compute as the reflections through spheres are involved.

Exercise 8.14

Write $y \in \mathbb{R}^{n+1}$ as $y = (x, x_{n+1})$ with $x \in \mathbb{R}^n$. Define $U(y) = x_{n+1} - u(x)$. Trivially, the function U is harmonic in $\mathbb{R}^{n+1}$. Suppose $y^0 = (x^0, x_{n+1}^0)$ be such that $U(y^0) = 0$. If $U(y) \leqslant x_{n+1} - x_{n+1}^0 + \nabla u(x^0) \cdot (x - x^0)$ for all $y \in \mathbb{R}^{n+1}$, then, using the result in Exercise 8.11, we conclude that $u(x) = u(x^0) - \nabla u(x^0) \cdot (x - x^0)$ for all $x \in \mathbb{R}^n$, modulo a constant term. Note that the set $\{U = 0\}$ in $\mathbb{R}^{n+1}$ represents the graph of the function u. Thus, the conclusion above is that the graph of u intersects any of its tangent planes, at two or more points.

Exercise 8.15

Recall that $u \in C^\infty$ and each derivative $D^\alpha u$ is also harmonic. Thus, by MVP, we have

$$\frac{\partial u}{\partial x_i}(x) = \frac{1}{\omega_n r^n} \int_{|x-y|<r} \frac{\partial u}{\partial x_i}(y)\,dy = \frac{1}{\omega_n r^n} \int_{|x-y|=r} u(y)\nu_i\,d\sigma(y),$$

for any $x \in \mathbb{R}^n$ and $r > 0$. Thus, using the hypothesis, we arrive at

$$\left|\frac{\partial u}{\partial x_i}(x)\right| \leqslant \frac{Cn}{r}(1 + (r + |x|)^s). \tag{8.2.1}$$

We claim that

$$|D^\alpha u(x)| \leqslant C\frac{n^k}{r^k}(1 + (rk + |x|)^s), \tag{8.2.2}$$

for any multi-index α with $|\alpha| = k \geqslant 1$. The case $k = 1$ is nothing, but equations (8.2.1) and (8.2.2) will be established by an inductive argument. Let equation (8.2.2) hold for any multi-index α with $|\alpha| = k \geqslant 1$. We show that it also holds for $k+1$.

Let α be any multi-index with $|\alpha| = k$. Then, by MVP,

$$\frac{\partial}{\partial x_i} D^\alpha u(x) = \frac{1}{\omega_n r^n} \int_{|x-y|<r} \frac{\partial}{\partial x_i} D^\alpha u(y)\,dy = \frac{1}{\omega_n r^n}\int_{|x-y|=r} D^\alpha u(y)\nu_i\,d\sigma(y).$$

For y satisfying $|x - y| = r$, we have $|y| \leqslant |x| + r$, and therefore, by induction hypothesis,

$$|D^\alpha u(y)| \leqslant C\frac{n^k}{r^k}(1 + (rk + |x| + r)^s) = C\frac{n^k}{r^k}(1 + (r(k+1) + |x|)^s).$$

Thus,

$$\left|\frac{\partial}{\partial x_i} D^\alpha u(x)\right| \leqslant C\frac{n^k}{r^k}(1 + (r(k+1) + |x|)^s)\frac{1}{\omega_n r^n}\int_{|x-y|=r} d\sigma(y)$$

$$= C\frac{n^k}{r^k}(1 + (r(k+1) + |x|)^s)\frac{1}{\omega_n r^n}n\omega_n r^{n-1}$$

$$= C\frac{n^{k+1}}{r^{k+1}}(1 + (r(k+1) + |x|)^s).$$

This shows that equation (8.2.2) holds good with k replaced by $k+1$ and the induction is complete.

We now take $k = [s] + 1$ in equation (8.2.2) and observe that the power of r in the numerator is s, whereas it is $[s] + 1$ in the denominator. Hence, letting $r \to \infty$, we conclude that $D^\alpha u(x) = 0$ for all $x \in \mathbb{R}^n$ and multi-indexes α with $|\alpha| = [s] + 1$. Thus, u is a polynomial of degree at most $[s]$.

Remark: Similar result holds for analytic functions of one complex variable. In that case, the required estimate easily follows from Cauchy integral formula, which is not available in the present situation. Instead, we need to use MVP. The remark is that the present result in the exercise is a result due to harmonicity. Recall that the real and imaginary parts of an analytic function of one complex variable are harmonic.

Exercise 8.16

Fix $x_0 \in \Omega$, and consider the function $v(x) = \frac{1}{2n} d(x, x_0)^2 = \frac{1}{2n} |x - x_0|^2$, $x \in \Omega$, which satisfies the equation $\Delta v = 1$ in Ω. Thus, the function $u(x) + v(x) = u(x) + \frac{1}{2n}|x - x_0|^2$ is harmonic in Ω. By the minimum principle, we have

$$u(x) + \frac{1}{2n}|x - x_0|^2 \geq \min_{x \in \partial\Omega} \left[u(x) + \frac{1}{2n}|x - x_0|^2 \right] = \frac{1}{2n}(d(x_0, \partial\Omega))^2.$$

Taking $x = x_0$, and noting that x_0 is arbitrary, the result follows.

Exercise 8.17

This exercise is exactly in the spirit of Exercise 8.7. As in that exercise, define v in $B_1(0)$ using the values u in the upper part of $B_1(0)$. That is,

$$v(x', x_n) = \begin{cases} u(x', x_n) & \text{if } x_n > 0, \\ -u(x', -x_n) & \text{if } x_n < 0. \end{cases}$$

Again, as in Exercise 8.7, we see that v is harmonic in $B_1(0)$ and satisfies the boundary condition $v = \phi$ on $\partial B_1(0)$, thanks to the condition that $\phi(x', x_n) = -\phi(x', -x_n)$. By uniqueness, $v = u$. Hence the result.

Exercise 8.18

This is again similar to Exercises 8.7 and 8.17. Define $u(x', -x_n) = -u(x', x_n)$, $x_n > 0$. Then, u is continuous in $B_1(0)$ and harmonic in B_1^+, B_1^-. Let $x^0 \in B_1(0)$ with $x_n^0 = 0$ and $0 < r < 1$. Then,

$$\int_{|x - x^0| = r} u(x) \, d\sigma(x) = \int_{x_n > 0} + \int_{x_n < 0} = 0.$$

Thus, u satisfies the MVP.

Exercise 8.19

Here, $u \in C(\Omega)$ is sub-harmonic if it satisfies the mean value inequality $u(x) \leqslant \frac{1}{n\omega_n r^{n-1}} \int_{|x-y|=r} u(y)\, d\sigma(y)$, for all $B_r(x) \Subset \Omega$.

If u is sub-harmonic in Ω and v is harmonic in Ω' (as stated in the question) with $v = u$ on $\partial \Omega'$, then $u - v$ is sub-harmonic in Ω' and is zero on $\partial \Omega'$. By maximum principle, $u \leqslant v$ in Ω', as required.

Conversely, suppose u satisfies the stated condition. Fix $x \in \Omega$, and choose $r > 0$ such that $B_r(x) \Subset \Omega$. Then, by the stated condition, if v is any harmonic function in $B_r(x)$ satisfying $v = u$ on $\partial B_r(x)$, then $u \leqslant v$ in $B_r(x)$. Now, by MVP, we have

$$u(x) \leqslant v(x) = \frac{1}{n\omega_n r^{n-1}} \int_{|x-y|=r} v(y)\, d\sigma(y) = \frac{1}{n\omega_n r^{n-1}} \int_{|x-y|=r} u(y)\, d\sigma(y),$$

as $u = v$ on $\{|x - y| = r\}$. This means that u satisfies the mean value inequality, which in turn proves that u is sub-harmonic.

Exercise 8.20

Let $A = \{x \in \Omega : u(x) > 0\}$ and $B = \{x \in \Omega : u(x) = 0\}$. The set A is open in Ω by the continuity of u. If $x_0 \in B$, then, by MVP,

$$0 = u(x_0) = \frac{1}{n\omega_n r^{n-1}} \int_{|x_0-y|=r} u(y)\, d\sigma(y),$$

for some $r > 0$ such that $B_r(x_0) \Subset \Omega$. Since $u \geqslant 0$, by assumption, it follows, again by the continuity of u, that $u = 0$ in $B_r(x_0)$. This shows that the set B is also open in Ω. Since Ω is connected, one of them must be an empty set, proving the assertion.

Exercise 8.21

If u were not identically constant, then it would have a strict maximum or minimum at a boundary point x_0. By Hopf's lemma, applied to either u or $-u$, we see that $\frac{\partial u}{\partial \nu} \neq 0$, a contradiction to the assumption.

The result of this exercise immediately provides a uniqueness (up to a constant) result for the Laplace equation with Neumann boundary condition. Note that $u \equiv$ constant is always a solution of the Neumann boundary value problem. Thus, the uniqueness is in the sense that if u and v are two solutions, then $u - v$ is a constant. Quite often, the constant is fixed requiring that the average of the solution is zero, that is, $\int_\Omega u(x)\, dx = 0$.

Exercise 8.22

Let $\Omega^- = \{x \in \Omega : u(x) < 0\}$. Then, Ω^- is an open subset in Ω, by the continuity of u, and u satisfies the inequality $\Delta u \leqslant 0$ in Ω^-. Since $u = 0$ on $\partial\Omega^- \cap \overline{\Omega}$, it follows from (weak) minimum principle (applied to Ω^-) that $\min_{\overline{\Omega}} \geqslant \min_{\partial\Omega} u_-$, where $u_- = \min\{0, u\}$. Since, by assumption, $u \geqslant 0$ on $\partial\Omega$, it follows that Ω^- is empty. This proves that $u \geqslant 0$ in Ω.

Suppose $u(x_0) = 0$ for some $x_0 \in \Omega$. Then, $\Delta u(x_0) \geqslant 0$. However, the given equation implies that $\Delta u(x_0) = f(x_0) < 0$, a contradiction. Therefore, $u > 0$ in Ω.

The assumption $f \leqslant 0$ may not suffice to claim that $u > 0$ in Ω. The example $\Delta u = u^3$, which has $u = 0$ as a solution shows.

Exercise 8.23

Consider the *spherical mean* of u:

$$M_u(x, r) = \frac{1}{n\omega_n r^n} \int_{|x-y|=r} u(y)\, d\sigma(y),\ r > 0,$$

for $x \in \Omega$ and all $r > 0$ such that $B_r(x) \Subset \Omega$. Then, by hypothesis, we have

$$0 = \int_{|x-y|=r} \frac{\partial u}{\partial \nu}\, d\sigma(y) = r^{n-1} \int_{|y|=1} \frac{\partial u}{\partial r}(x + ry)\, d\sigma(y)$$

$$= r^{n-1} \frac{\partial}{\partial r}\left(\int_{|y|=1} u(x + ry)\, d\sigma(y)\right) = r^{n-1} n\omega_n \frac{\partial}{\partial r} M_u(x, r).$$

This shows that M_u as a function of r is a constant. However, $M_u(x, r) \to u(x)$ as $r \to 0$, and thus the constant is $u(x)$, that is, $M_u(x, r) = u(x)$ for $r > 0$. This proves that u satisfies MVP, and thus u is harmonic.

Exercise 8.24

Put $\Gamma_1 = \partial B_1(0) \cap \{x_n \geqslant 0\}$ and $\Gamma_2 = \partial B_1(0) \cap \{x_n < 0\}$ so that $\partial B_1(0) = \Gamma_1 \cup \Gamma_2$. Also, by continuity, $u = 0$ on Γ_1. Observe that u cannot take a minimum value at a point in $B_1(0)$, as at such a point $\Delta u \geqslant 0$, but by hypothesis, $\Delta u = -1 < 0$. Thus, the minimum of u is attained on Γ_1 or Γ_2. If it is attained on Γ_1, it immediately follows that $u \geqslant 0$ in $\overline{B_1(0)}$ as $u = 0$ on Γ_1 and hence $u \geqslant 0$ on $\partial\Omega$.

Suppose, on the other hand, the minimum of u is attained at some point $x_* \in \Gamma_2$. Since $u = 0$ on Γ_1, it follows that $u(x_*) \leqslant 0$. But then, by Hopf's lemma, we have $-\frac{\partial u}{\partial \nu}(x_*) > 0$, and the boundary condition in the hypothesis implies that $u(x_*) > 0$, a contradiction. This contradiction in fact shows that $u > 0$ on Γ_2, as required. This completes the proof.

Exercise 8.25

The function $\cosh(a|x|)$ with $a \geqslant 1$ provides a required example.[3] However, this function is unbounded (not bounded above). Note that this function is infinitely differentiable everywhere in $\mathbb{R}^n$, including the origin, as can be easily checked.

We now show that if u is any function, bounded above and satisfying the hypothesis, it is necessarily non-positive. For, if possible, assume u is positive somewhere. If u assumes a local maximum at some point, it is easy to see a contradiction. In general, consider the auxiliary function $v(x) = u(x)/w(x)$ where $w(x) = \cosh(\varepsilon |x|)$ for some $\varepsilon > 0$ to be chosen. Then, the function v is also C^2, positive somewhere and $\to 0$ as $|x| \to \infty$. Thus, v attains a positive local maximum at, say x_0: $v(x_0) > 0$, $\nabla v(x_0) = 0$ and $\Delta v(x_0) \leqslant 0$.

Observe that $\Delta u = w \Delta v + 2 \nabla w \cdot \nabla v + v \Delta w$. In particular, $\Delta u(x_0) = w(x_0) \Delta v(x_0) + v(x_0) \Delta w(x_0)$. Now, using the hypothesis, we have

$$\lambda v(x_0) w(x_0) = \lambda u(x_0) \leqslant \Delta u(x_0)$$
$$= w(x_0) \Delta v(x_0) + v(x_0) \Delta w(x_0) \leqslant v(x_0) \Delta w(x_0).$$

Dividing throughout by $v(x_0) w(x_0)$, we therefore obtain $\lambda \leqslant \dfrac{\Delta w(x_0)}{w(x_0)}$. A simple computation shows that $\dfrac{\Delta w(x_0)}{w(x_0)} = \varepsilon^2 + \varepsilon(n-1) \dfrac{\tanh(\varepsilon |x_0|)}{|x_0|}$, taking some care at $x_0 = 0$. Using the inequality $|\tanh x| \leqslant |x|$ for all $x \in \mathbb{R}$, we thus obtain $\lambda \leqslant n \varepsilon^2$. This immediately gives a contradiction by choosing ε sufficiently small. This completes the proof.

An interesting consequence is the following: If u is a bounded (both from below and above) C^2 function satisfying $\Delta u = \lambda u$ in $\mathbb{R}^n$, for some $\lambda > 0$, then we have $\pm u \leqslant 0$ and hence $u \equiv 0$.

Exercise 8.26

Put $m = \sup\limits_{\overline{\Omega}} u$. If $m \leqslant 0$, there is nothing to prove. Assume $m > 0$. It suffices to show that u cannot attain a positive local maximum in Ω. Suppose not and assume that there is an $x_0 \in \Omega$, where u attains a local maximum and $u(x_0) > 0$. Then, $\Delta u(x_0) \leqslant 0$ and therefore $0 < \lambda u(x_0) \leqslant \Delta u(x_0) \leqslant 0$, which is a contradiction. Thus, the positive supremum of u can attain only on $\partial \Omega$ and $u \leqslant \sup\limits_{\partial \Omega} u_+$, proving the assertion.

Exercise 8.27

Consider the function v defined by $v(x) = u(x) - u(x_0) \exp(-\sqrt{\lambda} x_n)$, $x \in \Omega$. Then, $\Delta v - \lambda v \geqslant 0$, $v \leqslant 0$ and $v(x_0) = 0$. See Exercise 8.26. Also note that the outward normal ν is in

---

[3] A simpler function is $\exp(x_1 + \cdots + x_n)$. The function cosh is used in the solution to *pull down* the function u.

the direction of $(0,\ldots,0,-1)$. Therefore, $\dfrac{\partial v}{\partial \nu}(x_0) \leqslant 0$, which immediately gives $\dfrac{\partial u}{\partial \nu}(x_0) \leqslant -\sqrt{\lambda} u(x_0) < 0$.

Exercise 8.28

It is a bit lengthy, but straightforward computation to show that U is harmonic in $\{|x| > R\}$. Since $n \geqslant 3$, it easily follows that $U(x) \to 0$ as $|x| \to \infty$. Also, we have

$$u(x) = \left(\frac{R}{|x|}\right)^{n-2} U\left(\frac{R^2 x}{|x|^2}\right), \ |x| < R.$$

Exercise 8.29

The function $x \mapsto |x-y|^{-1}$ is harmonic in $B_r(x_0)$, as $y \notin \overline{B_r(x_0)}$. Thus, by MVP, the given integral equals $\dfrac{4\pi r^3}{3|x_0 - y|}$ (remember, $n = 3$).

(Try to compute the integral directly using spherical co-ordinates; may assume $x_0 = 0$. Since the solution involves only $|x_0 - y|$ in the denominator, is it sufficient to assume $y \neq x_0$?)

Exercise 8.30

Assume that $u \in C(\overline{\Omega})$ is subharmonic and put $M = \max\limits_{\overline{\Omega}} u$. If there is an $x_0 \in \Omega$ such that $u(x_0) = M$, we show that $u \equiv M$.

Put $\Omega_M = \{x \in \Omega : u(x) = M\}$, and assume that $x_0 \in \Omega_M$. By the continuity of u, Ω_M is a closed subset of Ω. If $x \in \Omega_M$, then using the hypothesis that u is subharmonic, we get

$$0 = u(x) - M \leqslant \frac{1}{|B_r(x)|} \int_{B_r(x)} u(y)\, dy - M = \frac{1}{|B_r(x)|} \int_{B_r(x)} (u(y) - M)\, dy,$$

for all $r > 0$ such that $B_r(x) \Subset \Omega$. Since $u \leqslant M$ in Ω, we conclude that $u(y) = M$ for all $y \in B_r(x)$ for some $r > 0$. This proves that Ω_M is an open subset of Ω as well. By connectedness, we conclude that $\Omega_M = \Omega$, thereby proving the weak maximum principle. The arguments for the weak minimum principle are the same.

Exercise 8.31

Let $x_0 \in \Omega$, and choose $r > 0$ such that $B_r(x_0) \Subset \Omega$. Let v be the harmonic function in $B_r(x_0)$ satisfying the boundary condition $v = u$ on $\partial B_r(x_0)$. The Poisson formula (8.1.15), by shifting the origin to x_0, provides such a v. Applying the result in Exercise 8.30 to the functions $\pm(v - u)$ in $B_r(x_0)$, we conclude that $u \equiv v$ in $B_r(x_0)$. In particular, $\Delta u(x_0) = 0$.

Exercise 8.32

Using the Taylor's theorem, we have

$$u(x-h, y-k) = u(x,y) - hu_x(x,y) - ku_y(x,y) + \frac{h^2}{2}u_{xx}(\xi_{1x}, \eta_{1y})$$
$$+ \frac{k^2}{2}u_{yy}(\xi_{1x}, \eta_{1y}) + hk u_{xy}(\xi_{1x}, \eta_{1y}),$$

where (ξ_{1x}, η_{1y}) is a point lying on the line joining the points (x, y) and $(x - h, y - k)$. Similar expressions can be written down at the other three points. Summing all these expressions and doing some simplification, we obtain

$$u(A) + u(B) + u(C) + u(D) = 4u(x,y) + \frac{h^2}{2}[u_{xx}(\xi_{1x}, \eta_{1y})$$
$$+ u_{xx}(\xi_{2x}, \eta_{2y}) + u_{xx}(\xi_{3x}, \eta_{3y})$$
$$+ u_{xx}(\xi_{4x}, \eta_{4y})] + \frac{k^2}{2}[u_{yy}(\xi_{1x}, \eta_{1y})$$
$$+ u_{yy}(\xi_{2x}, \eta_{2y}) + u_{yy}(\xi_{3x}, \eta_{3y}) + u_{yy}(\xi_{4x}, \eta_{4y})].$$

Now, use the hypothesis by taking $k = h$, and then let $h \to 0$ to see that $(u_{xx} + u_{yy})(x,y) = 0$. This proves the first assertion.

The harmonic function $e^x \cos y$ provides a counterexample, as can be easily checked.

Exercise 8.33

We do not need the stated smoothness on the function f; it is sufficient to assume f is continuous and has compact support. However, recall that if $f \in C_c^2(\mathbb{R}^n)$, then u solves the Poisson equation $-\Delta u = f$.

Recall that $\phi(x) = -(2\pi)^{-1} \log |x|$ if $n = 2$ and $\phi(x) = (n(n-2)\omega_n)^{-1} |x|^{2-n}$ if $n \geq 3$. Observe that ϕ is locally integrable and $|\phi(x)| \to \infty$ if $n = 2$ and $\phi(x) \to 0$ if $n \geq 3$ as $|x| \to \infty$. First, consider the case $n = 2$. Choose a function $f \in C_c^\infty(\mathbb{R}^2)$ such that $f(x) \leq -\alpha < 0$ for all $x \in \bar{B}$, for some ball B; it is not difficult to see such a function exists. Now, let $M > 0$ be arbitrary. Choose x so large that $\log |x - y| > M$ for all $y \in \bar{B}$. Then,

$$u(x) = (2\pi)^{-1} \int_{\mathbb{R}^2} \log |x-y|(-f(y))\, dy$$
$$\geq (2\pi)^{-1} \int_{\bar{B}} \log |x-y|(-f(y))\, dy \geq (2\pi)^{-1} \alpha M |\bar{B}|,$$

proving that $u(x) \to \infty$ as $|x| \to \infty$. Similarly, if f is chosen such that $f(x) \geq \alpha > 0$ for all x, then we infer that $u(x) \to \infty$ as $|x| \to \infty$. This proves (1).

For $n \geq 3$, let $f \in C_c^\infty(\mathbb{R}^n)$ be such that its support is in $\overline{B_R(0)}$ for some $R > 0$. Put $M = \max_{\overline{B_R(0)}} |f(x)|$. Then,

$$|u(x)| \leq c \int_{\mathbb{R}^n} |x-y|^{2-n} |f(y)| \, dy \leq cM \int_{B_R(0)} |x-y|^{2-n} \, dy,$$

where $c = (n(n-2)\omega_n)^{-1}$. If $x \notin \overline{B_R(0)}$, the function $y \mapsto |x-y|^{2-n}$ is harmonic in $B_R(0)$, and therefore, by MVP, we have $\int_{B_R(0)} |x-y|^{2-n} \, dy = c(R)|x|^{2-n}$, where the constant $c(R)$ depends on R. On the other hand, for $x \in \overline{B_R(0)}$, the function $y \mapsto |x-y|^{2-n}$ is locally integrable in $B_R(0)$, and therefore, $|u(x)|$ remains bounded for $x \in \overline{B_R(0)}$. This proves (2).

Exercise 8.34

Let $x_0 \in \mathbb{R}^n \setminus \partial\Omega$ be an arbitrary point, and put $d = d(x_0, \partial\Omega) > 0$. Note that the integrand functions in the definition of u_1 and u_2 belong to $C(\partial\Omega)$, as functions of ξ, for all $x \in B_{d/2}(x_0)$, and, as functions of x, they belong to $C^\infty(B_{d/2}(x_0))$, for all $\xi \in \partial\Omega$. Furthermore, the following statement holds: For any multi-index α and $x \in B_{d/2}(x_0)$, $|D_x^\alpha \phi(x-\xi)|$ and $|D_x^\alpha \frac{\partial}{\partial \nu_\xi} \phi(x-\xi)|$ are bounded by a constant, depending only on α and d. Thus, we can differentiate under the integral sign any number of times. This shows that the functions u_1 and u_2 are infinitely differentiable in $B_{d/2}(x_0)$, and $D^\alpha u_1(x) = \int_{\partial\Omega} \rho_1(\xi) D_x^\alpha \phi(x-\xi) \, d\sigma(\xi)$

and $D^\alpha u_2(x) = \int_{\partial\Omega} \rho_2(\xi) \frac{\partial}{\partial \nu_\xi} D_x^\alpha \phi(x-\xi) \, d\sigma(\xi)$, for any multi-index α. In particular, we have

$$\Delta u_1(x) = \int_{\partial\Omega} \rho_1(\xi) \Delta_x \phi(x-\xi) \, d\sigma(\xi) = 0 \text{ and } \Delta u_2(x) = \int_{\partial\Omega} \rho_2(\xi) \frac{\partial}{\partial \nu_\xi} \Delta_x \phi(x-\xi) \, d\sigma(\xi) = 0 \text{ as}$$

$\Delta_x \phi(x-\xi) = 0$ for $x \neq \xi$. This proves that the functions u_1 and u_2 are harmonic in $\mathbb{R}^n \setminus \partial\Omega$.

Exercise 8.35

The proof is somewhat technical and involves many algebraic manipulations. Since ϕ and its first derivatives are locally integrable, it follows that v is a C^1 function for bounded, measurable function f. In fact, we have

$$\frac{\partial v}{\partial x_i}(x) = \int_\Omega \frac{\partial \phi}{\partial x_i}(x-y) f(y) \, dy,$$

for $i = 1, 2, \ldots, n$ and $x \in \Omega$. Also, note that v is defined in $\mathbb{R}^n$. However, the second derivatives of ϕ are not locally integrable, and this implies that v, in general, does not possess second derivatives, even for a continuous function f.

If the density $f \in C^1(\overline{\Omega})$, then we can apply divergence theorem one more time and show that $-\Delta u = f$ in Ω; see, for instance, Ref.[45]. In particular, we can use this result when f is a constant function.

We begin with the following observation. Choose $R > 0$ small so that $\overline{B_R(x_0)} \subset \Omega$, and write $v = v_1 + v_2$, where $v_1(x) = \int_{B_R(x_0)} \phi(x-y) f(y) \, dy$ and $v_2(x) = \int_{\Omega \setminus B_R(x_0)} \phi(x-y) f(y) \, dy$.

Since the integrand in the expression for v_2 does not possess any singularity, it follows that v_2 possesses all the second derivatives at x_0 and $\Delta v_2(x_0) = 0$. Thus, it suffices to show that v_1 possesses all the second derivatives at x_0 and $\Delta v_1(x_0) = -f(x_0)$.

We now write $v_1(x) = v_{11}(x) + v_{12}(x)$, where

$$v_{11}(x) = \int_{|x_0 - y| < R} \phi(x-y)(f(y) - f(x_0)) \, dy,$$

$$v_{12}(x) = f(x_0) \int_{|x_0 - y| < R} \phi(x-y) \, dy,$$

for $|x_0 - x| < R$. Since v_{12} has constant density $f(x_0)$, it readily follows that $\Delta v_{12}(x_0) = -f(x_0)$. In v_{11}, the density vanishes at x_0 and satisfies the local Hölder continuity around x_0. Below, we show that $\Delta v_{11}(x_0) = 0$, and this would then complete the proof.

Technical Computations: By translation, we may assume that $x_0 = 0$. We only need to analyse the following Newtonian potential:

$$u(x) = \int_{|y| < R} \phi(x-y) f(y) \, dy, \ |x| < R.$$

Here, we assume that $f(0) = 0$ and $|f(x)| \leq A|x|^\alpha$ for all $|x| < R$ and some $\alpha \in (0,1)$ and a positive constant A. We need to show that u possesses all the second derivatives at 0 and $\Delta u(0) = 0$.

We have $\dfrac{\partial \phi}{\partial x_i} = \dfrac{1}{\sigma_n} \dfrac{y_i - x_i}{|x-y|^n}$ and

$$\frac{\partial^2 \phi}{\partial x_i^2}(x-y) = \frac{1}{\sigma_n} \left[\frac{n(y_i - x_i)^2}{|x-y|^{n+2}} - \frac{1}{|x-y|^n} \right],$$

for $i = 1, 2, \ldots, n$. Put

$$J_i = \frac{1}{\sigma_n} \int_{|y| < R} \left[\frac{n y_i^2}{|y|^{n+2}} - \frac{1}{|y|^n} \right] f(y) \, dy,$$

for $i = 1, 2, \ldots, n$. Here, the term in the parenthesis in the integrand is $\dfrac{\partial^2 \phi}{\partial x_i^2}(x-y)$ evaluated at $x = 0$, up to a constant. Because of the assumption of the local Hölder continuity of f, each of the terms in the integrand of J_i each behaves like $|y|^{\alpha-n}$ and so integrable. Therefore J_i is well-defined. Also, observe that $J_1 + \cdots + J_n = 0$.

Let
$$w_i(x) = \frac{\partial u}{\partial x_i}(x) = \int_{|y|<R} \frac{\partial \phi}{\partial x_i}(x-y) f(y)\, dy,$$

for $i = 1, 2, \ldots, n$. In particular,
$$w_i(0) = \frac{\partial u}{\partial x_i}(0) = \frac{1}{\sigma_n} \int_{|y|<R} \frac{y_i}{|y|^n} f(y)\, dy.$$

We next show that the function w_i is differentiable at $x = 0$ and $\dfrac{\partial^2 u}{\partial x_i^2}(0) = \dfrac{\partial w_i}{\partial x_i}(0) = J_i$, for $i = 1, 2, \ldots$. We begin by showing that $\lim_{h \to 0} \dfrac{w_i(he_i) - w_i(0)}{h} = J_i$, where $e_1, \ldots, e_n$ are the standard unit vectors in $\mathbb{R}^n$. We write
$$I(h) = \frac{w_i(he_i) - w_i(0)}{h} - J_i.$$

Thus,
$$I(h) = \frac{1}{\sigma_n} \int_{|y|<R} \left[\frac{1}{h}\left(\frac{y_i - h}{|he_i - y|^n} - \frac{y_i}{|y|^n} \right) - \left(\frac{n y_i^2}{|y|^{n+2}} - \frac{1}{|y|^n} \right) \right] f(y)\, dy.$$

Put $r = |he_i - y|$ and $r_0 = |y|$. Then,
$$\frac{1}{r^n} - \frac{1}{r_0^n} = \frac{r_0^n - r^n}{r_0^n r^n}$$
$$= \frac{(r_0 - r)(r_0^{n-1} + r_0^{n-2} r + \cdots + r^{n-1})}{r_0^n r^n}$$
$$= \frac{(r_0^2 - r^2)(r_0^{n-1} + r_0^{n-2} r + \cdots + r^{n-1})}{r_0^n r^n (r_0 + r)}$$
$$= \frac{h(2y_i - h)}{r_0 r (r_0 + r)} \sum_{k=0}^{n-1} r^{k+1-n} r_0^{-k}.$$

Thus,
$$\frac{1}{h}\left(\frac{y_i - h}{|he_i - y|^n} - \frac{y_i}{|y|^n} \right) = -\frac{1}{r^n} + \frac{y_i}{h}\left(\frac{1}{|he_i - y|^n} - \frac{1}{|y|^n} \right)$$
$$= -\frac{1}{r^n} + \frac{y_i(2y_i - h))}{r_0 r(r_0 + r)} \sum_{k=0}^{n-1} r^{k+1-n} r_0^{-k}$$

and

$$I(h) = \frac{1}{\sigma_n} \times$$

$$\int_{|y|<R} \left[\left(-\frac{1}{r^n} + \frac{y_i(2y_i - h)}{r_0 r(r_0 + r)} \sum_{k=0}^{n-1} r^{k+1-n} r_0^{-k} \right) - \left(\frac{ny_i^2}{|y|^{n+2}} - \frac{1}{|y|^n} \right) \right] f(y) \, dy.$$

We now show that $I(h) \to 0$ as $h \to 0$. This shows that $I(h)$ is continuous at $h = 0$ as $I(0) = 0$.

Let $0 < \delta < R$ be small, and assume $|h| < \delta$. We write $I(h)$ as

$$I(h) = \int_{|y|<\delta} + \int_{\delta<|y|<R} = I_1 + I_2, \text{ say.}$$

It is readily seen that I_2 is a continuous function of h and tends to 0 as $h \to 0$, uniformly for $|h| < \delta$. As far as I_1 is concerned, there are two infinities in its integrand, namely the denominators containing r_0 and the denominators containing r. The terms containing r_0 are taken care of by the assumption of local Hölder continuity of f. For the terms containing r, we proceed as follows. We have

$$\left| -\frac{1}{r^n} + \frac{y_i(2y_i - h)}{r_0 r(r_0 + r)} \frac{1}{r^{n-1}} \right| = \left| \frac{2y_i^2 - y_i h - r_0^2 - r_0 r}{r_0 r^n (r_0 + r)} \right|$$

$$= \left| \frac{2y_i^2 - y_i h - r_0^2}{r_0 r^n (r_0 + r)} - \frac{1}{r_0 r^{n-1} (r_0 + r)} \right|$$

$$\leqslant \frac{2}{r_0 r^{n-1}},$$

using the simple estimates

$$|2y_i^2 - y_i h - r_0^2| \leqslant r(r_0 + r) \text{ and } r_0 + r \geqslant r_0.$$

The remaining terms in the integrand of I_1 are easily estimated as

$$\left| \frac{y_i(2y_i - h)}{r_0 r(r_0 + r)} \sum_{k=1}^{n-1} r^{k+1-n} r_0^{-k} - \frac{ny_i^2}{|y|^{n+2}} + \frac{1}{|y|^n} \right| \leqslant \sum_{k=0}^{n-1} r^{k-n} r_0^{-k} + \frac{n+1}{r_0^n}.$$

For the density f, we have the estimate $|f(y)| \leqslant A r_0^\alpha$ for all y in $B_\delta(0)$. Therefore, the integrand of I_1 is dominated by

$$A r_0^\alpha \left[\sum_{k=1}^{n-1} r_0^{-k} r^{k-n} + (n+1) r_0^{-n} \right] \leqslant C \left[\sum_{k=1}^{n-1} r_0^{-k+\alpha} r^{k-n} + (n+1) r_0^{-n+\alpha} \right],$$

for some constant $C > 0$. Now, the last term can be estimated as $\leqslant C_1/r^{n-\alpha}$ for $r \leqslant r_0$, within $B_\delta(0)$, and as $\leqslant C_1/r_0^{n-\alpha}$ for $r_0 \leqslant r$. Therefore,

$$|I_1| \leqslant C_1 \left(\int_{|y|<\delta} \frac{dy}{r_0^{n-\alpha}} + \int_{|y|<\delta} \frac{dy}{r^{n-\alpha}} \right).$$

Put $\beta = n - \alpha$. Then, $0 < \beta < n$. Let

$$K_1 = \{y \in B_R(0) : |y| < \delta,\ |he_i - y| > \delta\}$$

and

$$K_2 = \{y \in B_R(0) : |y| > \delta,\ |he_i - y| < \delta\}.$$

See Figure 8.2.

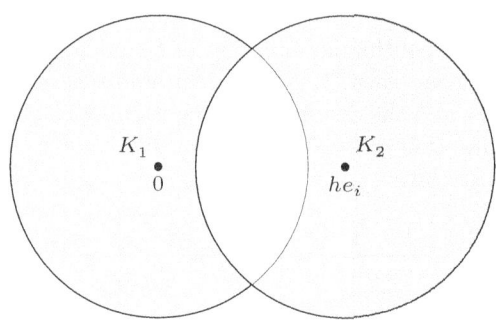

Figure 8.2 | The sets K_1 and K_2 in Exercise 8.35.

Then, we have

$$\int_{|y|<\delta} \frac{dy}{r^\beta} = \int_{|y|<\delta} \frac{dy}{|he_i - y|^\beta}$$

$$= \int_{|he_i - y|<\delta} \frac{dy}{|he_i - y|^\beta} + \int_{K_1} \frac{dy}{|he_i - y|^\beta} - \int_{K_2} \frac{dy}{|he_i - y|^\beta}$$

$$= \int_{|y|<\delta} \frac{dy}{|y|^\beta} + \int_{K_1} \frac{dy}{|he_i - y|^\beta} - \int_{K_2} \frac{dy}{|he_i - y|^\beta}$$

$$= \int_{|y|<\delta} \frac{dy}{r_0^\beta} + \int_{K_1} \frac{dy}{|he_i - y|^\beta} - \int_{K_2} \frac{dy}{|he_i - y|^\beta}.$$

Now, it is easily seen that both K_1 and K_2 have the same volume.[4] Therefore,

$$\int_{K_2} \frac{dy}{|he_i - y|^\beta} > \frac{\text{vol}(K_2)}{\delta^\beta} = \frac{\text{vol}(K_1)}{\delta^\beta} > \int_{K_1} \frac{dy}{|he_i - y|^\beta}.$$

Therefore,

$$\int_{|y|<\delta} \frac{dy}{r^{n-\alpha}} \leqslant \int_{|y|<\delta} \frac{dy}{r_0^{n-\alpha}}.$$

Taking into account all these estimates, we finally see that

$$|I_1| \leqslant 2C_1 \int_{|y|<\delta} \frac{dy}{|y|^{n-\alpha}} = \tilde{C}\delta^\alpha \to 0,$$

as $\delta \to 0$. Similar expressions hold true for the mixed second derivatives of $u(x)$ evaluated at $x = 0$. We now see that u possesses all the second derivatives at 0 and that $\Delta u(0) = 0$, by summing the expressions from $i = 1$ through n. This completes the proof. ∎

---

[4] More generally, if $x_1, x_2 \in \Omega$, $x_1 \neq x_2$ and the balls $B_\delta(x_1)$, $B_\delta(x_2)$ are contained in Ω along with their closures, for some $\delta > 0$, then

$$\text{vol}(\{y \in \Omega : |y - x_1| < \delta, \, |y - x_2| > \delta\}) = \text{vol}(\{y \in \Omega : |y - x_2| < \delta, \, |y - x_1| > \delta\}).$$

9
Heat Equation

9.1 Homogeneous Equation: Fourier–Poisson Formula

Consider the initial value problem (IVP) for the heat equation in space dimension $n \geqslant 1$:

$$u_t = a^2 \Delta u, \ t > 0, \ x \in \mathbb{R}^n, \tag{9.1.1}$$
$$u(x,0) = g(x), \ x \in \mathbb{R}^n.$$

The solution of the IVP (9.1.1) is given by the Fourier–Poisson formula:

$$u(x,t) = \frac{1}{(4\pi a^2 t)^{n/2}} \int_{\mathbb{R}^n} g(y) e^{-\frac{|y-x|^2}{4a^2 t}} \, dy, \tag{9.1.2}$$

for $x \in \mathbb{R}^n$ and $t > 0$. The Fourier–Poisson formula (9.1.2) can be written as a *convolution*. For this purpose, we define the *heat kernel* or *fundamental solution of the heat equation* by

$$K(x,t) = \begin{cases} (4\pi a^2 t)^{-n/2} e^{-\frac{|x|^2}{4a^2 t}} & \text{for } x \in \mathbb{R}^n, \ t > 0, \\ 0 & \text{for } x \in \mathbb{R}^n, \ t < 0. \end{cases} \tag{9.1.3}$$

We can write the Fourier–Poisson integral (9.1.2) as the *convolution* of K and g as $u(x,t) = (K(\cdot,t) * g)(x)$. Furthermore, the heat kernel K enjoys the following properties, which are frequently used:

- $K(x-y,t) = K(y-x,t)$ for all $x, y \in \mathbb{R}^n$ and $t > 0$.
- $\int_{\mathbb{R}^n} K(x-y,t) \, dx = 1$ for all $y \in \mathbb{R}^n$ and $t > 0$.
- $(\partial_t - a^2 \Delta_x) K(x-y,t) = 0$ for all $y \in \mathbb{R}^n$ and $t > 0$.

9.1.1 Inhomogeneous Equation: Duhamel's Principle

We now consider the *inhomogeneous heat equation*:

$$u_t - a^2 \Delta u = f(x,t), \ x \in \mathbb{R}^n, t > 0,$$
$$u(x,0) = g(x), \ x \in \mathbb{R}^n. \tag{9.1.4}$$

Assume that the function f and its partial derivatives f_{x_j} are continuous in $x \in \mathbb{R}^n, t > 0$ and the function g is continuous and bounded. Owing to the linearity in the problem, the required solution can be written as some of two functions, namely the solution of the homogeneous equation with initial condition g and the solution of the inhomogeneous equation with zero initial condition. Thus, it suffices to consider equation (9.1.4) with $g \equiv 0$ therein. A solution of this problem can be obtained via the *Duhamel's principle*: Fix $s \geqslant 0$, and consider the IVP for the heat equation:

$$v_t - a^2 \Delta v = 0, \ x \in \mathbb{R}^n, t > s,$$
$$v(x,s) = f(x,s), \ x \in \mathbb{R}^n. \tag{9.1.5}$$

Denoting the solution by $v(x,t;s)$, we get

$$v(x,t;s) = \int_{\mathbb{R}^n} K(x-y, t-s) f(y,s) \, dy, \ t > s, \tag{9.1.6}$$

using equation (9.1.2) and the change of variable $t \mapsto t-s$. We can now write down the expression for a solution to equation (9.1.4) with $g \equiv 0$, as:

$$u(x,t) = \int_0^t v(x,t;s) \, ds = \int_0^t \int_{\mathbb{R}^n} K(x-y, t-s) f(y,s) \, dy ds. \tag{9.1.7}$$

9.1.2 Heat Equation in a Finite Interval: Fourier Method

In this subsection, we consider initial boundary value problem for the one-dimensional heat equation and see how the Fourier method can be used to obtain the solution. Consider IVP for the heat equation in a finite interval $[0, L]$ on the real line

$$u_t - a^2 u_{xx} = 0, \ 0 < x < L, t > 0,$$
$$u(x,0) = g(x), \ 0 < x < L. \tag{9.1.8}$$

The initial condition $u(x,0) = g(x)$ in equation (9.1.8) represents the initial temperature distribution at all points of the rod at the initial instant of time $t = 0$. At the end points, $x = 0$ and $x = L$, different boundary conditions may be given. For example, consider the following conditions:

$$u(0,t) = h_1(t), \quad u(L,t) = h_2(t), \qquad (9.1.9)$$

for $t > 0$. These conditions imply that the rod is maintained at the prescribed temperatures at the end points. The resulting initial boundary value problem is termed as the *Dirichlet problem*. If, instead, the heat flux density at the end points is supplied, then u is replaced by $\dfrac{\partial u}{\partial x}$ in equation (9.1.9). The resulting problem is termed as the *Neumann problem*. If at one end of the rod temperature is prescribed and at the other end heat flux density, we arrive at a situation where the resulting problem is known as a *mixed problem*.

The general references are [45], [24], [25], [42], [35], [55] and [57].

9.2 Exercises

Exercise 9.1

Let $u \in C^2(\mathbb{R} \times (0, \infty))$ be a solution of the equation $u_t = a^2 u_{xx} + b u_x + cu + f(x,t)$, where a, b and c are real constants and f is a given function. Define the function v by $v(x,t) = e^{-ct} u(x - bt, t)$ for $x \in \mathbb{R}$ and $t > 0$. Show that v satisfies the non-homogeneous equation $v_t = a^2 v_{xx} + g$, where $g(x,t) = e^{-ct} f(x - bt, t)$ for $x \in \mathbb{R}$ and $t > 0$.

Exercise 9.2

Solve the IVP: $u_t = a^2 u_{xx} + b u_x + cu + f(x,t)$, $x \in \mathbb{R}, t > 0$ and $u(x,0) = u_0(x)$, $x \in \mathbb{R}$, with the following data:

(i) $f(x,t) = t \sin x$, $u_0 \equiv 1$, $a = c > 0$, $b = 0$.

(ii) $f(x,t) = h(t) \in C^1([0, \infty))$ and u_0 is a bounded continuous function.

Exercise 9.3

For an arbitrary $s \geq 0$, let $u(x,t;s)$ be a C^2 solution of the IVP

$$u_t = a^2 \Delta u, \ x \in \mathbb{R}^n, t > s \text{ and } u(x,s;s) = f(x,s), \ x \in \mathbb{R}^n.$$

Define v by $v(x,t;s) = \displaystyle\int_s^t u(x,t;\tau)\,d\tau$. Show that v satisfies the IVP

$$v_t = a^2 \Delta v + f(x,t), \ x \in \mathbb{R}^n, t > s \text{ and } v(x,s;s) = 0, \ x \in \mathbb{R}^n.$$

Furthermore, show that the converse also holds.

Thus, the homogeneous problem with non-homogeneous initial data can be transformed into a non-homogeneous problem with zero initial data and vice versa.

Exercise 9.4

Let $u_0 : \mathbb{R}^n \to \mathbb{R}$ be such that $u_0(x) = \prod_{j=1}^{n} u_{0j}(x_j)$, where for each $j = 1, 2, \ldots, n$, $u_{0j} : \mathbb{R} \to \mathbb{R}$ is a bounded continuous function. If u_j solves the *one*-dimensional heat equation $\dfrac{\partial u_j}{\partial t} = a^2 \dfrac{\partial^2 u_j}{\partial x_j^2}$ with initial condition $u_j(x_j, 0) = u_{0j}(x_j)$ for $j = 1, 2, \ldots, n$, show that the solution u of the IVP $u_t = a^2 \Delta u$, $u(x, 0) = u_0(x)$ is given by $u(x, t) = \prod_{j=1}^{n} u_j(x_j, t)$.

Exercise 9.5

Let $\mathcal{M}$ denote the class of functions $f(x, t)$, $x \in \mathbb{R}^n$ and $t \geq 0$, that are bounded in each strip $\{(x, t) : x \in \mathbb{R}^n, 0 \leq t \leq T\}$. Assume that the function $f \in C^2$ in $\mathbb{R}^n \times [0, \infty)$ such that f and all its derivatives up to the second order belong to $\mathcal{M}$. Show that the function u given by the expression in equation (9.1.7) belongs to $C^2(\mathbb{R}^n \times (0, \infty)) \cap C^1(\mathbb{R}^n \times [0, \infty))$ and satisfies the inhomogeneous equation (9.1.4). Furthermore, show that $u(x, 0) = 0$ for $x \in \mathbb{R}^n$.

Exercise 9.6

Show that
$$\iint_{E(1)} \frac{|y|^2}{s^2} \, dy\, ds = 4.$$

Here, $E(1) = E(0, 0; 1)$ with
$$E(x, t; r) = \left\{ (y, s) \in \mathbb{R}^{n+1} : s \leq t, \; K(x - y, t - s) \geq \frac{1}{r^n} \right\},$$
where K is the fundamental solution of the heat equation.

Exercise 9.7

(i) Show that the function u defined by
$$u(x, t) = \left(1 - \frac{x}{L}\right) u_1 + \frac{x}{L} u_2 + \sum_{n=1}^{\infty} g_n \exp\left(-\frac{n^2 \pi^2 a^2}{L^2} t\right) \sin\left(\frac{n\pi}{L} x\right)$$

satisfies the equation (9.1.8) and the boundary conditions

$$u(0,t) = u_1, \ u(L,t) = u_2, \ t > 0,$$

where u_1 and u_2 are given constants and g_n are the Fourier coefficients of the function g:

$$g_n = \frac{2}{L} \int_0^L g(x) \sin\left(\frac{n\pi}{L}x\right) dx.$$

Physically, these boundary conditions represent that the ends of the rod are maintained at the given temperatures.

(ii) Show that the function u defined by

$$u(x,t) = \sum_{n=1}^{\infty} g_n \exp\left(-\frac{a^2 \lambda_n^2}{L^2} t\right) X_n(x)$$

satisfies the equation (9.1.8) and the boundary conditions

$$u_x(0,t) - hu(0,t) = 0, \ u_x(L,t) + hu(L,t) = 0, \ t > 0,$$

for some $h \geqslant 0$. Here, $\lambda_n = \frac{\mu_n}{L}$ with μ_n are the roots of the algebraic equation

$$2 \cot \mu = \frac{\mu}{hL} - \frac{hL}{\mu}, \ h > 0.$$

The functions X_n are the non-trivial solutions of the boundary value problem

$$\left. \begin{array}{l} X_n'' + \lambda_n X_n = 0 \\ X_n'(0) - hX_n(0) = 0, \ X_n'(L) + hX_n(L) = 0, \end{array} \right\}$$

for $n = 1, 2, ...$, and satisfy the *orthogonality conditions*:

$$\int_0^L X_n(x) X_m(x) \, dx = 0, \ \text{for } n \neq m.$$

Furthermore, the functions X_n are normalized in the sense that

$$\int_0^L X_n^2(x) \, dx = 1,$$

and g_n are the *Fourier coefficients* of a function g with respect to the orthonormal family $\{X_n\}$, which are given by

$$g_n = \int_0^L g(x) X_n(x)\, dx.$$

Exercise 9.8

Describe the solution of the one-dimensional boundary value problem $u_t = (1/2)u_{xx}$, $x > 0$, $t > 0$ satisfying the condition $u(0,t) = f(t)$, where f is a smooth periodic function defined on $\mathbb{R}$, of period 1.

This problem models the description of the temperature profile beneath the earth's surface ($x > 0$) given the temperature profile on the surface ($x = 0$). The period 1 of the initial profile means 1 year, after choosing the appropriate physical units. The mean temperature on the surface may be set to 0; temperature will have a maximum during the summer and a minimum during the winter.

Exercise 9.9

Suppose $u_0 \in C(\mathbb{R}^n)$ satisfies the condition that $|u_0(x)| \leqslant M e^{-\delta|x|^2}$ for all $x \in \mathbb{R}^n$ and for some constants $M > 0$, $\delta \geqslant 0$. Show that the solution u of the heat equation $u_t = a^2 \Delta u$ with initial data u_0 satisfies the estimate

$$|u(x,t)| \leqslant M\left(1 + 4a^2 \delta t\right)^{-n/2} \exp\left(-\frac{\delta |x|^2}{1 + 4a^2 \delta t}\right),$$

for all $x \in \mathbb{R}^n$ and $t \geqslant 0$.

Exercise 9.10

This exercise is used in the proof of uniqueness of positive solutions to the heat equation. See Exercise 9.11.

Let K be the fundamental solution of the heat operator $\partial_t - \Delta$ and u be a solution of the heat equation in $Q = \Omega \times (a,b)$, where Ω is a bounded open set in $\mathbb{R}^n$ with smooth boundary. Assume that u is continuous in $\overline{Q}$ and that all the first derivatives $\dfrac{\partial u}{\partial x_i}$ are continuous in $\partial \Omega \times [a,b]$. Prove the following: For any $(x,t) \in Q$,

$$u(x,t) = \int_\Omega K(x,t;\xi,a)\, u(\xi,a)\, d\xi$$

$$- \int_a^t \int_{\partial \Omega} \left[u(\xi,\tau) \frac{\partial}{\partial \nu} K(x,t;\xi,\tau) - K(x,t;\xi,\tau) \frac{\partial u}{\partial \nu}(\xi,\tau) \right] d\sigma(\xi)\, d\tau.$$

Here, the following notation has been used:

$$K(x,t;\xi,\tau) = K(x-\xi, t-\tau) = (4\pi(t-\tau))^{-n/2} \exp\left(-\frac{|x-\xi|^2}{4(t-\tau)}\right),$$

for $x, \xi \in \mathbb{R}^n$ and $\tau < t$. It follows that solutions of the heat equation belong to C^∞.

Exercise 9.11

Let $Q = \mathbb{R}^n \times (0,T)$, $T > 0$ and $u \in C^{2,1}(Q)$. This means that the real-valued function u has continuous derivatives up to second order with respect to x variables and up to first order with respect to t variable. If u is a *non-negative* solution of the heat equation $u_t - \Delta u = 0$ in Q and is continuous in $\mathbb{R}^n \times [0,T)$, show that

$$u(x,t) = (4\pi t)^{-n/2} \int_{\mathbb{R}^n} \exp\left(-\frac{|x-y|^2}{4t}\right) u(y,0)\, dy,$$

for all $(x,t) \in Q$. In particular, if $u(x,0) = 0$ for all $x \in \mathbb{R}^n$, it follows that $u \equiv 0$ in Q. For a fixed $s \in [0,T)$, using a translation in the t variable, we obtain from the above expression

$$u(x,t) = (4\pi t)^{-n/2} \int_{\mathbb{R}^n} \exp\left(-\frac{|x-y|^2}{4(t-s)}\right) u(y,s)\, dy,$$

for $s < t < T$.

Through a translation, we may replace the interval $(0,T)$ by (a,b), $a < b$. For $n = 1$, this is proved in Ref. [57]. This proof extends to general n with minor changes. The proof is also sketched in Ref. [25] through several problems. Uniqueness results for more general parabolic equations in which the Laplacian is replaced by a uniformly elliptic operator are treated in Ref. [24]. There are also uniqueness results concerning the solutions of the heat equation in a Riemannian manifold [20]. See also Ref. [47].

Exercise 9.12

Show that the following IVP has no classical solution in any strip $\{0 < t < T\}$:

$$u_t - \Delta u = f(x),\ x \in \mathbb{R}^n,\ t > 0,$$
$$u(x,0) = \varphi(x),\ x \in \mathbb{R}^n,$$

where $\varphi(x) = (x_1^2 - x_2^2)\zeta(|x|)(-\log|x|)^{1/2}$ and

$$f(x) = \frac{x_1^2 - x_2^2}{2|x|^2} \zeta(|x|)\left[(n+2)(-\log|x|)^{-1/2} + (1/2)(-\log|x|)^{-3/2}\right]$$
$$- \frac{x_1^2 - x_2^2}{|x|} \zeta'(|x|)\left[(n+3)(-\log|x|)^{1/2} - (-\log|x|)^{-1/2}\right]$$
$$- (x_1^2 - x_2^2)\zeta''(|x|)(-\log|x|)^{1/2},$$

with $\zeta \in C_0^\infty(\mathbb{R}^n)$, which is identically equal to 1 for $|x| < 1/2$ and identically equal to 0 for $|x| > 3/4$. Here, $x = (x_1, x_2, \dots, x_n)$.

Remark: This example can also be used to show that the Poisson equation $-\Delta u = f$ has no C^2 solution in $B_1(0)$.

9.3 Solutions

Exercise 9.1

From the expression for v, it follows that

$$v_t(x, t) = e^{-ct}\left(-cu(x - bt, t) - bu_x(x - bt, t) + u_t(x - bt, t)\right)$$
$$v_{xx}(x, t) = e^{-ct} u_{xx}(x - bt, t).$$

Therefore,

$$\begin{aligned} v_t(x, t) - a^2 v_{xx}(x, t) &= e^{-ct}\left(-cu(x - bt, t) - bu_x(x - bt, t) \right.\\ &\quad \left. + u_t(x - bt, t) - a^2 u_{xx}(x - bt, t)\right)\\ &= e^{-ct} f(x - bt, t) = g(x, t). \end{aligned}$$

This exercise helps us solve the heat-type equations with lower-order terms. Note that $v(x, 0) = u(x, 0)$ and u can be obtained in terms of v as $u(x, t) = e^{ct} v(x + bt, t)$. The problems in Exercise 9.2 below make use of this observation. Also, the above procedure easily gets extended to the case of $\mathbb{R}^n$.

Exercise 9.2

We follow the notations in Exercise 9.1.

(i) Here, we have $g(x, t) = e^{-ct} f(x, t) = te^{-ct} \sin x$ and $v(x, 0) \equiv 1$. Therefore, using the Fourier–Poisson formula (9.1.2) and the solution formula (9.1.7) for the inhomogeneous heat equation, we have

$$v(x, t) = \int_{\mathbb{R}^n} K(x - y, t)\, dy + \int_0^t \int_{\mathbb{R}^n} K(x - y, t - s) g(y, s)\, dy ds$$

$$= 1 + \int_0^t \int_{\mathbb{R}^n} K(x - y, t - s) s e^{-cs} \sin y\, dy ds,$$

and we have $u(x, t) = e^{ct} v(x, t)$.

(ii) Here, we have $g(x,t) = e^{-ct}h(t)$ and $v(x,0) = u_0(x)$. Hence,

$$v(x,t) = \int_{\mathbb{R}^n} K(x-y,t) u_0(y)\, dy$$

$$+ \int_0^t \int_{\mathbb{R}^n} K(x-y, t-s) e^{-cs} h(s)\, dy ds,$$

and we have $u(x,t) = e^{ct} v(x+bt, t)$.

Exercise 9.3

Clearly, $v(x, s; s) = 0$. We have

$$v_t(x, t; s) = u(x, t; t) + \int_s^t u_t(x, t; \tau)\, d\tau$$

$$= f(x, t) + a^2 \int_s^t \Delta_x u(x, t; \tau)\, d\tau$$

$$= f(x, t) + a^2 \Delta_x \int_s^t u(x, t; \tau)\, d\tau$$

$$= f(x, t) + a^2 \Delta v(x, t; s).$$

For the converse, since $v(x, s; s) = 0$ for $s \geqslant 0$ arbitrary, we can write

$$v(x, t; s) = \int_s^t u(x, t; \tau)\, d\tau,$$

for some function u; u is an anti-derivative of v considered as a function of the variable s. Thus,

$$f(x, t) = v_t - a^2 \Delta_x v = u(x, t; t) + \int_s^t (u_t - \Delta_x u)(x, t; \tau)\, d\tau.$$

Taking $t = s$, we obtain $u(x, s; s) = f(x, s)$. Therefore, we have

$$\int_s^t (u_t - a^2 \Delta_x u)(x, t; \tau)\, d\tau = 0,$$

for all $t > s$ and $x \in \mathbb{R}^n$. It then follows that

$$\int_{t_1}^{t_2} (u_t - a^2 \Delta_x u)(x, t; \tau) \, d\tau = 0,$$

for all $t_2 > t_1 > s$ and $x \in \mathbb{R}^n$, from which we conclude that $u_t - a^2 \Delta u = 0$ for all $t > s$ and $x \in \mathbb{R}^n$, as required.

Exercise 9.4

Observe that u_0 is also a bounded continuous function. Thus, a solution of the IVP $u_t = a^2 \Delta u$, $u(x, 0) = u_0(x)$ is unique. We verify that the function u given by

$$u(x, t) = \prod_{j=1}^{n} u_j(x_j, t)$$

is a solution of this IVP, thereby concluding the proof.[1] Clearly, we have $u(x, 0) = u_0(x)$. Also,

$$\frac{\partial u}{\partial t} = \sum_{j=1}^{n} \left(\prod_{k=1, k \neq j}^{n} u_k \right) \frac{\partial u_j}{\partial t}.$$

Now, as u_j depends on x_j only, we obtain

$$\frac{\partial^2 u}{\partial x_j^2} = \left(\prod_{k=1, k \neq j}^{n} u_k \right) \frac{\partial^2 u_j}{\partial x_j^2},$$

for $j - 1, 2, \ldots, n$. Therefore,

$$\Delta u = \sum_{j=1}^{n} \frac{\partial^2 u}{\partial x_j^2} = \sum_{j=1}^{n} \left(\prod_{k=1, k \neq j}^{n} u_k \right) \frac{\partial^2 u_j}{\partial x_j^2}.$$

Putting these expressions together and using the hypotheses, we see that u satisfies the heat equation $u_t = a^2 \Delta u$.

Exercise 9.5

Note that since the function f and the fundamental solution K are locally integrable in $\mathbb{R}^{n+1}$, the convolution defining the function u exists. Furthermore, u is also locally integrable in $\mathbb{R}^{n+1}$. If we put

$$h(x, t) = \int_0^t \int_{\mathbb{R}^n} K(x - y, t - s) |f(y, s)| \, dy \, ds,$$

---

[1] In any case, the function u as defined is a solution of the IVP.

then

$$h(x,t) \leqslant \sup_{\substack{y\in\mathbb{R}^n \\ 0\leqslant s\leqslant t}} |f(y,s)| \int_0^t\!\!\int_{\mathbb{R}^n} K(x-y,t-s)\,dyds$$

$$\leqslant t \sup_{\substack{y\in\mathbb{R}^n \\ 0\leqslant s\leqslant t}} |f(y,s)|,\ t>0.$$

This shows that $u \in \mathcal{M}$ and $u(x,0) = 0$. At this stage, it should be remarked that the existence of the convolution in $\mathbb{R}^{n+1}$ is sufficient to show that u satisfies equation (9.1.4) *in the sense of distributions*.

To prove the stated smoothness of u, introduce the change of variables in the integral in equation (9.1.7):

$$y = x - 2a\sqrt{s}\,\xi,\ \tau = t - s.$$

Thus,

$$u(x,t) = \pi^{-n/2}\int_0^t\!\!\int_{\mathbb{R}^n} f(x - 2a\sqrt{s}\,\xi, t-\tau)e^{-|\xi|^2}\,d\xi d\tau.$$

Therefore, using the smoothness hypothesis on f, we infer that

$$\frac{\partial u}{\partial t}(x,t) = \pi^{-n/2}\int_0^t\!\!\int_{\mathbb{R}^n} \frac{\partial f}{\partial t}(x - 2a\sqrt{s}\,\xi, t-\tau)e^{-|\xi|^2}\,d\xi d\tau$$

$$+ \pi^{-n/2}\int_{\mathbb{R}^n} f(x - 2a\sqrt{s}\,\xi, 0)e^{-|\xi|^2}\,d\xi.$$

Similar expressions hold for the derivatives with respect to the x variables (up to the second order) of u. This proves the stated smoothness of u. Furthermore, we can directly prove that u satisfies the inhomogeneous equation (9.1.4), using the fact that K satisfies the homogeneous heat equation as mentioned in Section 9.1. ∎

Exercise 9.6

Put

$$I = \iint_{E(1)} \frac{|y|^2}{s^2}\,dyds. \tag{9.2.1}$$

Note that the domain of integration $E(1)$ is given by[2]

$$E(1) = \{(y,s) : s < 0,\ K(-y,-s) \geqslant 1\},$$

---

[2] Since division by s is involved, the value $s = 0$ is avoided.

with the fundamental solution K given by equation (9.1.3) (with $a = 1$). Thus,

$$1 \leqslant (4\pi(-s))^{-n/2} \exp(-|y|^2/(-4s)),$$

for $(y, s) \in E(1)$. This further restricts the s variable to the interval $[-1/4\pi, 0)$ and y to the ball $|y| \leqslant \rho$, where $\rho = \rho(s) = \sqrt{2ns\log(-4\pi s)}$. Thus, first, performing the integration with respect to the y variables in equation (9.2.1), we obtain

$$\int_{|y|\leqslant \rho} |y|^2 \, dy = \sigma_n \int_0^\rho r^2 r^{n-1} \, dr = \frac{\sigma_n}{n+2} \rho^{n+2}.$$

Here, $\sigma_n = \dfrac{2\pi^{n/2}}{\Gamma(n/2)}$ is the surface area of the unit ball in $\mathbb{R}^n$ and Γ is the Euler gamma function. Thus,

$$I = \frac{\sigma_n}{n+2} \int_{-1/4\pi}^0 s^{-2} (2ns\log(-4\pi s))^{n/2+1} \, ds.$$

Now, change the variable of integration: $\log(-4\pi s) = -\xi$. Then,

$$I = \frac{\sigma_n}{(n+2)\pi^{n/2}} \frac{n^{n/2+1}}{2^{n/2-1}} \int_0^\infty \xi^{n/2+1} e^{-n\xi/2} \, d\xi.$$

Now, changing the last integral to the standard gamma integral, we conclude that $I = 4$.

Exercise 9.7

(i) Since $0 < \exp\left(-\dfrac{n^2\pi^2 a^2}{L^2} t\right) \leqslant 1$ for all $t \geqslant 0$, it easily follows that the given series in the expression for u converges absolutely and uniformly. Also, the term $n^k \exp\left(-\dfrac{n^2\pi^2 a^2}{L^2} t\right)$ is bounded, for all $t > 0$ and $k = 0, 1, 2, ...$, if we choose n large enough. These observations show that term-by-term differentiations once with respect to t and twice with respect to x, in the given series, are easily justified. It is now straightforward to perform these term-by-term differentiations and verify that u satisfies the heat equation $u_t = a^2 u_{xx}$ for $t > 0$. Because of the presence of $\sin\left(\dfrac{n\pi}{L} x\right)$ term in the series, the given boundary conditions are easily seen to be satisfied.

(ii) The arguments regarding the term-by-term differentiations are exactly similar to those in (i) above. Thus, u satisfies the heat equation. That u satisfies the boundary conditions follows since X_n's satisfy the same boundary conditions. These boundary conditions physically represent free heat exchange at the end points of the rod.

Exercise 9.8

For $x > 0$, it is natural to expect that the solution $u(x,t)$ is also a periodic function of t of period 1. Assume $u(x,t) = \sum_{n=-\infty}^{\infty} c_n(x) \exp(2\pi i n t)$, where the Fourier coefficients c_n are functions of x. Hence, $c_n(x) = \int_0^1 u(x,t) \exp(-2\pi i n t)\, dt$. Differentiating twice with respect to x, we obtain

$$c_n''(x) = \int_0^1 u_{xx}(x,t) \exp(-2\pi i n t)\, dt = 2 \int_0^1 u_t(x,t) \exp(-2\pi i n t)\, dt$$

$$= 4\pi i n c_n = \left(\sqrt{2\pi |n|}\,(1 \pm i)\right)^2 c_n,$$

where the positive sign is chosen if $n > 0$ and the negative sign if $n < 0$. However,

$$c_n(0) = \int_0^1 u(0,t) \exp(-2\pi i n t)\, dt = \int_0^1 f(t) \exp(-2\pi i n t)\, dt = \hat{f}(n),$$

where $\hat{f}(n)$ denote the Fourier coefficients of the periodic function f. Solving the second-order ODE satisfied by c_n, we thus obtain

$$u(x,t) = \sum_{n=-\infty}^{\infty} \hat{f}(n) \exp\left(-\sqrt{2\pi|n|}\,x\right) \exp\left(2\pi i n t \mp \sqrt{2\pi|n|}\,i\,x\right).$$

In particular, if we take the simple harmonic $f(t) = \exp(2\pi i n t)$, $n > 0$ fixed, we get $u(x,t) = \exp\left(-\sqrt{2\pi|n|}\,x\right) \exp\left(2\pi i n t - \sqrt{2\pi|n|}\,i\,x\right)$. Thus, the temperature response at depth x is *damped* by $\exp\left(-\sqrt{2\pi|n|}\,x\right)$ and suffers a *phase shift* by an amount $\sqrt{2\pi|n|}\,i\,x$.

To get a better understanding of this phenomenon, let us take $f(t) = \sin(2\pi t)$. Then, $u(x,t) = \exp\left(-\sqrt{2\pi}\,x\right) \sin\left(2\pi t - \sqrt{2\pi}\,x\right)$. Choose $x = \sqrt{\pi/2}$. At this depth, the temperature profile is damped by a factor of $e^{-\pi} \approx 0.04$ and is *completely out of phase*, with the seasons; that is, at this depth, the summer feels like winter and vice versa (see Figure 9.1). This is the reason for the construction of vegetable cellars. At proper depth, not only is the 'local climate' much more nearly constant (damping effect), but it is cooler in the summer and warmer in the winter (phase shift effect). In his famous work on the heat equation, Fourier suggested that a depth of about 4 m, after putting in the proper physical constants, is good enough for the construction of such cellars (see Ref. [22]).

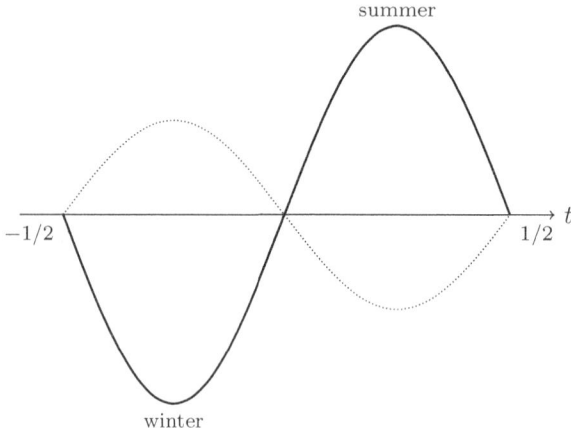

Figure 9.1 | Temperature profiles in Exercise 9.8.

Exercise 9.9

Using the Fourier–Poisson formula and the hypothesis, we obtain, for $t > 0$,

$$|u(x,t)| \leqslant M \left(4\pi a^2 t\right)^{-n/2} \int_{\mathbb{R}^n} \exp\left(-\frac{|x-y|^2}{4a^2 t}\right) e^{-\delta|y|^2} \, dy.$$

Rewriting the exponents, we get

$$\frac{|x-y|^2}{4a^2 t} + \delta|y|^2 = \frac{|x|^2}{4a^2 t} - \frac{2x \cdot y}{4a^2 t} + \frac{1 + 4\delta a^2 t}{4a^2 t}|y|^2$$

$$= \frac{\delta|x|^2}{1 + 4\delta a^2 t} + |\alpha x - \beta y|^2,$$

with $\alpha^2 = \left(4a^2 t(1 + 4\delta a^2 t)\right)^{-1}$ and $\beta^2 = \dfrac{1 + 4\delta a^2 t}{4a^2 t}$. Furthermore, we have

$$\int_{\mathbb{R}^n} \exp\left(-|\alpha x - \beta y|^2\right) dy = (\pi/\beta^2)^{n/2}.$$

Substituting these expressions into the above estimate for $|u(x,t)|$, we get the desired inequality.

Exercise 9.10

We begin by deriving the following identity:

$$\int_a^t \int_\Omega [v(u_\tau - \Delta u) + u(v_\tau + \Delta v)] \, d\xi \, d\tau$$

$$= \int_a^t \int_\Omega (uv)_\tau \, d\xi \, d\tau - \int_a^t \int_{\partial\Omega} \left(v \frac{\partial u}{\partial \nu} - u \frac{\partial v}{\partial \nu} \right) d\sigma(\xi) \, d\tau, \quad (9.2.2)$$

for functions u and v defined on $\overline{Q}$, possessing continuous first derivative with respect to t and continuous derivatives up to second order with respect to x and being continuous in $\overline{Q}$. We have

$$\int_\Omega v(\xi,\tau)(u_\tau - \Delta_\xi u) \, d\xi = \int_\Omega v(\xi,\tau) u_\tau(\xi,\tau) \, d\xi - \int_\Omega v(\xi,\tau) \Delta_\xi u \, d\xi$$

$$= \int_\Omega v \, u_\tau \, d\xi - \int_\Omega u \Delta_\xi v \, d\xi$$

$$- \int_{\partial\Omega} \left(v \frac{\partial u}{\partial \nu} - v \frac{\partial u}{\partial \nu} \right) d\sigma(\xi).$$

Writing $v \, u_\tau = (vu)_\tau - u \, v_\tau$ and a further integration with respect to τ over $[a,t]$ gives equation (9.2.2).

Fix $(x,t) \in Q$. Let $u(\xi,\tau)$ be a solution of the heat equation $u_\tau - \Delta_\xi u = 0$ and $v(\xi,\tau) = K(x,t;\xi,\tau)$. Then, $v_\tau + \Delta_\xi v = 0$. Using these u and v in equation (9.2.2), we obtain

$$\int_a^t \int_\Omega (uv)_\tau \, d\xi \, d\tau = \int_a^t \int_{\partial\Omega} \left(v \frac{\partial u}{\partial \nu} - u \frac{\partial v}{\partial \nu} \right) d\sigma(\xi) \, d\tau. \quad (9.2.3)$$

As $\tau = t$ is a singularity for $K(x,t;\xi,\tau)$, care is needed to evaluate the limit as $\tau \to t$ in the integral on the left-hand side with respect to τ. We have

$$\int_\Omega u(\xi,\tau) K(x,t;\xi,\tau) \, d\xi = \int_\Omega u(\xi,\tau) K(x,t;\xi,\tau) \, d\xi - u(x,t) + u(x,t)$$

$$= u(x,t) + \int_\Omega (u(\xi,\tau) - u(x,t)) K(x,t;\xi,\tau) \, d\xi$$

$$- u(x,t) \int_{\Omega^c} K(x,t;\xi,\tau) \, d\xi.$$

Given an $\varepsilon > 0$, choose $\delta > 0$ such that $|u(\xi, \tau) - u(x, t)| < \varepsilon$ whenever $|x - \xi| < \delta$ and $t - \tau < \delta$, using the continuity of u. Write

$$\int_\Omega (u(\xi, \tau) - u(x, t)) \, K(x, t; \xi, \tau) \, d\xi \leqslant$$

$$\int_{\Omega \cap \{|x-\xi|<\delta\}} |u(\xi, \tau) - u(x, t)| \, K(x, t; \xi, \tau) \, d\xi$$

$$+ \int_{\Omega \cap \{|x-\xi|\geqslant\delta\}} (u(\xi, \tau) - u(x, t)) \, K(x, t; \xi, \tau) \, d\xi.$$

The first integral is less than ε by the continuity of u. For the second integral, we have

$$\int_{\Omega \cap \{|x-\xi|\geqslant\delta\}} |u(\xi, \tau) - u(x, t)| \, K(x, t; \xi, \tau) \, d\xi \leqslant 2M \int_{|x-\xi|\geqslant\delta} K(x, t; \xi, \tau) \, d\xi,$$

where M is the supremum of $|u|$ over $\overline{Q}$, and the latter integral tends to 0 as $\tau \to t$. Similarly, as $x \in \Omega$, there is a $\tilde{\delta}$ such that $|x - \xi| \geqslant \tilde{\delta}$ for all $\xi \in \Omega^c$. Therefore,

$$\int_{\Omega^c} K(x, t; \xi, \tau) \, d\xi \leqslant \int_{\{|x-\xi|\geqslant\tilde{\delta}\}} K(x, t; \xi, \tau) \, d\xi \to 0 \quad \text{as} \quad \tau \to t.$$

We thus conclude that

$$\lim_{\tau \to t-} \int_\Omega u(\xi, \tau) \, K(x, t; \xi, \tau) \, d\xi = u(x, t).$$

Finally, from equation (9.2.3), we obtain

$$u(x, t) = \int_\Omega K(x, t; \xi, a) \, u(\xi, a) \, d\xi$$

$$- \int_a^t \int_{\partial\Omega} \left[u(\xi, \tau) \frac{\partial}{\partial \nu} K(x, t; \xi, \tau) - K(x, t; \xi, \tau) \frac{\partial u}{\partial \nu}(\xi, \tau) \right] d\sigma(\xi) \, d\tau,$$

as required.

Exercise 9.11

We begin by proving the following inequality:

$$\int_{\mathbb{R}^n} K(x - y, t) u(y, s) \, dy \leqslant u(x, t + s), \tag{9.2.4}$$

for all $x \in \mathbb{R}^n$ and $0 < t < T - s$. The proof of the convergence of the integral in the above inequality is part of the problem. Pick $R > 0$ and $s \in [0, T)$. Define the function v by

$$v(x, t) = u(x, t + s) - \int_{|y| < R} K(x - y, t) u(y, s) \, dy,$$

for (x, t) in the cylinder $\mathcal{C} = \{(x, t) : |x| \leq \widetilde{R}, \; t \in [0, T - s]\}$, $\widetilde{R} > R$. It is clear that v satisfies the heat equation in the interior of $\mathcal{C}$. We now show that $v \geq 0$.

As observed in Exercise 9.10, we have

$$u(x, s) = \lim_{t \to 0} \int_{|y| < R} K(x - y, t) u(y, s) \, dy,$$

uniformly on compact subsets of $\{|x| < R\}$, and the above limit is 0 uniformly on compact subsets of $\{|x| > R\}$. We claim that

$$\limsup_{(x,t) \to (\xi, 0)} \int_{|y| < R} K(x - y, t) u(y, s) \, dy \leq u(\xi, s),$$

when $|\xi| = R$. Let $\varepsilon > 0$ be arbitrary. By continuity, there is a $\delta > 0$ such that $|u(y, s) - u(\xi, s)| < \varepsilon/2$ whenever $|y - \xi| < \delta$. If $|x - \xi| < \delta/2$, we have $|y - \xi| < \delta$ for $|y| < R$ and $|x - y| < \delta/2$. Therefore,

$$\int_{\{|x-y|<\delta/2\} \cap \{|y|<R\}} K(x - y, t) u(y, s) \, dy < (u(\xi, s) + \varepsilon/2)$$

$$\times \int_{|y|<R} K(x - y, t) \, dy < u(\xi, s) + \varepsilon/2.$$

On the other hand, we have

$$\int_{\{|x-y|\geq\delta/2\} \cap \{|y|<R\}} K(x - y, t) u(y, s) \, dy < \varepsilon/2,$$

if we choose t very small. Thus,

$$\int_{|y|<R} K(x - y, t) u(y, s) \, dy < u(\xi, s) + \varepsilon,$$

if $|x - \xi| < \delta/2$ and t sufficiently small. Since $\varepsilon > 0$ is arbitrary, we conclude that

$$\limsup_{(x,t) \to (\xi, 0)} \int_{|y|<R} K(x - y, t) u(y, s) \, dy \leq u(\xi, s),$$

where $|\xi| = R$.

Thus, the function v, which satisfies the heat equation, is non-negative at the base of the cylinder $\mathcal{C}$, namely $|x| \leqslant \widetilde{R}$ and $t = 0$. We next estimate v on the boundary $|x| = \widetilde{R}$, $0 < t < T - s$. Using the elementary fact that $r^k e^{-r^2}$ is bounded for $r \geqslant 0$ and any $k \geqslant 0$, we obtain that $K(x,t) < c|x|^{-n}$ for some constant c and all $t > 0$ and $x \neq 0$. Therefore, if $|x| = \widetilde{R}$, we have

$$\int_{|y|<R} K(x-y,t)u(y,s)\,dy < \frac{c}{(\widetilde{R}-R)^n} \int_{|y|<R} u(y,s)\,dy.$$

By choosing $\widetilde{R}$ sufficiently large, we can make the right-hand side less than any preassigned $\varepsilon > 0$. Thus, we conclude that on the lateral boundary of $\mathcal{C}$, v satisfies the inequality $v(x,t) = u(x,t+s) - \varepsilon \geqslant -\varepsilon$, as u is assumed to be non-negative. Since ε is arbitrary, we conclude $v \geqslant 0$ on the lateral boundary of $\mathcal{C}$. By maximum principle, it follows that $v(x,t) \geqslant 0$ in the interior of $\mathcal{C}$ as well. Therefore,

$$\int_{|y|<R} K(x-y,t)u(y,s)\,dy \leqslant u(x,t+s).$$

Since R is arbitrary, we arrive at the desired inequality (9.2.4). This also proves the convergence of the integral therein. We now reset v as

$$v(x,t) = u(x,t+s) - \int_{\mathbb{R}^n} K(x-y,t)u(y,s)\,dy,$$

for $x \in \mathbb{R}^n$ and $t \in [0, T-s)$. Our final task is to show $v \equiv 0$, thereby proving equality in equation (9.2.4). Note that v satisfies the heat equation in the strip $\{0 < t < T - s\}$, is non-negative and satisfies $v(x,0) = 0$ for all $x \in \mathbb{R}^n$. In particular, whatever we have shown above regarding the function u, including equation (9.2.4), applies to the function v as well. Put $w(x,t) = \int_0^t v(x,s)\,ds$. Then, $w \geqslant 0$ satisfies the heat equation and $w_t = \Delta w = v \geqslant 0$, as $v(x,0) = 0$. Furthermore, $w \equiv 0$ implies $v \equiv 0$ as $v \geqslant 0$. However, w is a non-decreasing function of t and is a subharmonic function of x. Therefore, without loss of generality, we may as well assume that v itself is a non-decreasing function of t and is a subharmonic function of x.

Let $\tau \in (0, T-s)$ and $0 < t_0 < T - s - \tau$. Set

$$M = (4\pi t_0)^{n/2} \int_{\mathbb{R}^n} K(y,t_0) v(y,\tau)\,dy.$$

From equation (9.2.4), we see that $M \leqslant (4\pi t_0)^{n/2} v(0, t_0 + \tau)$. Let $x \neq 0$. Now,

$$(4\pi t_0)^{n/2} K(y,t_0) = \exp\left(-\frac{|y|^2}{4t_0}\right) \geqslant \exp\left(-\frac{|x|^2}{t_0}\right),$$

if $|y| \leqslant 2|x|$. Therefore,

$$M \geqslant \int_{|y|<2|x|} \exp\left(-\frac{|y|^2}{4t_0}\right) v(y,\tau)\, dy$$

$$\geqslant \exp\left(-\frac{|x|^2}{t_0}\right) \int_{|y|<2|x|} v(y,\tau)\, dy$$

$$\geqslant \exp\left(-\frac{|x|^2}{t_0}\right) \int_{|y-x|<|x|} v(y,\tau)\, dy$$

$$\geqslant \exp\left(-\frac{|x|^2}{t_0}\right) \omega_n |x|^n v(x,\tau),$$

where ω_n is the volume of the unit ball in $\mathbb{R}^n$, and, in the last line, we have used the mean value inequality enjoyed by a subharmonic function. Thus,

$$v(x,\tau) \leqslant \frac{M \exp\left(\frac{|x|^2}{t_0}\right)}{\omega_n |x|^n} \leqslant M \exp\left(\frac{|x|^2}{t_0}\right),$$

for $|x| \geqslant \rho$ for some $\rho > 0$. For $|x| \leqslant \rho$, v is bounded by continuity. Also, since $v_t \geqslant 0$, we obtain that $v(x,t) \leqslant v(x,\tau)$ for all $t \in [0,\tau]$. Thus, we conclude that

$$0 \leqslant v(x,t) \leqslant M_1 \exp\left(\frac{|x|^2}{t_0}\right),$$

for some constant $M_1 > 0$ and all $x \in \mathbb{R}^n$, $t \in [0,\tau]$. However, we have already obtained a uniqueness result for solutions of the heat equation having the above-mentioned growth. Since $v(x,0) = 0$ for all x, it follows that $v(x,t) = 0$ for all x and $t \in [0,\tau]$. Since $\tau < T-s$ is arbitrary, this proves that $v(x,t) = 0$ for all $x \in \mathbb{R}^n$ and $t \in [0, T-s)$, which in turn establishes the equality in equation (9.2.4).

Exercise 9.12

Recall the IVP:

$$\left.\begin{array}{l} u_t - \Delta u = f(x),\ x \in \mathbb{R}^n,\ t > 0, \\ u(x,0) = \varphi(x),\ x \in \mathbb{R}^n. \end{array}\right\} \quad (9.2.5)$$

Observe that the function[3] $\varphi \in C^1(\mathbb{R}^n) \cap C^\infty(\mathbb{R}^n \setminus \{0\})$, vanishes for $|x| > 3/4$ and hence, is bounded in $\mathbb{R}^n$. Similarly, the function $f \in C(\mathbb{R}^n) \cap C^\infty(\mathbb{R}^n \setminus \{0\})$, also vanishes for

---

[3] The factor $x_1^2 - x_2^2$ is for convenience and easy computation. In its place, we can use a general quadratic function $q(x) = \sum_{i,j=1}^{n} a_{ij} x_i x_j$, where the constants a_{ij} are not all zero and satisfy the condition $a_{11} + a_{22} + \cdots + a_{nn} = 0$. Thus, q is a harmonic function and satisfies the relation $x \cdot \nabla q(x) = 2q(x)$.

$|x| > 3/4$ and hence, is bounded in $\mathbb{R}^n$. A lengthy calculation shows that the function $u(x,t) = \varphi(x)$, which is independent of t, satisfies equation (9.2.5) for $x \neq 0$. However, the function u is not a classical solution of equation (9.2.5) as, for example, $\lim_{|x|\to 0} u_{x_1 x_1}(x,t) = \infty$.

We now show that IVP does not possess a classical solution in any strip $\{0 < t < T\}$. Suppose, on the contrary, $v(x,t)$ is a classical solution in some strip $\{0 < t < T\}$. Then, the function
$$w(x,t) = u(x,t) - v(x,t) = \varphi(x) - v(x,t)$$
satisfies the homogeneous heat equation and belongs to the function class $C^{2,1}(\mathbb{R}^n \setminus \{0\} \times (0,T))$. Furthermore, since $\varphi \in C^1(\mathbb{R}^n)$, it follows $w \in C^1$ in the strip $\{T/2 \leq t < T\}$. In particular,
$$w(x,t) \in C^{2,1}(\{0 < |x| \leq 1\} \times [T/2, T)) \cap C^1(\{|x| \leq 1\} \times [T/2, T))$$
and $w_t - \Delta w = 0$ for all $(x,t) \in \{0 < |x| < 1\} \times (T/2, T)$.

We now show that $w \in C^{2,1}(\{|x| < 1\} \times (T/2, T))$, which is impossible as it would imply that the function $u(x,t) = \varphi(x)$ belongs to $C^{2,1}(\{|x| < 1\} \times (T/2, T))$, which is not true. To establish the claim, we derive the representation formula for w, similar to the one obtained in Exercise 9.10.

Let $(x,t) \in \{0 < |x| < 1\} \times (T/2, T)$ and $\varepsilon \in (0, t - T/2)$ be arbitrary. We have the following identity
$$\frac{\partial}{\partial \tau}(w(\xi, \tau) K(\xi - x, t - \tau)) + \sum_{i=1}^n \frac{\partial}{\partial \xi_i}\left(w \frac{\partial K}{\partial \xi_i} - K \frac{\partial w}{\partial \xi_i}\right)$$
$$= K(\xi - x, t - \tau)(w_\tau - \Delta_\xi w) - w(K_\tau + \Delta_\xi K) = 0,$$
which is easy to verify. Integrating this equality over the region
$$\{(\xi, \tau) : \delta < |\xi| < 1, \ T/2 < \tau < t - \varepsilon\},$$
where $\delta \in (0, |x|)$ is arbitrary, we obtain, using the divergence theorem,
$$\int_{\delta < |\xi| < 1} w(\xi, t - \varepsilon) K(x - \xi, \varepsilon) \, d\xi$$
$$= \int_{\delta < |\xi| < 1} w(\xi, T/2) K(x - \xi, t - T/2) \, d\xi$$
$$- \int_{T/2}^{t-\varepsilon} \int_{|\xi|=1} \left(w(\xi, \tau) \frac{\partial}{\partial \nu_\xi} K(x - \xi, t - \tau) - K(x - \xi, t - \tau) \frac{\partial w}{\partial \nu_\xi}\right) d\sigma(\xi) \, d\tau$$
$$- \int_{T/2}^{t-\varepsilon} \int_{|\xi|=\delta} \left(w(\xi, \tau) \frac{\partial}{\partial \nu_\xi} K(x - \xi, t - \tau) - K(x - \xi, t - \tau) \frac{\partial w}{\partial \nu_\xi}\right) d\sigma(\xi) \, d\tau$$
$$= I_1 + I_2 + I_3, \text{ say.}$$

We now take the limits as $\varepsilon \to 0$ first and then as $\delta \to 0$ in the above equation. Choose δ_0 arbitrary such that $0 < \delta_0 < \min(1 - |x|, |x| - \delta)$. Then,

$$\int_{\delta < |\xi| < 1} w(\xi, t - \varepsilon) K(x - \xi, \varepsilon) \, d\xi = \int_{|x - \xi| < \delta_0} w(\xi, t - \varepsilon) K(x - \xi, \varepsilon) \, d\xi$$

$$+ \int_{\{\delta < |\xi| < 1\} \cap \{|x - \xi| \geq \delta_0\}} w(\xi, t - \varepsilon) K(x - \xi, \varepsilon) \, d\xi.$$

On the set $\{\delta < |\xi| < 1\} \cap \{|x - \xi| \geq \delta_0\}$, we use the obvious estimate $|w(\xi, t - \varepsilon)| \leq \max |w(\xi, \tau)|$ to obtain

$$\lim_{\varepsilon \to 0} \int_{\{\delta < |\xi| < 1\} \cap \{|x - \xi| \geq \delta_0\}} w(\xi, t - \varepsilon) K(x - \xi, \varepsilon) \, d\xi = 0.$$

On the other hand, changing the variable of integration, we have

$$\int_{|x - \xi| < \delta_0} w(\xi, t - \varepsilon) K(x - \xi, \varepsilon) \, d\xi = \pi^{-n/2} \int_{|\eta| < \frac{\delta_0}{2\varepsilon}} w(x + 2\eta\sqrt{\varepsilon}, t - \varepsilon) e^{-|\eta|^2} \, d\eta.$$

The continuity of w gives $\lim_{\varepsilon \to 0} w(x + 2\eta\sqrt{\varepsilon}, t - \varepsilon) = w(x, t)$. Hence,

$$\lim_{\varepsilon \to 0} \int_{|x - \xi| < \delta_0} w(\xi, t - \varepsilon) K(x - \xi, \varepsilon) \, d\xi = w(x, t) \, \pi^{-n/2} \int_{\mathbb{R}^n} e^{-|\eta|^2} \, d\eta$$

$$= w(x, t).$$

Consequently,

$$\lim_{\delta \to 0} \lim_{\varepsilon \to 0} \int_{\delta < |\zeta| < 1} w(\xi, t - \varepsilon) K(x - \xi, \varepsilon) \, d\xi = w(x, t).$$

Next, it is evident that

$$\lim_{\delta \to 0} I_1 = \int_{|\xi| < 1} w(\xi, T/2) K(x - \xi, t - T/2) \, d\xi$$

and $\lim_{\varepsilon \to 0} I_2 =$

$$-\int_{T/2}^{t} \int_{|\xi| = 1} \left(w(\xi, \tau) \frac{\partial}{\partial \nu_\xi} K(x - \xi, t - \tau) - K(x - \xi, t - \tau) \frac{\partial w}{\partial \nu_\xi} \right) d\sigma(\xi) \, d\tau.$$

Since $w \in C^1(\{|x| \leqslant 1\} \times [T/2, T])$, it follows that $\lim_{\delta \to 0} \lim_{\varepsilon \to 0} I_3 = 0$. Thus, we obtain the representation

$$w(x,t) = \int_{|\xi|<1} w(\xi, T/2) \, K(x-\xi, t-T/2) \, d\xi$$

$$- \int_{T/2}^{T} \int_{|\xi|=1} \left(w(\xi, \tau) \frac{\partial}{\partial \nu} K(x-\xi, t-\tau) - K(x-\xi, t-\tau) \frac{\partial w}{\partial \nu} \right) d\sigma(\xi) \, d\tau,$$

for $0 < |x| < 1$ and $T/2 < t < T$. However, looking at the integral expressions on the right-hand side, it immediately follows that $w \in C^\infty(\{|x| < 1\} \times (T/2, T))$, in particular, $w \in C^{2,1}(\{|x| < 1\} \times (T/2, T))$. This completes the proof. ∎

10
One-Dimensional Wave Equation

10.1 Homogeneous Equation: D'Alembert's Formula

The Cauchy problem or the initial value problem (IVP) for the one-dimensional wave equation is

$$\left.\begin{array}{l} \Box_c u \equiv u_{tt} - c^2 u_{xx} = 0,\ t > 0,\ x \in \mathbb{R}, \\ u(x,0) = u_0(x),\ u_t(x,0) = u_1(x),\ x \in \mathbb{R}. \end{array}\right\} \quad (10.1.1)$$

Here, $c > 0$ is a constant, called the *speed of propagation.*

This equation models many real-world problems: small transversal vibrations of a string, longitudinal vibrations of a rod, electrical oscillations in a wire, torsional oscillations of shafts, oscillations in gases and so on. The wave equation is a prototype of hyperbolic equations. It has two real and distinct characteristics, given by $x \pm ct =$ constant. Using the characteristic variables $\xi = x + ct$ and $\tau = x - ct$, we have $u_{tt} - c^2 u_{xx} = -4c^2 u_{\xi\tau}$, and hence, $u_{\xi\tau} = 0$. Its general solution is therefore given by

$$u(x,t) = F(\xi) + G(\tau) = F(x+ct) + G(x-ct).$$

Using the initial conditions in equation (10.1.1) to determine F and G results in the *D'Alembert's formula* for the solution u:

$$u(x,t) = \frac{1}{2}\left(u_0(x+ct) + u_0(x-ct)\right) + \frac{1}{2c}\int_{x-ct}^{x+ct} u_1(s)\,ds. \quad (10.1.2)$$

More generally, data can be prescribed on a non-characteristic curve: $t = \varphi(x)$, with some conditions on φ, but an explicit formula for the solution may not be found. We state the foregoing in the following theorem:

Theorem 10.1. Suppose the initial conditions satisfy that $u_0 \in C^2(\mathbb{R})$ and $u_1 \in C^1(\mathbb{R})$. Then, the function u given by the D'Alembert's formula (10.1.2) is a C^2 function in $\mathbb{R} \times [0, \infty)$ and satisfies equation (10.1.1) and hence unique.

10.2 Domain of Dependence and Other Concepts

The following observations based on the D'Alembert's formula (10.1.2) may be made. The value of the solution u at (x, t), $t > 0$ depends on the values of the initial data only in the interval $[x - ct, x + ct]$ on the initial line $t = 0$, that is, the x-axis. This is referred to as the *domain of dependence* (of the solution) at (x, t). Similarly, a point y on the initial line can *influence* the value of u for some $t > 0$, only in a line segment $\{(x, t) : x \in [y - ct, y + ct]\}$. This is referred to as the *range of influence* of the point $(y, 0)$. These are illustrated in Figure 10.1.

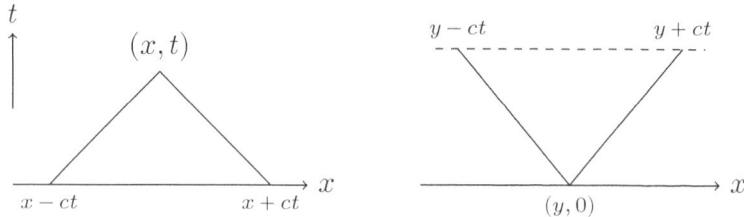

Figure 10.1 | Domain of dependence and range of influence.

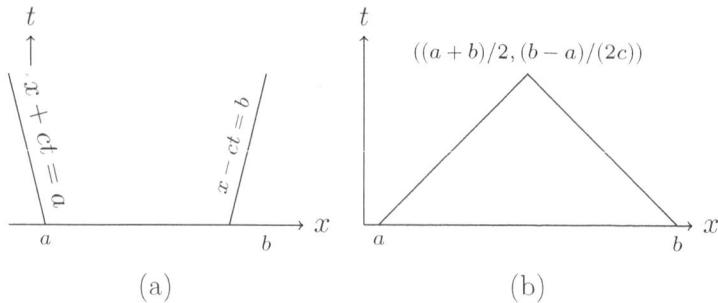

Figure 10.2 | (a) Range of influence and (b) domain of determinacy of $[a, b]$.

The initial values in an interval $[a, b]$ on $t = 0$ will influence the values of the solution u in the region $\{(x, t) : x \in [a - ct, b + ct], t > 0\}$. This is the region between the diverging characteristics $x + ct = a$ and $x - ct = b$ drawn from the points $(a, 0)$ and $(b, 0)$, $a < b$, respectively; see Figure 10.2(a). On the other hand, the converging characteristics emanating from $(a, 0)$ and $(b, 0)$ meet when $t = \dfrac{1}{2c}(b-a)$. Thus, the values of the solution u in the triangle with vertexes at $(a, 0)$, $(b, 0)$ and $\left(\dfrac{a+b}{2}, \dfrac{b-a}{2c}\right)$ are determined by the initial values within the interval $[a, b]$ at $t = 0$; see Figure 10.2(b). This triangular region is referred to as the *domain of determinacy* of the interval $[a, b]$.

10.3 Discontinuous Initial Data: Propagation of Singularities

The D'Alembert's formula gives a C^2 solution of the wave equation (10.1.1), provided that the initial values $u_0 \in C^2(\mathbb{R})$ and $u_1 \in C^1(\mathbb{R})$. What happens if the initial values have discontinuities or do not possess derivatives at some points? For example, suppose $u_0(x) = -1$ if $x < 0$ and $u_0(x) = 1$ if $x > 0$ and $u_1 \equiv 0$. The D'Alembert's formula gives the solution u, for $t > 0$, as $u(x, t) = -1$ if $x < -ct$; $u(x, t) = 1$ if $x > ct$ and $u(x, t) = 0$ if $-ct < x < ct$. This is schematically depicted in Figure 10.3(a).

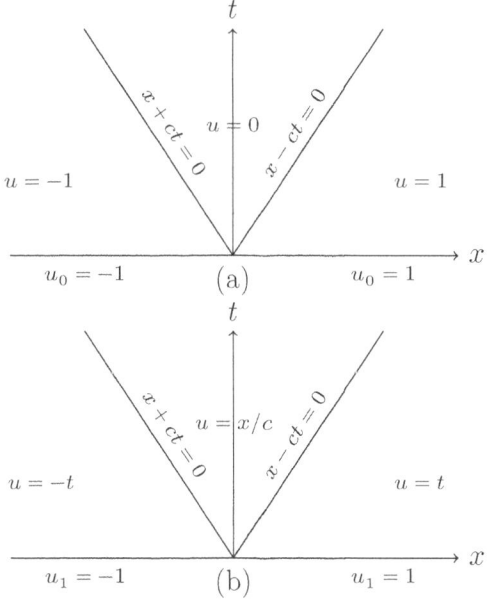

Figure 10.3 | Discontinuous data: for u_0 and u_1.

On the other hand, if we take the initial values as $u_0 \equiv 0$ and $u_1(x) = -1$ if $x < 0$ and $u_1(x) = 1$ if $x > 0$, the D'Alembert's formula gives the solution u, for $t > 0$, as $u(x,t) = -t$ if $x < -ct$; $u(x,t) = t$ if $x > ct$ and $u(x,t) = x/c$ if $-ct < x < ct$. This is schematically depicted in Figure 10.3(b).

In either case, we observe that the solution u fails to be continuous or do not possess continuous derivatives across the characteristics, emanating from the point where the initial values have discontinuities. This phenomenon is called *propagation of singularities* and is an important topic in the theory of general hyperbolic equations. This study was initiated by Sato in the analytic case and Hörmander in the C^∞ case. The subject has renewed interest in the last couple of decades due to important applications in seismic exploration, remote sensing and tomography.

10.4 Inhomogeneous Equation: Duhamel's Principle

Consider the inhomogeneous wave equation

$$u_{tt} - c^2 u_{xx} = f(x,t), \ x \in \mathbb{R}, t > 0. \tag{10.4.1}$$

Fix $s \geqslant 0$ and consider the following initial value problem:

$$\left.\begin{array}{l} v_{tt} - c^2 v_{xx} = 0, \ x \in \mathbb{R}, t > s, \\ v(x,s) = 0, \ v_t(x,s) = f(x,s). \end{array}\right\} \tag{10.4.2}$$

Using the translation invariance of the wave equation and changing the variable t to $t - s$, the solution of equation (10.4.2) is given by the D'Alembert's formula

$$v(x,t;s) = \frac{1}{2c} \int_{x-c(t-s)}^{x+c(t-s)} f(\xi, s) \, d\xi. \tag{10.4.3}$$

The solution u of the inhomogeneous equation (10.4.1) with zero initial conditions is given by

$$u(x,t) = \int_0^t v(x,t;s) \, ds = \frac{1}{2c} \int_0^t \int_{x-c(t-s)}^{x+c(t-s)} f(\xi, s) \, d\xi \, ds, \tag{10.4.4}$$

provided that f is a C^1 function.

10.5 Equation with Two Distinct Speeds

The wave operator $\Box_c$ can be written as $\Box_c = \partial_t^2 - c^2\partial_x^2 = (\partial_t - c\partial_x)(\partial_t + c\partial_x)$. We can consider a more general second-order hyperbolic operator of the form $(\partial_t - c_1\partial_x)(\partial_t - c_2\partial_x)$ with $c_1 \neq c_2$. If we use the characteristic variables: $\xi = x + c_1 t$ and $\tau = x + c_2 t$, the equation $(\partial_t - c_1\partial_x)(\partial_t - c_2\partial_x)u = 0$ reduces to $u_{\xi\tau} = 0$. Again, the general solution is given by $u(x,t) = F(x + c_1 t) + G(x + c_2 t)$. If we prescribe the initial values as in equation (10.1.1), we obtain the following formula for the solution, generalizing the D'Alembert's formula:

$$u(x,t) = \left[\frac{c_1}{c_1 - c_2}u_0(x + c_2 t) - \frac{c_2}{c_1 - c_2}u_0(x + c_1 t)\right]$$
$$+ \frac{1}{c_1 - c_2}\int_{x+c_2 t}^{x+c_1 t} u_1(s)\,ds. \qquad (10.5.1)$$

With $c_2 = -c_1 = -c$, formula (10.5.1) reduces to equation (10.1.2). The above procedure can now be generalized to higher-order equations of the form $Lu = 0$, where $L = (\partial_t - c_1\partial_x)\cdots(\partial_t - c_k\partial_x)$, with $c_1, \ldots, c_k$ are real and distinct. An inductive argument shows that the general solution of $Lu = 0$ is given by

$$u(x,t) = F_1(x + c_1 t) + \cdots + F_k(x + c_k t).$$

The functions F_j can be determined by providing k initial conditions.

What happens when $c_1 = c_2$? Consider the second-order equation $(\partial_t - c\partial_x)^2 u = 0$. Now, there is only one real characteristic. Should we call this equation parabolic or not? The equation $(\partial_t - c\partial_x)^2 u = 0$, though possesses only one real characteristic, should not be classified as *parabolic*, for the following reasons. If we change the independent variables as $\xi = x + ct$ (this is a characteristic variable) and $\tau = x - ct$, then the equation $(\partial_t - c\partial_x)^2 u = 0$ reduces to $\partial_\tau^2 u = 0$, which has no resemblance with the heat equation $u_t = a^2 u_{xx}$. Also, the equation $(\partial_t - c\partial_x)^2 u = 0$ requires two initial conditions, for its solution, whereas the heat equation $u_t = a^2 u_{xx}$ requires only one initial condition. This is the reason for the requirement of the presence of a specific first-order term for the equation to be called parabolic.

The equation $(\partial_t - c\partial_x)^2 u = 0$ is classified as *weakly hyperbolic*. Its general solution is given by (integrate $\partial_\tau^2 u = 0$) $u(x,t) = F(x + ct) + G(x + ct)\tau$. Since $\tau = x - ct = (x + ct) - 2ct$, the general solution can be written as $u(x,t) = F(x + ct) + tG(x + ct)$. If we now prescribe the initial data $u(x,0) = u_0(x)$, $u_t(x,0) = u_1(x)$, then the solution is given by

$$u(x,t) = u_0(x + ct) - ctu_0'(x + ct) + tu_1(x + ct).$$

Notice the *loss of regularity* in the solution. We require u_0 to be a C^3 function and u_1 to be a C^2 function, in order that u is a C^2 solution. Again, by an induction argument, the general solution of the equation $(\partial_t - c\partial_x)^k u = 0$, where $k \geqslant 1$, is given by $u(x,t) = \sum_{j=0}^{k-1} t^j F_j(x+ct)$. The functions F_j are determined by prescribing the initial conditions with appropriate smoothness conditions. Combining the two operators above, we can write down the general solution of the equation $Lu = 0$, where $L = (\partial_t - c_1\partial_x)^{n_1} \cdots (\partial_t - c_k\partial_x)^{n_k}$, with c_i distinct real numbers and n_i positive integers.

This has an immediate application to the first-order systems: $u_t + Au_x = 0$. Here, A is a real $N \times N$ matrix with *real* eigenvalues and u an N- vector. Such a system is called *hyperbolic*. By Jordan decomposition theorem, there is a non-singular real matrix C such that $C^{-1}AC$ is in the Jordan form. If we put $v = C^{-1}u$, then each component of v satisfies the first-order homogeneous or inhomogeneous equation and hence can be explicitly written down. Notice that the eigenvalues of A play the roles of speeds of propagation of the given system.

10.6 Characteristic Parallelogram Property

We return to the wave equation: $(\partial_t^2 - c^2\partial_x^2)u = 0$. Consider a parallelogram $ABCD$ in the upper half-plane $(t \geqslant 0)$, whose sides are the characteristics. It is a straightforward to prove the relation $u(A) + u(C) = u(B) + u(D)$, using the D'Alembert's formula. This relation is termed as *characteristic parallelogram property* (CPP). The converse is also true: Any C^2 function u satisfying the CPP (for every characteristic parallelogram) is a solution of the wave equation. The CPP is quite useful in deriving a formula for the solution of the wave equation in the first quadrant and in a finite interval.

Consider the case of the first quadrant:

$$\left.\begin{array}{l} u_{tt} - c^2 u_{xx} = 0,\ t > 0,\ x > 0, \\ u(x,0) = u_0(x),\ u_t(x,0) = u_1(x),\ x > 0, \\ u(0,t) = h(t),\ t > 0. \end{array}\right\} \quad (10.6.1)$$

In the region $x > ct$ of the quarter plane, the D'Alembert's formula applies (look at the domain of dependence) and the solution is given by

$$u(x,t) = (1/2)\left(u_0(x+ct) + u_0(x-ct)\right) + (1/(2c)) \int_{x-ct}^{x+ct} u_1(s)\,ds,\ x > ct. \quad (10.6.2)$$

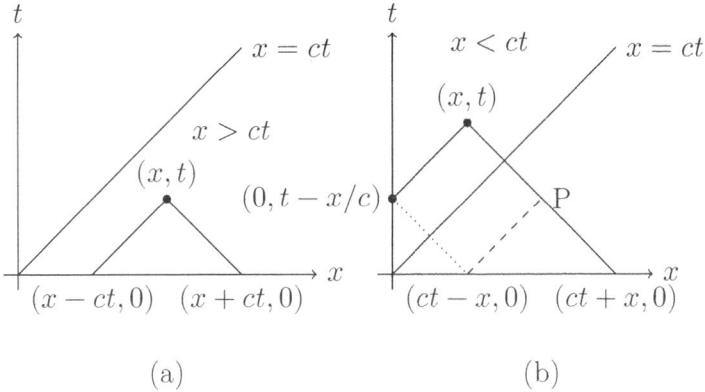

Figure 10.4 | Characteristics of the wave equation in the first quadrant.

In the region $x < ct$, we use CPP to write down the solution. Referring to Figure 10.4(b), the point P is the intersection of the two characteristics indicated there; its co-ordinates are not needed in the computation, but its domain of dependence $[ct - x, ct + x]$ is used. Using D'Alembert's formula, we have

$$u(\text{P}) = (1/2) \left[u_0(ct + x) + u_0(ct - x) \right] + (1/(2c)) \int_{ct-x}^{ct+x} u_1(\xi) \, d\xi. \tag{10.6.3}$$

On the other hand, considering the characteristic parallelogram with vertices P, (x, t), $(0, t - x/c)$ and $(ct - x, 0)$ and using CPP, we get

$$u(\text{P}) + u(0, t - x/c) = u(x, t) + u(ct - x, 0) \quad \text{or}$$
$$u(\text{P}) = -h(t - x/c) + u(x, t) + u_0(ct - x). \tag{10.6.4}$$

Combining equations (10.6.3) and (10.6.4), we get the required solution as

$$u(x, t) = h(t - x/c) + (1/2) \left[u_0(ct + x) - u_0(ct - x) \right]$$
$$+ (1/(2c)) \int_{ct-x}^{ct+x} u_1(\xi) \, d\xi. \tag{10.6.5}$$

Notice a negative sign resulting from the reflected characteristic. The characteristics in this case are schematically depicted in Figure 10.4. Note that the point $(ct - x, 0)$ is the point where the *reflected* characteristic (shown as dotted line) emanating from the point $(0, (ct-x)/c)$ on the positive t-axis meets the positive x-axis. In order that the solution u is C^2 in the first quadrant (including the characteristic line $x = ct$), the initial conditions u_0 and u_1 and the boundary data h need to satisfy certain *compatibility conditions*. These are given by

$$u_0(0) = h(0), \; u_1(0) = h'(0), \; c^2 u_0''(0) = h''(0), \qquad (10.6.6)$$

and it is straightforward to verify now that the solution given by equations (10.6.2) and (10.6.5) is indeed a C^2 function in the first quadrant $\{x > 0, \, t > 0\}$ and a solution of equations (10.6.1).

We next consider the case of finite interval. The initial boundary value problem in this case is given by

$$\left. \begin{array}{l} u_{tt} - c^2 u_{xx} = 0, \; t > 0, \; 0 < x < L, \\ u(x,0) = u_0(x), \; u_t(x,0) = u_1(x), \; 0 < x < L, \\ u(0,t) = h_1(t), \; u(L,t) = h_2(t), \; t > 0. \end{array} \right\} \qquad (10.6.7)$$

The solution u of equation (10.6.7) is determined using the D'Alembert's formula and CPP, in each of the regions in the infinite rectangle $[0, L] \times [0, \infty)$, formed by the characteristics. The details will be provided in an exercise.

10.7 Telegraph Equation

The telegraph equation is given by $u_{tt} - c^2 u_{xx} + \lambda u$, $\lambda \neq 0$. The second-order equation $u_{tt} - c^2 u_{xx} + a u_x + b u_t + \alpha u = 0$, where a, b and α are constants, can be transformed into the telegraph equation, by absorbing the first-order terms $a u_x$ and $b u_t$ in the second-order derivative terms by a simple change of variable. However, this procedure does not apply to the term containing u. A formula will be derived using the solution of the two-dimensional wave equation, using the method of descent.

The general references are [45], [31], [16], [35], [55] and [49].

10.8 Exercises

Exercise 10.1

Let u be a C^2 solution of the wave equation $u_{tt} - c^2 u_{xx} = f(x,t)$ in the upper half-plane $x \in \mathbb{R}, \, t > 0$ satisfying the initial conditions $u(x, 0) = u_0(x)$, $u_t(x,0) = u_1(x)$, $x \in \mathbb{R}$. Derive the formula for the solution u, by integrating the wave equation over the characteristic triangle with vertexes (x,t), $(x-ct,0)$ and $(x+ct,0)$ and using the Green's theorem. This gives another proof of the uniqueness of the solution to the IVP.

Exercise 10.2

If $c_1, \ldots, c_k$ are distinct real numbers, show that the general solution of the one-dimensional equation $\prod_{j=1}^{k}(\partial_t - c_j \partial_x)u = 0$ is given by $u(x,t) = \sum_{j=1}^{k} F_j(x + c_j t)$, where F_j's are smooth functions.

Exercise 10.3

If c is a real number and $k \geqslant 1$ is an integer, show that the general solution of the one-dimensional equation $(\partial_t - c\partial_x)^k u = 0$ is given by $u(x,t) = \sum_{j=1}^{k} t^{j-1} F_j(x + c_j t)$, where F_j's are smooth functions.

Exercise 10.4

Consider the wave equation $u_{tt} - c^2 u_{xx} = 0$ in the first quadrant $x > 0$, $t > 0$, and impose the Neumann boundary condition on the boundary $x = 0$: $u_x(0,t) = g(t)$, $t > 0$ and the initial conditions at $t = 0$, $x > 0$.

$$u(x,0) = u_0(x),\ u_t(x,0) = u_1(x),\ x > 0.$$

Here, $u_x(0,t) = \lim_{x \to 0+} x^{-1}(u(x,t) - u(0,t))$. Thus, u_x denotes the normal derivative to the boundary line $x = 0$, with respect to the *inward* unit normal; $-u_x$ will be the normal derivative with respect to the *outward* unit normal. Assume that u_0 is a C^2 function and u_2 and g are C^1 functions. Derive the formula for the solution using the CPP. Furthermore, discuss the conditions under which the solution would be C^2 in the first quadrant.

Exercise 10.5

Write down the solution of the initial boundary value problem (10.6.7) for the wave equation in a finite interval $[0, L]$.

Exercise 10.6

Write the solution of equation (10.1.1), where $u_1 \equiv 0$ and u_0 is given by

$$u_0(x) = \begin{cases} 0 & \text{if } |x| \geqslant a \\ (b/a)(a - |x|) & \text{if } |x| \leqslant a, \end{cases}$$

where a and b are positive real numbers. Determine the points in the upper half-plane ($t > 0$), where u is discontinuous or not differentiable.

Exercise 10.7

Solve the first-order system $u_t + Au_x = 0$, where A is the 2×2 real matrix $\begin{bmatrix} 3 & -1 \\ 1 & 1 \end{bmatrix}$, with initial data $u_1(x,0) = u_{10}(x)$ and $u_2(x,0) = u_{20}(x)$.

Exercise 10.8

Consider the mixed initial boundary value problem stated in equation (10.6.1), with $h \equiv 0$, that is zero Dirichlet boundary condition. Derive the formula for the solution using a *reflection* in the x-variable.

Exercise 10.9

Consider the mixed boundary value problem for the wave equation $u_{tt} - c^2 u_{xx} = 0$ in the first quadrant $x > 0$, $t > 0$ by imposing the following mixed boundary condition on the boundary $x = 0$: $u_t + \alpha u_x = 0$, $x = 0$, $t > 0$ and the initial conditions at $t = 0$, $x > 0$.

(i) If $\alpha \neq c$, derive a formula for the solution.

(ii) If $\alpha = c$, show that a solution in general does not exist but exists if the initial conditions satisfy some additional conditions. Interpret the boundary condition in this case geometrically.

Exercise 10.10

Using the Fourier method (separation of variables), construct the solution of the following initial boundary value problem:

$$u_{tt} - c^2 u_{xx} = 0, \ 0 < x < L, \ t > 0,$$
$$u(x,0) = \varphi(x), \ u_t(x,0) = 0, \ 0 \leqslant x \leqslant L,$$
$$u(0,t) = u(L,t) = 0, \ t > 0,$$

where L is a positive real number, and φ is given by

$$\varphi(x) = \frac{u_0}{L}(L - |L - 2x|), \ x \in [0, L],$$

with u_0 being a constant. Physically, the problem describes the vibrations of a string tied at the ends $x = 0$ and $x = L$ and, initially, plucked in the middle to u_0.

10.9 Solutions

Exercise 10.1

Let (x,t), $t > 0$ be an arbitrary point and T be the (interior) of the triangle with sides S_1, S_2 and S_3, where S_1 is the line segment joining the points (x,t) and $(x-ct,0)$, S_2 is the line segment on the x-axis joining the points $(x-ct,0)$ and $(x+ct,0)$ and S_3 is the line segment joining the points $(x+ct,0)$ and (x,t) (see Figure 10.5).

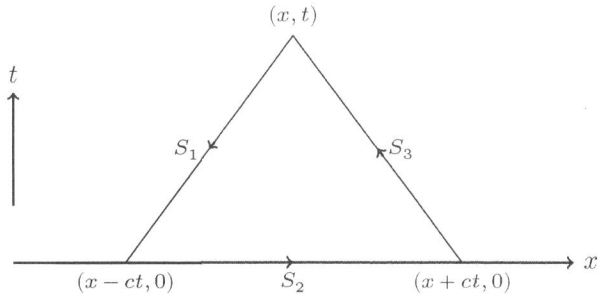

Figure 10.5 | Domain of integration in Exercise 10.1.

Integrating the equation $u_{\tau\tau} - c^2 u_{\xi\xi} = f(\xi,\tau)$ over T, we obtain

$$\int_T f(\xi,\tau)\,d\xi d\tau = \int_T (u_{\tau\tau} - c^2 u_{\xi\xi})\,d\xi d\tau = \int_T ((u_\tau)_\tau - c^2(u_\xi)_\xi)\,d\xi d\tau.$$

Applying Green's formula to the last integral, we get

$$\int_T ((u_\tau)_\tau - c^2(u_\xi)_\xi)\,d\xi d\tau = -\left(\int_{S_1} + \int_{S_2} + \int_{S_3}\right)(u_\tau\,d\xi + c^2 u_\xi\,d\tau).$$

On S_1, $\xi = x - c(t-\tau)$, $0 \leqslant \tau \leqslant t$ and therefore $d\xi = c\,d\tau$. Thus, if we put $g(\tau) = cu(\xi,\tau) = cu(x-c(t-\tau),\tau)$, then

$$\frac{dg}{d\tau} = cu_\tau(x - c(t-\tau),\tau) + c^2 u_\xi(x - c(t-\tau),\tau).$$

Therefore (the orientation of S_1 is such that a point traverses from the point (x,t) to the point $(x-ct, 0)$),

$$-\int_{S_1} (u_\tau \, d\xi + c^2 u_\xi \, d\tau) = -\int_t^0 [cu_\tau(x - c(t-\tau), \tau) + c^2 u_\xi(x - c(t-\tau), \tau)] \, d\tau$$

$$= \int_0^t \frac{dg}{d\tau} \, d\tau = g(t) - g(0) = cu(x,t) - cu(x-ct, 0).$$

On S_2, $\tau = 0$, and therefore there is no contribution from the integral with respect to τ and ξ varies in the interval $[x-ct, x+ct]$. Hence,

$$-\int_{S_2} (u_\tau \, d\xi + c^2 u_\xi \, d\tau) = -\int_{x-ct}^{x+ct} u_\tau(\xi, 0) \, d\xi.$$

The computation of the line integral over S_3 is similar to that of integrating over S_1, and we obtain

$$-\int_{S_3} (u_\tau \, d\xi + c^2 u_\xi \, d\tau) = cu(x,t) - cu(x+ct, 0).$$

Note that the orientation of S_3 is such that the point moves from $(x+ct, 0)$ to (x,t). Taking into account that $u(x, 0) = u_0(x)$ and $u_t(x, 0) = u_1(x)$ for $x \in \mathbb{R}$, we finally obtain the formula

$$u(x,t) = \frac{1}{2}(u_0(x+ct) + u_0(x-ct)) + \frac{1}{2c} \int_{x-ct}^{x+ct} u_1(\xi) \, d\xi$$

$$+ \frac{1}{2c} \int_0^t \int_{x-c\tau}^{x+c\tau} f(\xi, \tau) \, d\xi d\tau.$$

The term in the second line on the right-hand side of the above expression is nothing but $\iint_T f(\xi, \tau) \, d\xi d\tau$.

Exercise 10.2

The assertion is proved by induction on k. For $k=1$, the statement follows from the method of characteristics applied to the first-order transport equation. Suppose the assertion holds for $k-1$ for some $k > 1$. First, recall that the general solution of the equation $u_t - cu_x = f(x, t)$, where f is a C^1 function, is given by $u(x,t) = u_0(x+ct) + \int_0^t f(x + c(t-s), s) \, ds$, with u_0 being an arbitrary C^1 function. In particular, if $f(x,t) = F(x + \tilde{c}t)$ and $\tilde{c} \neq c$, then the general solution of $u_t - cu_x = F(x + \tilde{c}t)$ can be written in the form

$$u(x,t) = F_1(x + ct) + F_2(x + \tilde{c}t), \qquad (10.8.1)$$

for appropriate C^1 functions F_1 and F_2.

Let $\tilde{L} = \prod_{j=1}^{k-1}(\partial_t - c_j\partial_x)$ and $L = \prod_{j=1}^{k}(\partial_t - c_j\partial_x)$. If $Lu = 0$, put $v = (\partial_t - c_k\partial_x)u$. Then, we have $\tilde{L}v = 0$, and therefore, by induction hypothesis, v can be written as

$$v(x,t) = \sum_{j=1}^{k-1} \tilde{F}_j(x + c_j t),$$

for appropriate C^1 functions $\tilde{F}_j$. Since $(\partial_t - c_k\partial_x)u = v$ and c_j's are distinct, the induction step now follows from the observation made in equation (10.8.1).

Exercise 10.3

We again use an induction argument. Using a similar argument as in the previous exercise, we note that the general solution of the equation $u_t - cu_x = t^j f(x + ct)$, $j \geq 0$, is given by $u(x,t) = u_0(x + ct) + \dfrac{t^{j+1}}{j+1} f(x + ct)$ with an arbitrary C^1 function u_0.

When $k = 1$, the assertion follows trivially. Suppose the assertion holds for $k - 1$ for $k > 1$, and let u be a solution of $(\partial_t - c\partial_x)^k u = 0$. Put $v = (\partial_t - c\partial_x)u$. Then, $(\partial_t - c\partial_x)^{k-1}v = 0$, and thus, by induction hypothesis, we have

$$v(x,t) = \sum_{j=1}^{k-1} t^{j-1} g_j(x + ct),$$

for arbitrary C^1 functions g_j. Since $u_t - cu_x = v$, we see that the assertion is true for k, using the observation made earlier.

Another argument can be provided as follows. Introduce the characteristic variables $\xi = x + ct$, $\eta = x - ct$. Then, $(\partial_t - c\partial_x)^k u = 0$ implies that $(\partial_\eta)^k u = 0$. This is an ordinary differential equation (ODE) with ξ playing the role of a parameter. The general solution of this ODE is given by $u = \sum_{j=1}^{k} \eta^{j-1} g_j(\xi)$, a polynomial in η of degree at most $k - 1$, with coefficients, which can be functions of ξ alone. Since we can write $\eta = x - ct = x + ct - 2ct$, and similarly, for the powers of η, we see that u can be written in the form

$$u(x,t) = \sum_{j=1}^{k} t^{j-1} F_j(x + ct),$$

320 | Notes, Problems and Solutions in Differential Equations

with arbitrary C^1 functions F_j, as claimed.

Remark: Combining the solutions of Exercises 10.2 and 10.3, we can write down the general solution of the equation $Lu = 0$, where

$$L = (\partial_t - c_1 \partial_x)^{n_1} \cdots (\partial_t - c_k \partial_x)^{n_k},$$

with real and distinct constants $c_1, \ldots, c_k$ and positive integers $n_1, \ldots, n_k$.

Exercise 10.4

The case of the Dirichlet boundary condition was considered in Section 10.6. In the region $x > ct$, we obtain the solution using the D'Alembert's formula (10.1.2). For the region $x < ct$, we attempt to find the Dirichlet boundary condition from the given Neumann boundary condition so that we can use equation (10.6.5) to obtain the expression for the solution. This process is termed as the *Neumann condition to Dirichlet condition map*. There is similarly the notion of the *Dirichlet condition to Neumann condition map*. These concepts are important and arise in the analysis of partial deferential equation (PDE). See, for instance, Ref. [18].

Assume that $u(0, t) = h(t)$, $t > 0$. We now find relation between g and h. Using equation (10.6.5), we have

$$u(x, t) = h(t - x/c) + (1/2)\left(u_0(ct + x) - u_0(ct - x)\right)$$

$$+ (1/(2c)) \int_{ct-x}^{ct+x} u_1(\xi)\, d\xi,$$

for $x < ct$. Using this expression, we now compute the limit $\lim_{x \to 0+} x^{-1}(u(x, t) - u(0, t))$, for $t > 0$. This will then relate the Dirichlet data h to the Neumann data g. Assume $0 < x < ct$. Then,

$$u(x, t) - u(0, t) = (h(t - x/c) - h(t)) + (1/2)\left(u_0(ct + x) - u_0(ct - x)\right)$$

$$+ (1/(2c)) \int_{ct-x}^{ct+x} u_1(\xi)\, d\xi.$$

Dividing throughout by x and taking the limit as $x \to 0$, we obtain

$$g(t) = -(1/c)h'(t)) + u_0'(ct) + (1/c)u_1(ct).$$

An integration now gives

$$h(t) = -c \int_0^t g(s)\, ds + u_0(ct) + (1/c) \int_0^{ct} u_1(s)\, ds + k,$$

where k is an arbitrary constant. Substituting this expression into equation (10.6.5), we obtain the following expression for the solution of the Neumann problem:

$$u(x,t) = -c \int_0^{t-x/c} g(s)\, ds + (1/2)\left(u_0(ct+x) + u_0(ct-x)\right)$$

$$+ (1/(2c))\left(\int_0^{ct+x} + \int_0^{ct-x}\right) u_1(\xi)\, d\xi + k, \tag{10.8.2}$$

for $x < ct$.

In the region $x > ct$, the D'Alembert's formula is applicable:

$$u(x,t) = (1/2)\left(u_0(x+ct) + u_0(x-ct)\right) + (1/(2c))\int_{x-ct}^{x+ct} u_1(\xi)\, d\xi. \tag{10.8.3}$$

Thus, we obtain a C^2 solution u in the regions $x < ct$ and $x > ct$. Let us now dwell into the conditions on the data so that u would be a C^2 function throughout the first quadrant. Looking at the formulas in equations (10.8.2) and (10.8.3), we conclude that $k = 0$ for u to be continuous on $x = ct$. Similarly, by taking the first and second derivatives (with respect to t will do) of the expressions in these formulas, we infer that u would be a C^2 function in the first quadrant if $g(0) = u_0'(0)$ and $g'(0) = u_1'(0)$. These are the compatibility conditions in the present case. Note that all the compatibility conditions are required only at the *interface point* $(0,0)$. Thus, the required C^2 solution of the initial Neumann boundary value problem is given by the formulas in equations (10.8.2) (take $k = 0$) and (10.8.3), provided the above compatibility conditions are satisfied.

Exercise 10.5

This is algebraically much more involved compared to the quarter plane problem. There would be an infinite number of expressions of the solution corresponding to an infinite number of regions of the infinite rectangle $[0, L] \times [0, \infty)$, formed by the characteristics. These are schematically depicted in Figure 10.6. We consider the four regions depicted in Figure 10.6(b), in which we would be writing the expressions for the solution. For other regions, we can use an inductive argument, but writing a general expression is very difficult, if not impossible! The expression for the solution at points on the characteristic lines is not written. These can be easily written down by the requirement that the solution is a C^2 function, which requires certain compatibility conditions that need to be satisfied by u_0, u_1, h_1 and h_2, similar to the ones in equation (10.6.6) and are given by

$$u_0(0) = h_1(0),\ u_1(0) = h_1'(0),\ c^2 u_0''(0) = h_1''(0),$$
$$u_0(L) = h_2(0),\ u_1(L) = h_2'(0),\ c^2 u_0''(L) = h_2''(0).$$

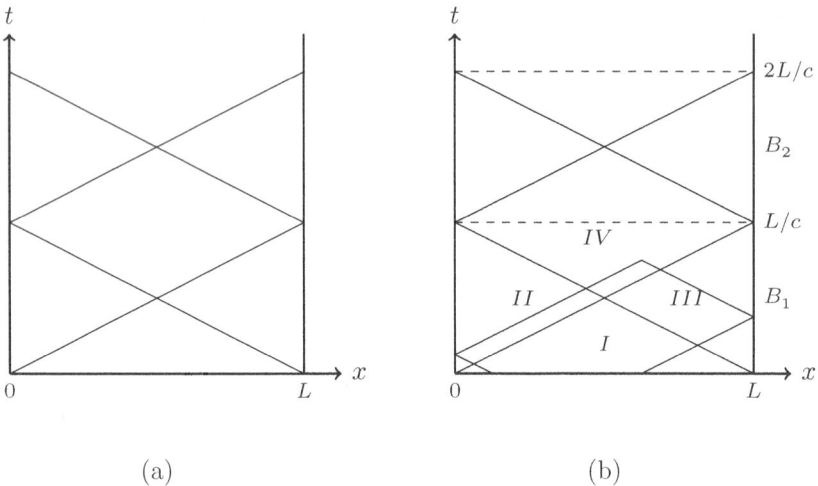

(a) (b)

Figure 10.6 | Characteristics of the wave equation in a finite interval.

Consider a point (x, t) below the characteristic $x = ct$, which meets the boundary line $x = L$ at $(L, L/c)$. Then, $x > ct$. If, in addition, $x + ct < L$, then (x, t) lies in the region (I) and the D'Alembert's formula (10.1.2) applies. If, on the other hand, $x + ct > L$, then (x, t) lies in the region (III) and one backward characteristic through it meets the boundary line $x = L$ at $(L, (x + ct - L)/c)$ and its reflected characteristic meets the line $t = 0$ at $(2L - (x + ct), 0)$. The situation is therefore similar to the quarter plane case, and the required formula for the solution is given by

$$u(x,t) = h_2\left((x+ct-L)/c\right) + (1/2)\left[u_0(x-ct) - u_0(2L-x-ct)\right]$$
$$+ (1/(2c)) \int_{x-ct}^{2L-x-ct} u_1(\xi)\, d\xi.$$

Observe the negative sign coming from the reflected characteristic, as in the case of the quarter plane problem. If the point (x, t) satisfies the relations $x < ct$ and $x + ct < L$, then it belongs to the region (II) and the expression for the solution is the same as in the quarter plane case and is given by equation (10.6.5) with h replaced by h_1.

Next, consider that the point (x, t) satisfies the relations $x < ct$ and $x + ct > L$. This point belongs to the region (IV). The backward characteristics now meet both the boundary lines $x = 0$, $x = L$ at $(0, t - x/c)$ and $(L, (x + ct - L)/c)$, respectively. The corresponding reflected characteristics meet the line $t = 0$ at $(ct - x, 0)$ and $(2L - (x + ct), 0)$. Assume, in addition, that $ct < L$. This implies that $ct - x < 2L - (x + ct) < L$. We now draw a suitable characteristic through one of the points $(ct - x, 0)$ and $(2L - (x + ct), 0)$ to obtain a characteristic parallelogram. If $ct - x = 2L - (x + ct)$, then we already have a characteristic parallelogram (see Figure 10.6(a)). The characteristic through $(ct - x, 0)$ meets the backward characteristic through (x, t) at the point $(ct, x/c)$ which lies in the region (III), and we have the characteristic parallelogram with vertices at (x, t), $(0, t - x/c)$, $(ct - x, 0)$ and $(ct, x/c)$. Now, apply the solution formula for the region (III), use CPP and do some computations to get the following formula:

$$u(x, t) = h_1(t - x/c) + h_2((x + ct - L)/c)$$
$$- (1/2) [u_0(ct - x) + u_0(2L - x - ct)] + (1/(2c)) \int_{ct-x}^{2L-x-ct} u_1(\xi) \, d\xi.$$

Observe the negative signs coming from the two reflected characteristics and the terms appearing from both the boundary conditions. With the above-mentioned compatibility conditions, it is straightforward to verify that the solution u given by the these different formulas is actually a C^2 function in the region $\{(x, t) \in (0, L) \times (0, L/c)\}$.

We now proceed to find solution formulas for all (x, t), using an inductive argument. For this purpose, put $B_n = \{(x, t) \in (0, L) \times (((n-1)L)/c, (nL)/c)\}$, $n = 1, 2, \ldots$ (see Figure 10.6(b)). We have found solution formulas in the block B_1. Assume that we have solution formulas in the block B_n for some $n \geqslant 1$. We show how to obtain solution formulas in the block B_{n+1}.

Let $(x, t) \in B_{n+1}$. The backward characteristics through (x, t) meet the boundary lines $x = 0$, $x = L$ at $(0, t - x/c)$ and $(L, (x + ct - L)/c)$, respectively. The reflected characteristics through these latter points intersect at $(L - x, (ct - L)/c)$, and we have a characteristic parallelogram with vertices at (x, t), $(0, t - x/c)$, $(L - x, (ct - L)/c))$ and $(L, (x + ct - L)/c)$. Now, apply CPP to get

$$u(x, t) = -u(L - x, (ct - L)/c)) + u(0, t - x/c) + u(L, (x + ct - L)/c)$$
$$= h_1(t - x/c) + h_2((x + ct - L)/c) - u(L - x, (ct - L)/c)).$$

However, the point $(L - x, (ct - L)/c)) \in B_n$, and we have a solution formula there. Thus, we also have solution formula for the block B_{n+1}, and this completes the induction argument. Since we have constructed solution formulas in B_1, we conclude that we have solution formulas at all points in $(0, L) \times (0, \infty)$. In passing, we also observe that even the geometry of straight lines can be quite complex!

Exercise 10.6

Observe that u_0 is continuous in $\mathbb{R}$, but not differentiable at the points $x = 0$, $\pm a$ (see Figure 10.7). We therefore expect the solution to be not smooth across various characteristics emanating from these points $x = 0$, $\pm a$. The following schematic diagram showing the various characteristics through the points $(0,0)$ and $(\pm a, 0)$ will help in determining the solution. These characteristics divide the upper half-plane into 10 regions, in each of which u has a different expression. For a point (x,t), $t > 0$, first we need to identify the (open) regions it belongs and then determine the points $(x - ct, 0)$ and $(x + ct, 0)$ on the x-axis, to apply the D'Alembert's formula. Except the regions (5) and (6), we observe that the regions are pairwise symmetric about the line $x = 0$. For example, regions (1) and (10) are symmetric. After a careful look at the geometry of the characteristics (see Figure 10.8), we obtain the solution as follows:

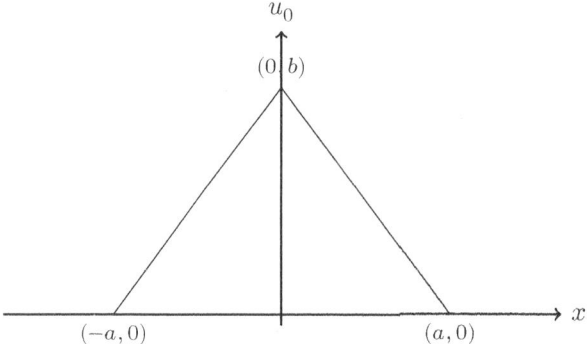

Figure 10.7 | The function u_0 in Exercise 10.6.

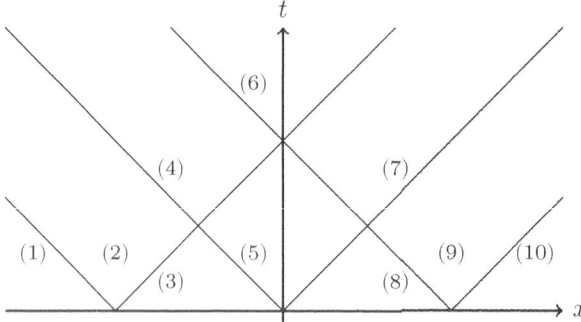

Figure 10.8 | The characteristics for Exercise 10.6.

$$u(x,t) = \begin{cases} 0 & \text{if } |x| > a+ct \text{ (regions (1) and (10))}, \\ \dfrac{b}{2a}(a+ct-|x|) & \text{if } \max\{ct, a-ct\} < |x| < ct \\ & \text{(regions (2) and (9))}, \\ \dfrac{b}{a}(a-|x|) & \text{if } ct < |x| < a-ct \\ & \text{(regions (3) and (8))}, \\ \dfrac{b}{2a}(a-ct+|x|) & \text{if } \max\{ct-a, a-ct\} < |x| < ct \\ & \text{(regions (4) and (7))}, \\ \dfrac{b}{a}(a-ct) & \text{if } \max\{-ct, ct-a\} < x < \min\{ct, a-ct\} \\ & \text{(region (5))}, \\ 0 & \text{if } a-ct < x < ct-a \text{ (region (6))}. \end{cases}$$

Using the above expressions, we can find out the smoothness of the solution across various characteristics. ∎

Exercise 10.7

The matrix A has only one eigenvalue 2, whose algebraic multiplicity is 2 and the geometric multiplicity is 1. Let $C = \begin{bmatrix} 1 & 1 \\ 1 & 0 \end{bmatrix}$. Then,

$$C^{-1} = \begin{bmatrix} 0 & 1 \\ 1 & -1 \end{bmatrix}, \text{ and } C^{-1}AC = \begin{bmatrix} 2 & 1 \\ 0 & 2 \end{bmatrix}.$$

If we put $v = C^{-1}u$, then $v_1 = u_2$, $v_2 = u_1 - u_2$ and the given system reduces to $v_{1t} + 2v_{1x} + v_{2x} = 0$ and $v_{2t} + 2v_{2x} = 0$. Therefore, $v_2(x,t) = v_{20}(x-2t) = u_{10}(x-2t) - u_{20}(x-2t)$. Substituting this expression into the inhomogeneous equation satisfied by v_1 and solving the same, we obtain

$$u_2(x,t) = v_1(x,t) = v_{10}(x-2t) - t[u'_{10}(x-2t) - u'_{20}(x-2t)].$$

Reverting to the original variables, the solution to the given system is given by

$$u_1(x,t) = u_{10}(x-2t) - t[u'_{10}(x-2t) - u'_{20}(x-2t)]$$
$$u_2(x,t) = u_{20}(x-2t) - t[u'_{10}(x-2t) - u'_{20}(x-2t)].$$

Note that we require both u_{10} and u_{20} to be C^2 functions to obtain a C^1 solution, exhibiting loss of regularity. ∎

Exercise 10.8

Assume $h \equiv 0$ in equation (10.6.1). Extend the initial data u_0, u_1 to $\mathbb{R}$ as odd functions: $u_j(-x) = -u_j(x)$, $x > 0$ for $j = 0, 1$. We also assume that the extended functions denoted by $\tilde{u}_0$ and $\tilde{u}_1$ are, respectively, C^2 and C^1 functions. By D'Alembert's formula, we have

$$\tilde{u}(x,t) = \frac{1}{2}\left(\tilde{u}_0(x+ct) + \tilde{u}_0(x-ct)\right) + \frac{1}{2c}\int_{x-ct}^{x+ct}\tilde{u}_1(\xi)\,d\xi,$$

denoting the solution of the wave equation with initial data $\tilde{u}_0$ and $\tilde{u}_1$ in $\mathbb{R}$. As these functions are odd, it follows that $\tilde{u}(0,t) = 0$ for all $t > 0$; $\tilde{u}_0(x-ct) = u_0(ct-x)$ and $\int_{x-ct}^{x+ct}\tilde{u}_1(\xi)\,d\xi = \int_{ct-x}^{x+ct} u_1(\xi)\,d\xi$ for $x < ct$. Thus, if we restrict $\tilde{u}$ to the first quadrant and denote it by u, we see that u is the required solution, as it is now straightforward to see that this formula is the same as the one given by equation (10.6.5), with $h \equiv 0$ in the region $x < ct$. In the region $x > ct$, the above formula is the same as the D'Alembert's formula.

Exercise 10.9

We follow the procedure used in Exercise 10.4 and attempt to find that Dirichlet data from the given boundary condition. Assume that $u(0,t) = h(t)$ for $t > 0$. Using equation (10.6.5), we have

$$u(x,t) = h(t - x/c) + (1/2)\left(u_0(ct+x) - u_0(ct-x)\right)$$
$$+ (1/(2c))\int_{ct-x}^{ct+x} u_1(\xi)\,d\xi,$$

for $x < ct$. Using this expression, we now compute $u_t + \alpha u_x$ evaluated on the line $x = 0$. We have, for $x < ct$,

$$u_t(x,t) = h'(t - x/c) + (c/2)(u_0'(ct+x) - u_0'(ct-x))$$
$$+ (1/2)(u_1(ct+x) - u_1(ct-x)),$$

$$u_x(x,t) = (-1/c)h'(t - x/c) + (1/2)(u_0'(ct+x) + u_0'(ct-x))$$
$$+ (1/(2c))(u_1(ct+x) + u_1(ct-x)).$$

Therefore, using the given boundary condition, we obtain

$$0 = (u_t + \alpha u_x)\Big|_{x=0} = (1 - \alpha/c)h'(t) + \alpha u_0'(ct) + (\alpha/c)u_1(ct).$$

This immediately gives

$$(\alpha - c)h'(t) = \alpha c u_0'(ct) + \alpha u_1(ct). \qquad (10.8.4)$$

If $\alpha \neq c$, we then obtain

$$h(t) = \frac{\alpha}{\alpha - c} u_0(ct) + \frac{\alpha}{c} \int_0^{ct} u_1(\xi)\, d\xi + k,$$

where k is a constant of integration. Substituting this into the above expression of the solution, we obtain

$$u(x,t) = \frac{u_0(ct+x)}{2} + \frac{1}{2c} \int_0^{ct+x} u_1(\xi)\, d\xi$$
$$+ \frac{\alpha + c}{\alpha - c} \left[\frac{u_0(ct-x)}{2} + \frac{1}{2c} \int_0^{ct-x} u_1(\xi)\, d\xi \right] + k,$$

for $x < ct$. In the region $x > ct$, the solution is given by the D'Alembert's formula. The compatibility conditions for the solution to be a C^2 function in the first quadrant can now be derived in a straightforward manner. These conditions now include α and the constant of integration k as well.

On the other hand, if $\alpha = c$, equation (10.8.4) gives no information about h but shows that u_0 and u_1 satisfy the relation $c u_0'(t) + u_1(t) = 0$ for $t > 0$. Hence, the initial conditions are no longer arbitrary. This has a curious effect on the solution. In fact, a solution is given by

$$u(x,t) = \begin{cases} h(t - x/c) & \text{if } x < ct, \\ u_0(x - ct) & \text{if } x > ct, \end{cases}$$

for arbitrary C^2 function h. For u to be a C^2 function in the first quadrant, it suffices to assume the compatibility conditions $u_0(0) = h(0)$, $c u_0'(0) = -h'(0)$ and $c^2 u_0''(0) = h''(0)$. We do not have uniqueness of the solution either.

If $\alpha = c$, then $u_t + \alpha u_x$ is the directional derivative of u along a family of the characteristics, and we have seen above its effect in obtaining a solution.

Exercise 10.10

Let $u(x,t) = X(x)T(t)$. Then, $XT'' - c^2 X''T = 0$. Therefore, $\dfrac{X''}{X} = \dfrac{T''}{c^2 T} = \lambda$, a constant. Thus, $X'' - \lambda X = 0$ and $T'' - \lambda c^2 T = 0$. Furthermore, the given initial and boundary conditions imply that

$$X(0) = X(L) = 0 \text{ and } T(0) = 1, T'(0) = 0,$$

as we seek non-trivial solutions X and T. It is easily checked that if $\lambda \geq 0$, any solution X satisfying the boundary conditions is the trivial solution, that is $X \equiv 0$. When $\lambda < 0$, the general solution X is given by $X(x) = c_1 \sin\sqrt{-\lambda}x + c_2 \cos\sqrt{-\lambda}x$. The boundary condition $X(0) = 0$ gives $c_2 = 0$. The boundary condition $X(L) = 0$ gives, for X to be non-trivial, $\lambda = -\dfrac{n^2\pi^2}{L^2}$ for $n = 1, 2, \dots$. Let $X_n(x) = c_n \sin\left(\dfrac{n\pi x}{L}\right)$ for $n = 1, 2, \dots$ for some constants c_n. We then see that the corresponding functions T are given by $T_n(t) = \cos\left(\dfrac{cn\pi t}{L}\right)$. By linearity, therefore,

$$u(x,t) = \sum_{n=1}^{\infty} c_n \sin\left(\frac{n\pi x}{L}\right) \cos\left(\frac{cn\pi t}{L}\right).$$

We need to determine the constants c_n. The initial condition at $t = 0$ gives that

$$\varphi(x) = \sum_{n=1}^{\infty} c_n \sin\left(\frac{n\pi x}{L}\right).$$

Thus, c_n are the Fourier coefficients of the function φ when it is expressed as a Fourier sine series. To do so, we extend φ to $\mathbb{R}$ as an odd periodic function of period $2L$. We then obtain that

$$c_n = \frac{2}{L} \int_0^L \varphi(x) \sin\left(\frac{n\pi x}{L}\right) dx.$$

Using the given expression for φ, we find that

$$c_n = \begin{cases} 0 & \text{if } n = 2k, \\ \dfrac{4u_0}{\pi^2} \dfrac{(-1)^k}{(2k+1)^2} & \text{if } n = 2k+1, \end{cases}$$

for $k = 0, 1, 2, \dots$. Thus,

$$u(x,t) = \frac{4u_0}{\pi^2} \sum_{k=0}^{\infty} \frac{(-1)^k}{(2k+1)^2} \sin\left(\frac{(2k+1)\pi x}{L}\right) \cos\left(\frac{(2k+1)c\pi t}{L}\right).$$

Note that the justification of term-by-term differentiation for the second derivatives is not possible (in the resulting series, we lose the term $(2k+1)^2$ in the denominator), and thus we cannot claim that u is a C^2 solution. In fact, if we require that u is a C^2 solution, then sufficient conditions on the initial conditions $u\big|_{t=0} = \phi$ and $u_t\big|_{t=0} = \psi$ are that $\phi \in C^3$ and $\psi \in C^2$. In the present case, ϕ is not differentiable at $x = L/2$, and its second derivative is a Dirac function.

However, if we truncate the above trigonometric series to any finite number of terms, we do obtain a solution of the wave equation, satisfying the initial conditions approximately. Though not in explicit form, we do obtain a solution with minimum requirements on the initial conditions, by the method of characteristics using the CPP. See Exercise 10.5.

11
Wave Equation in Higher Dimensions

11.1 Introduction

The Cauchy problem or initial value problem (IVP) for the homogeneous wave equation in the free space $\mathbb{R}^n$ is given by

$$\Box_c u \equiv u_{tt} - c^2 \Delta u = 0,\ x \in \mathbb{R}^n, t > 0, \tag{11.1.1}$$
$$u(x,0) = u_0(x),\ u_t(x,0) = u_1(x),\ x \in \mathbb{R}^n. \tag{11.1.2}$$

Here, $n \geqslant 2$ is an integer, the *(spatial) dimension*; $c > 0$ is a constant, the *speed of propagation* and u_0 and u_1 are given smooth functions, the initial values.

Spherical Mean Function: Given a C^2 function h defined on $\mathbb{R}^n$, define its *spherical mean function*, denoted by M_h, by

$$M_h(x,r) = \frac{1}{\sigma_n r^{n-1}} \int_{|x-y|=r} h(y)\, d\sigma(y), \tag{11.1.3}$$

for $x \in \mathbb{R}^n$ and $r > 0$. The integration is over the sphere of radius r, centred at x and $\sigma_n r^{n-1}$ is the surface measure of this sphere with $\sigma_n = 2\pi^{n/2}/\Gamma(n/2)$ denoting the surface measure of the unit sphere in $\mathbb{R}^n$; Γ is the Euler gamma function. By a change of variable, equation (11.1.3) can be written as

$$M_h(x,r) = \frac{1}{\sigma_n} \int_{|\xi|=1} h(x+r\xi)\, d\sigma(\xi). \tag{11.1.4}$$

The form of equation (11.1.4) enables us to define M_h for all real r, and it is readily seen that $M_h(x,-r) = M_h(x,r)$, that is, M_h is an even function of r. This property is used repeatedly in the computations below.

329

A computation using the divergence theorem yields the *Darboux equation*:

$$\left(\frac{\partial^2}{\partial r^2} + \frac{n-1}{r}\frac{\partial}{\partial r}\right) M_h(x,r) = \Delta_x M_h(x,r). \tag{11.1.5}$$

The notation Δ_x in the above expressions means the Laplacian taken with respect to the x variables. Note that in the above equation, x is a parameter, and the equation (11.1.5) is a second-order ordinary differential equation (ODE) in the variable r.

Using the Darboux equation and some manipulation gives us the solution of IVP (11.1.1) and (11.1.2), for $n = 3$:

$$u(x,t) = tM_{u_1}(x,ct) + \frac{\partial}{\partial t}[tM_{u_0}(x,ct)]$$

$$= \frac{1}{4\pi c^2 t}\int_{|y-x|=ct} u_1(y)\,d\sigma(y) + \frac{\partial}{\partial t}\left(\frac{1}{4\pi c^2 t}\int_{|y-x|=ct} u_0(y)\,d\sigma(y)\right). \tag{11.1.6}$$

The representation (11.1.6) is known as the *Kirchhoff's formula*. By carrying out the t differentiation, we can also write the Kirchoff's formula as follows:

$$\frac{\partial}{\partial t}\left[tM_{u_0}(x,ct)\right] = M_{u_0}(x,ct) + t\frac{\partial}{\partial t}M_{u_0}(x,ct)$$

and

$$\frac{\partial}{\partial t}M_{u_0}(x,ct) = \frac{1}{4\pi c^2 t^2}\int_{|y-x|=ct} \nabla u_0(y)\cdot(y-x)\,d\sigma(y).$$

Thus, the Kirchoff's formula (11.1.6) is rewritten as

$$u(x,t) = \frac{1}{4\pi c^2 t^2}\int_{|y-x|=ct} [tu_1(y) + u_0(y) + \nabla u_0(y)\cdot(y-x)]\,d\sigma(y). \tag{11.1.7}$$

The above formula brings out the essential features of the solution in the case $n = 3$. Thus, any C^2 solution of the Cauchy problems (11.1.1) and (11.1.2) is given by equation (11.1.6) and hence unique.

For $n \geqslant 3$ odd, we now write down a formula for the solution of the homogeneous wave equation, similar to the Kirchhoff's formula for $n = 3$; the formula for the solution for n even is obtained by the method of descent from dimension $n+1$, which is discussed in the next section. We introduce the following notation: For sufficiently smooth function h defined on $\mathbb{R}^n$, set

$$Q_h(x,t) = \frac{1}{\Gamma(k)} \left(\frac{1}{2t}\frac{\partial}{\partial t}\right)^{k-1} \left(\int_0^t \frac{r}{\sqrt{t^2-r^2}} \left(r^{2k-2} M_h(x,r)\right) dr\right), \tag{11.1.8}$$

for $n = 2k$, $k \geq 1$, and

$$Q_h(x,t) = \frac{\sqrt{\pi}}{2\Gamma(k+\frac{1}{2})} \left(\frac{1}{2t}\frac{\partial}{\partial t}\right)^{k-1} \left(t^{2k-1} M_h(x,t)\right), \tag{11.1.9}$$

for $n = 2k+1$, $k \geq 1$.

Theorem 11.1. *Consider the IVP for the homogeneous wave equation*

$$\Box_c u \equiv u_{tt} - c^2 \Delta u = 0, \ x \in \mathbb{R}^n, t > 0,$$
$$u(x,0) = \psi(x), \ u_t(x,0) = \phi(x), \ x \in \mathbb{R}^n.$$

Then, the solution u has the following representation:

$$u(x,t) = \frac{\partial}{\partial t} Q_\psi(x,t) + Q_\phi(x,t).$$

Here, $\psi \in C^{[n/2]+2}(\mathbb{R}^n)$ and $\phi \in C^{[n/2]+1}(\mathbb{R}^n)$.

11.1.1 Two-Dimensional Wave Equation: Method of Descent

Using the *method of descent*, due to Hadamard, we are now going to obtain the solution of two-dimensional (2D) wave equation using that of three-dimensional equation (*descending* from $n=3$ to $n=2$). In fact, the procedure can easily be extended to n-dimensional situation.

Consider the Cauchy problem for the homogeneous 2D wave equation, and let $u \in C^2(\mathbb{R}^2 \times [0,\infty))$ be the solution of

$$u_{tt} - \Delta u = 0, \ x \in \mathbb{R}^2, \ t > 0, \tag{11.1.10}$$
$$u(x,0) = u_0(x), \ u_t(x,0) = u_1(x), \ x \in \mathbb{R}^2. \tag{11.1.11}$$

Define $\tilde{u}(\tilde{x},t) = u(x,t)$, where $\tilde{x} = (x_1, x_2, x_3) \in \mathbb{R}^3$ and $x = (x_1, x_2) \in \mathbb{R}^2$, and similarly define $\tilde{u}_0$ and $\tilde{u}_1$. Then, $\tilde{u}$ solves the Cauchy problem for the three-dimensional wave equation:

$$\tilde{u}_{tt} - \Delta \tilde{u} = 0, \ \tilde{x} \in \mathbb{R}^3, \ t > 0,$$
$$\tilde{u}(\tilde{x},0) = \tilde{u}_0(\tilde{x}), \ \tilde{u}_t(\tilde{x},0) = \tilde{u}_1(\tilde{x}), \ \tilde{x} \in \mathbb{R}^3.$$

Therefore, from equation (11.1.6), we obtain (taking $c = 1$)

$$u(x,t) = \tilde{u}(x,0,t) = \frac{1}{4\pi t} \int_{|\tilde{y}-\tilde{x}|=t} \tilde{u}_1(\tilde{y}) \, d\sigma(\tilde{y})$$

$$+ \frac{\partial}{\partial t}\left(\frac{1}{4\pi t} \int_{|\tilde{y}-\tilde{x}|=t} \tilde{u}_0(\tilde{y}) \, d\sigma(\tilde{y})\right), \qquad (11.1.12)$$

By performing the integration with respect to y_3 variable, we obtain the solution for $n = 2$:

$$u(x_1, x_2, t) = \frac{1}{2\pi} \iint_{B_t(x)} \frac{u_1(y_1, y_2)}{\sqrt{t^2 - r^2}} \, dy_1 dy_2$$

$$+ \frac{\partial}{\partial t}\left(\frac{1}{2\pi} \iint_{B_t(x)} \frac{u_0(y_1, y_2)}{\sqrt{t^2 - r^2}} \, dy_1 dy_2\right), \qquad (11.1.13)$$

where $B_t(x)$ denotes the 2D open ball centred at $x = (x_1, x_2)$ and radius t: $B_t(x) = \{y = (y_1, y_2) : (x_1 - y_1)^2 + (x_2 - y_2)^2 < t^2\}$ and $r = |y - x|$.

Remark: Observe that the formula obtained for the solution for odd n is in terms of integrations over the sphere, the boundary of the ball, whereas, the solution formula for even n is in terms of integrations over the entire ball, in respective spaces. This has far-reaching consequences in the wave propagation problems, leading to the concept of *strong* and *weak Huygens principles*. See Ref. [45] for details.

11.1.2 Telegraph Equation

Consider the one-dimensional wave equation with lower-order terms present:

$$w_{tt} - w_{xx} + \alpha w_t + \beta w_x + \gamma w = 0, \ x \in \mathbb{R}, t > 0. \qquad (11.1.14)$$

Here, the coefficients α, β and γ are real constants. It is always possible to absorb the terms containing w_t and w_x by a simple change of variable: change $w(x,t)$ to $w(x,t) \exp\left((\alpha t - \beta x)/2\right)$. However, it is *not* possible to absorb the term containing w. This is the case we want to analyse now. The *telegraph equation* is given by

$$w_{tt} - w_{xx} - \lambda^2 w = 0, \ x \in \mathbb{R}, \ t > 0. \qquad (11.1.15)$$

Here, the constant $\lambda > 0$. We impose the initial conditions

$$w(x,0) = 0, \ w_t(x,0) = \psi(x), \ x \in \mathbb{R}. \qquad (11.1.16)$$

Put $x_1 = x$, and consider the function $u(x_1, x_2, t)$ defined by

$$u(x_1, x_2, t) = w(x_1, t) \cos(\lambda x_2).$$

Using equation (11.1.15), we see that u satisfies the 2D wave equation

$$u_{tt} - u_{x_1 x_1} - u_{x_2 x_2} = 0,$$

with initial conditions

$$u(x_1, x_2, 0) = 0, \ u_t(x_1, x_2, 0) = \psi(x_1) \cos(\lambda x_2).$$

Using the representation (11.1.13) for $u(x_1, x_2, t)$ and taking $x_2 = 0$ (*descending* from $n = 2$ to $n = 1$), we get

$$w(x, t) = w(x_1, t) = u(x_1, 0, t) = \frac{1}{2\pi} \iint_{B_t(x)} \frac{\cos(\lambda y_2) \psi(y_1)}{\sqrt{t^2 - r^2}} \, dy_1 dy_2,$$

where $B_t(x)$ is the 2D open ball centred at $(x_1, 0)$ and radius t: $B_t(x) = \{y = (y_1, y_2) : (x_1 - y_1)^2 + y_2^2 < t^2\}$ and $r^2 = (x_1 - y_1)^2 + y_2^2$. Performing the integration with respect to y_2, first, we obtain, by changing x_1 to x,

$$w(x, t) = \frac{1}{2} \int_{x-t}^{x+t} J_0(\lambda s) \psi(y) \, dy, \tag{11.1.17}$$

where $s^2 = t^2 - (x - y)^2$, and J_0 is the Bessel's function of the first kind of order 0 and is given by

$$J_0(z) = \frac{2}{\pi} \int_0^{\pi/2} \cos(z \sin \theta) \, d\theta.$$

We remark that if we include the speed of propagation c in equation (11.1.15), that is, replacing w_{xx} by $c^2 w_{xx}$, then the formula (11.1.17) changes to

$$w(x, t) = \frac{1}{2c} \int_{x-ct}^{x+ct} J_0(\lambda s/c) \psi(y) \, dy,$$

where $s^2 = c^2 t^2 - (x - y)^2$. From this representation, it is not difficult to write down a formula for the solution of the inhomogeneous Cauchy problem:

$$\left. \begin{array}{l} u_{tt} - c^2 u_{xx} - \lambda^2 u = f(x, t), \ x \in \mathbb{R}, \ t > 0, \\ u(x, 0) = \phi(x), \ u_t(x, 0) = \psi(x), \ x \in \mathbb{R}. \end{array} \right\} \tag{11.1.18}$$

Here, c and λ are positive constants. We remark that if the term $-\lambda^2 u$ is changed to $\lambda^2 u$ ($\lambda > 0$) in equation (11.1.18), then the cosh function needs to be used in place of the cos function. This results in replacing the Bessel function J_0 by the modified Bessel function I_0. For a quick reference to Bessel functions and their properties, see Ref. [9].

Exercises 11.1 and 11.2 deal with the fourth-order equations. For that purpose, we first consider these equations in one dimension. Consider the IVP

$$\left.\begin{array}{l}(\partial_t^2 - c_1^2 \partial_x^2)(\partial_t^2 - c_2^2 \partial_x^2)u = 0, \ x \in \mathbb{R}, \ t > 0, \\ \partial_t^j u(x,0) = u_j(x), \ x \in \mathbb{R}, \ j = 0,1,2,3.\end{array}\right\} \quad (11.1.19)$$

Here, $c_1 > c_2 > 0$ are constants, and u_j are given smooth functions. The general solution of equation (11.1.19) is given by

$$u(x,t) = F_1(x + c_1 t) + F_2(x - c_1 t) + F_3(x + c_2 t) + F_4(x - c_2 t),$$

where F_j are arbitrary functions in $C^4(\mathbb{R})$. Substituting the initial conditions in this expression, we get

$$\left.\begin{array}{l}F_1(x) + F_2(x) + F_3(x) + F_4(x) = u_0(x), \\ c_1(F_1'(x) - F_2'(x)) + c_2(F_3'(x) - F_4'(x)) = u_1(x), \\ c_1^2(F_1''(x) + F_2''(x)) + c_2^2(F_3''(x) + F_4''(x)) = u_2(x), \\ c_1^3(F_1'''(x) - F_2'''(x)) + c_2^3(F_3'''(x) - F_4'''(x)) = u_3(x).\end{array}\right\} \quad (11.1.20)$$

Solving these equations for F_j's, we obtain

$$u(x,t) = \frac{1}{2(c_1^2 - c_2^2)}(I_1 + I_2 + I_3 + I_4), \quad (11.1.21)$$

as the solution of IVP (11.1.19), where

$$\begin{aligned}I_1 &= c_1^2(u_0(x + c_2 t) + u_0(x - c_2 t)) \\ &\quad - c_2^2(u_0(x + c_1 t) + u_0(x - c_1 t)),\end{aligned}$$

$$I_2 = \frac{c_1^2}{c_2}\int_{x-c_2 t}^{x+c_2 t} u_1(y)\,dy - \frac{c_2^2}{c_1}\int_{x-c_1 t}^{x+c_1 t} u_1(y)\,dy,$$

$$I_3 = \int_0^{x+c_1t} (x+c_1t-y)u_2(y)\,dy + \int_0^{x-c_1t} (x-c_1t-y)u_2(y)\,dy$$
$$- \int_0^{x+c_2t} (x+c_2t-y)u_2(y)\,dy - \int_0^{x-c_2t} (x-c_2t-y)u_2(y)\,dy,$$

$$I_4 = \frac{1}{2c_2} \int_0^{x+c_2t} (x+c_2t-y)^2 u_3(y)\,dy - \frac{1}{2c_2} \int_0^{x-c_2t} (x-c_2t-y)^2 u_3(y)\,dy \quad (11.1.22)$$
$$- \frac{1}{2c_1} \int_0^{x+c_1t} (x+c_1t-y)^2 u_3(y)\,dy + \frac{1}{2c_1} \int_0^{x-c_1t} (x-c_1t-y)^2 u_3(y)\,dy.$$

Next, consider IVP for the *double characteristic* equation
$$(\partial_t^2 - c^2 \partial_x^2)^2 u = 0, \ x \in \mathbb{R}, \ t > 0$$
$$\partial_t^j u(x,0) = u_j(x), \ x \in \mathbb{R}, \ j = 0, 1, 2, 3, \quad (11.1.23)$$

where $c > 0$. The general solution of this equation is given by
$$u(x,t) = F_1(x+ct) + tF_2(x+ct) + F_3(x-ct) + tF_4(x-ct),$$

where F_j's are arbitrary functions in $C^4(\mathbb{R})$. As in the previous case, we can solve for F_j after substituting the given initial conditions. Alternatively, we can obtain the solution of equation (11.1.23) by letting $c_2 \to c_1 = c$ in equations (11.1.21) and (11.1.22), using L'Hospital's rule. We obtain
$$u(x,t) = I_1 + I_2 + I_3 + I_4, \quad (11.1.24)$$

as the solution of IVP (11.1.23). Here,

$$\left.\begin{aligned}
I_1 &= \frac{1}{2}(u_0(x+ct) + u_0(x-ct)) - \frac{ct}{4}(u_0'(x+ct) - u_0'(x-ct)), \\
I_2 &= \frac{3}{4c} \int_{x-ct}^{x+ct} u_1(y)\,dy - \frac{t}{4c}(u_1(x+ct) + u_1(x-ct)), \\
I_3 &= \frac{t}{4c} \int_{x-ct}^{x+ct} u_2(y)\,dy, \\
I_4 &= \frac{1}{8c^3} \int_{x-ct}^{x+ct} (x+ct-y)(x-ct-y) u_3(y)\,dy.
\end{aligned}\right\} \quad (11.1.25)$$

The general references are [45], [16], [31], [42] and [55].

11.2 Exercises

Exercise 11.1

Let $c_1, \ldots, c_k$ be positive and distinct real numbers. Show that the solution of the equation

$$\prod_{j=1}^{k} \Box_{c_j} u = \prod_{j=1}^{k} \left(\partial_t^2 - c_j^2 \Delta \right) u = 0$$

can be written as

$$u(x,t) = \sum_{j=1}^{k} u_j(x,t),$$

where u_j satisfies the equation $\Box_{c_j} u = 0$.

Exercise 11.2

Let $n = 3$ and consider the equation $\Box_c^2 u = 0$, where $c > 0$. Taking sufficiently smooth initial data $\partial_t^j u$ for $j = 0, 1, 2, 3$ at $t = 0$, write down the solution explicitly.

Exercise 11.3

From the formula for the solution of the homogeneous wave equation for general n mentioned in the Introduction, derive the formula for the solution of the inhomogeneous wave equation using Duhamel's principle.

Exercise 11.4

Consider the IVP for the wave equation

$$\Box_c u = u_{tt} - c^2 \Delta u = 0, \ x \in \mathbb{R}^n, \ t > 0,$$
$$u(x,0) = \psi(x), \ u_t(x,0) = \phi(x), \ x \in \mathbb{R}^n. \quad (11.2.7)$$

Let V^η denote the solution of the wave equation (11.2.7) satisfying $V^\eta(x,0) = \eta(x)$ and $V_t^\eta(x,0) = 0$, for $x \in \mathbb{R}^n$. Similarly, let U_η denote the solution of the wave equation (11.2.7) satisfying $U_\eta(x,0) = 0$ and $\partial_t U_\eta(x,0) = \eta(x)$, for $x \in \mathbb{R}^n$.

- Verify that the solution of the IVP (11.2.7) is given by $u(x,t) = V^\psi(x,t) + \int_0^t V^\phi(x,s) \, ds$.

- Verify that the solution of the IVP can also be represented by

$$u(x,t) = \frac{\partial}{\partial t} U_\psi(x,t) + U_\phi(x,t).$$

Exercise 11.5

Let L be linear partial differential operator defined by $L = \sum_{k=0}^{m} a_k r^{k+1} \partial_r^k$, where m is a non-negative integer and $a_0, \ldots, a_m$ are real numbers with $a_m = 1$. Assume that L satisfies the following commutator relation:

$$[\partial_r^2, L] \equiv \partial_r^2 L - L\partial_r^2 = (n-1)L\left(\frac{1}{r}\partial_r\right), \qquad (11.2.8)$$

where $n \geqslant 3$ is an integer.

(a) Show that the integer n is necessarily odd and $m = (n-3)/2$.

(b) Write down the recursive relation to determine the coefficients a_k.

(c) Show that the operator L can be written as $L = \left(\frac{1}{r}\partial_r\right)^m r^{2m+1}$, $r \neq 0$.

Exercise 11.6

Reduce equation (11.1.14) to an equation of the form (11.1.15), with $\pm \lambda^2 u$ term, by a suitable change of variables.

Exercise 11.7

Write down the formula for the solution of the inhomogeneous problem (11.1.18) for the telegraph equation.

Exercise 11.8

Show that the function u defined by $u(x,t) = \frac{\partial}{\partial t} Q_\psi(x,t) + Q_\phi(x,t)$, where Q_h is as defined in expressions (11.1.8) and (11.1.9), for sufficiently smooth function h, satisfies the IVP for the homogeneous wave equation ($c=1$) $u_{tt} - c^2 \Delta u = 0$, $x \in \mathbb{R}^n$, $t > 0$ satisfying the initial conditions $u(x,0) = \psi(x)$ and $u_t(x,0) = \phi(x)$, for $x \in \mathbb{R}^n$.

Exercise 11.9

If A is a tridiagonal matrix with *all* non-zero subdiagonal elements, show that the geometric multiplicity of any eigenvalue of A is 1. Use this result to establish that there are matrices A_1 and A_2 of arbitrary order such that the matrix $\xi_1 A_1 + \xi_2 A_2$ has real and distinct eigenvalues for all real ξ_1 and ξ_2, not both zero.

Exercise 11.10

Consider a system of second-order wave-like equations:

$$\mathbf{u}_{tt} + \mathbf{A}\Delta\mathbf{u} = \mathbf{0},\ x \in \mathbf{R}^n,\ t > 0. \tag{11.2.13}$$

Here, $\mathbf{A}$ is a real $N \times N$ matrix whose eigenvalues are all negative real numbers; $\mathbf{u}$ is N-(column) vector and $\Delta\mathbf{u}$ denotes the column vector $\Delta\mathbf{u} = \begin{bmatrix} \Delta u_1 & \cdots & \Delta u_N \end{bmatrix}^T$. Show how to obtain a partially decoupled system from equation (11.2.13), and indicate the procedure to obtain a solution with prescribed conditions at $t = 0$.

The system of Maxwell's equations of electrodynamics is a typical example of a system of wave-like equations ([30]). Another example is the system of equations of elastic waves ([31], [16]).

Exercise 11.11

Transform the following system of second-order equations

$$\mathbf{u}_{tt} + \mathbf{A}\Delta\mathbf{u} - \mathbf{B}^2\mathbf{u} = \mathbf{0},\ x \in \mathbf{R}^n,\ t > 0, \tag{11.2.14}$$

to a system of wave-like equations as in equation (11.2.13). Here, $\mathbf{A}$ is a real symmetric, negative definite $N \times N$ matrix and $\mathbf{B}$ is any real $N \times N$ matrix that commutes with $\mathbf{A}$; $\mathbf{u}$ is an N-vector (column vector) and $\Delta\mathbf{u}$ denotes the column vector $\Delta\mathbf{u} = \begin{bmatrix} \Delta u_1 & \cdots & \Delta u_N \end{bmatrix}^T$.

For $N = 1$ and $n = 3$, the equation $u_{tt} - c^2\Delta u + m^2 u = 0$ ($m > 0$) is called the *Klein–Gordon equation*. It is also called *Klein–Gordon–Fock equation* in the literature. This is a *relativistic wave equation*, related to the *Schrödinger equation*.

Exercise 11.12

Transform the following system of second-order equations

$$\mathbf{u}_{tt} + 2\mathbf{B}\mathbf{u}_t + \mathbf{A}\Delta\mathbf{u} = \mathbf{0},\ x \in \mathbf{R}^n,\ t > 0, \tag{11.2.15}$$

to a system of wave-like equations as in equation (11.2.13). Here, $\mathbf{A}$ is a real symmetric, negative definite $N \times N$ matrix, and $\mathbf{B}$ is any real $N \times N$ matrix that commutes with $\mathbf{A}$; $\mathbf{u}$ is N-(column) vector and $\Delta\mathbf{u}$ denotes the column vector $\Delta\mathbf{u} = \begin{bmatrix} \Delta u_1 & \cdots & \Delta u_N \end{bmatrix}^T$.

Exercise 11.13

Solve the following system of equations

$$\left.\begin{aligned}(u_1)_{tt} - (u_1)_{xx} + u_2 &= 0,\\ (u_2)_{tt} - (u_2)_{xx} - u_1 &= 0,\end{aligned}\right\} \tag{11.2.16}$$

subject to the initial conditions

$$u_i(x,0) = 0 \text{ and } (u_i)_t(x,0) = g_i(x),\ x \in \mathbb{R},\ i = 1.2,$$

where g_i are given smooth functions.

11.3 Solutions

Exercise 11.1

Let $n = 3$ and $k = 2$ with $c_1 > c_2 > 0$. The general case follows from an inductive argument using the operator L (see Exercise 11.5 and also Exercises 10.2 and 10.3 of Chapter 10), for n odd; for even n, we use the method of descent. Of course, computations are tedious. Thus, consider IVP

$$\left. \begin{array}{l} \Box_{c_1} \Box_{c_2} u = 0,\ x \in \mathbb{R}^3,\ t > 0 \\[4pt] \partial_t^j u(x,0) = \phi_j(x),\ x \in \mathbb{R}^3,\ j = 0,1,2,3. \end{array} \right\} \quad (11.2.1)$$

Using the Darboux equation (11.1.5), we easily verify that the function $v(x,r,t) = rM_u(x,r,t)$ satisfies the one-dimensional equation

$$\left. \begin{array}{l} (\partial_t^2 - c_1^2 \partial_r^2)(\partial_t^2 - c_2^2 \partial_r^2) v = 0, \\[4pt] \partial_t^j v(x,r,0) = M_{\phi_j}(x,r),\ x \in \mathbb{R}^3,\ r \in \mathbb{R},\ j = 0,1,2,3. \end{array} \right\} \quad (11.2.2)$$

Here, x is a parameter, and the general solution of the above equation is given by

$$v(x,r,t) = F_1(x, r - c_1 t) + F_2(x, r + c_1 t) \\ + F_3(x, r - c_2 t) + F_4(x, r + c_2 t),$$

for C^m functions F_i, $m \geq 4$. Using the initial conditions $\partial_t^j u(x,0) = \phi_j(x)$ for $j = 0,1,2,3$, we have

$$\partial_t^j v(x,r,0) = rM_{\phi_j}(x,r),\ r \in \mathbb{R},\ j = 0,1,2,3.$$

Using equation (11.1.21), we have

$$v(x,r,t) = \frac{1}{2(c_1^2 - c_2^2)} (I_1 + I_2 + I_3 + I_4), \quad (11.2.3)$$

as the solution of IVP (11.2.2), where

$$
\begin{aligned}
I_1 &= c_1^2[(r+c_2t)M_{\phi_0}(x, r+c_2t) + (r+c_2t)M_{\phi_0}(x, r-c_2t)] \\
&\quad - c_2^2[(r+c_1t)M_{\phi_0}(x, r+c_1t) + (r-c_1t)M_{\phi_0}(x, r-c_1t)], \\
I_2 &= \frac{c_1^2}{c_2}\int_{r-c_2t}^{r+c_2t} \rho M_{\phi_1}(x,\rho)\,d\rho - \frac{c_2^2}{c_1}\int_{r-c_2t}^{r+c_2t} \rho M_{\phi_1}(x,\rho)\,d\rho, \\
I_3 &= \int_0^{r+c_1t} (r+c_1t-\rho)\rho M_{\phi_2}(x,\rho)\,d\rho \\
&\quad + \int_0^{r-c_1t} (r-c_1t-\rho)\rho M_{\phi_2}(x,\rho)\,d\rho \\
&\quad - \int_0^{r+c_2t} (r+c_2t-\rho)\rho M_{\phi_2}(x,\rho)\,d\rho \\
&\quad - \int_0^{r-c_2t} (r-c_2t-\rho)\rho M_{\phi_2}(x,\rho)\,d\rho, \\
I_4 &= \frac{1}{2c_2}\int_0^{r+c_2t} (r+c_2t-\rho)^2 \rho M_{\phi_3}(x,\rho)\,d\rho \\
&\quad - \frac{1}{2c_2}\int_0^{r-c_2t} (r-c_2t-\rho)^2 \rho M_{\phi_3}(x,\rho)\,d\rho \\
&\quad - \frac{1}{2c_1}\int_0^{r+c_1t} (r+c_1t-\rho)^2 \rho M_{\phi_3}(x,\rho)\,d\rho \\
&\quad + \frac{1}{2c_1}\int_0^{r-c_1t} (r-c_1t-\rho)^2 \rho M_{\phi_3}(x,\rho)\,d\rho.
\end{aligned}
\qquad (11.2.4)
$$

The required solution u can be obtained from v as $u(x,t) = \lim_{r \to 0} \frac{v(x,r,t)}{r}$; we need to divide throughout the equation (11.2.3) by r and then take the limit as $r \to 0$. The following observation helps in the evaluation of these limits.

Let $h(x,t)$ be a C^2 function. The spherical mean function $M_h(x,r)$ is an even function of r. Put $\chi(t) = ctM_h(x,ct)$ (x plays the role of a parameter). Then,

$$\frac{d\chi}{dt} = \lim_{r \to 0} \frac{\chi(t+r) - \chi(t-r)}{2r}.$$

However, $\dfrac{d\chi}{dt} = c\dfrac{\partial}{\partial t}(tM_h(x,ct))$. Therefore, we conclude that

$$\lim_{r \to 0} \frac{1}{2r}[(ct+r)M_h(x,r+ct) + (r-ct)M_h(x,r-ct)] = \frac{\partial}{\partial t}(tM_h(x,ct)).$$

In a similar manner, the other limits can be evaluated. Now, a careful computation of the limits of each of the expressions in equation (11.2.4) gives

$$u(x,t) = \frac{1}{(c_1^2 - c_2^2)}(J_1 + J_2 + J_3 + J_4), \qquad (11.2.5)$$

as the required solution of equation (11.2.1), where now

$$\left.\begin{aligned}
J_1 &= c_1^2 \frac{\partial}{\partial t}\left(tM_{\phi_0}(x, c_2 t)\right) - c_2^2 \frac{\partial}{\partial t}\left(tM_{\phi_0}(x, c_1 t)\right), \\
J_2 &= c_1^2 tM_{\phi_1}(x, c_2 t) - c_2^2 tM_{\phi_1}(x, c_1 t), \\
J_3 &= \int_{c_2 t}^{c_1 t} \rho M_{\phi_2}(x, \rho)\, d\rho, \\
J_4 &= \int_0^{c_1 t}(c_1 t - \rho)\rho M_{\phi_3}(x, \rho)\, d\rho - \int_0^{c_2 t}(c_2 t - \rho)\rho M_{\phi_3}(x, \rho)\, d\rho.
\end{aligned}\right\} \qquad (11.2.6)$$

Exercise 11.2

As in the previous exercise, the function v now satisfies the equation $(\partial_t^2 - c^2 \partial_r^2)^2 v = 0$. Consider the operator $(\partial_t^2 - c^2 \partial_r^2)^2 (\partial_t^2 - c_1^2 \partial_r^2)^2$, with $c_1 \neq c$. The required solution is now obtained using equation (11.2.5), with c_2 replaced by c, and then taking the limit as $c_1 \to c$ in each of the expressions in equation (11.2.6).

The computations are bit hard, but straightforward. The interested reader can work them out and enjoy!

Exercise 11.3

By the linearity of the problem, it suffices to consider the following inhomogeneous problem:

$$\left.\begin{aligned}
\partial_t^2 u - c^2 \Delta u &= f(x,t), \ x \in \mathbb{R}^n,\ t > 0, \\
\partial_t^j u(x,0) &= 0,\ x \in \mathbb{R}^n,\ j = 0, 1.
\end{aligned}\right\}$$

We assume that f is a $C^{[n/2]+1}$ function. Now, consider the following IVP for the homogeneous wave equation

$$\left.\begin{array}{l} \partial_t^2 v - c^2 \Delta v = 0, \; x \in \mathbb{R}^n, \; t > s, \\ v(x,s) = 0, \; \partial_t v(x,s) = f(x,s), \; x \in \mathbb{R}^n, \end{array}\right\}$$

where $s \geqslant 0$ is a fixed number. Using the translation invariant property of the wave operator and the formula for the solution given in the Introduction, we obtain $v(x,t;s) = Q_\phi(x, t-s)$, $t > s$, where $\phi(x) = f(x,s)$, and the expression for Q_ϕ is as given in equations (11.1.8) and (11.1.9); the notation used to denote the solution is to stress the dependence on the variable s. Then, the required solution u of the inhomogeneous equation with zero initial conditions is given by

$$u(x,t) = \int_0^t v(x,t;s)\, ds = \int_0^t Q_\phi(x, t-s)\, ds, \text{ with } \phi(x) = f(x,s).$$

Exercise 11.4

The function $V^\psi(x,t)$ satisfies the equation $\Box_c V^\psi = 0$ and $V^\psi(x,0) = \psi(x)$, $V_t^\psi(x,0) = 0$.

Put $v(x,t) = \int_0^t V^\phi(x,s)\, ds$. Then, $v(x,0) = 0$ and $v_t(x,0) = \phi(x)$. Furthermore, $v_{tt} = u_t$

and $\Delta v = \int_0^t \Delta V^\phi(x,s)\, ds$.

Therefore,

$$v_{tt} - c^2 \Delta v = V_t^\phi - c^2 \int_0^t \Delta V^\phi(x,s)\, ds$$

$$= \int_0^t (V_{tt}^\phi - c^2 \Delta V^\phi)(x,s)\, ds, \text{ using } V_t^\phi(x,0) = 0$$

$$= 0.$$

Thus, $u(x,t) = V^\psi(x,t) + v(x,t)$ as required, proving the first statement. The proof of the second statement is exactly similar.

This exercise shows that the solution u can be obtained by taking the homogeneous initial condition for $u(x,0)$ or $u_t(x,0)$.

Exercise 11.5

We have $L\partial_r^2 = \sum_{k=0}^{m} a_k r^{k+1} \partial_r^{k+2}$ and $\partial_r^2(r\chi) = 2\partial_r \chi + r\partial_r^2 \chi$, for any function χ and for $k \geq 1$,

$$\partial_r^2(r^{k+1}\partial_r^k) = k(k+1)r^{k-1}\partial_r^k + 2(k+1)r^k \partial_r^{k+1} + r^{k+1}\partial_r^{k+2}.$$

Summing together the above expressions, we therefore obtain

$$\partial_r^2 L - L\partial_r^2 = \sum_{k=0}^{m}(k+1)\left(2a_k + (k+2)a_{k+1}\right) r^k \partial_r^{k+1}, \qquad (11.2.9)$$

with the convention that $a_{m+1} = 0$. Next, for $k \geq 0$, we have by Leibniz's rule,

$$\partial_r^k\left(\frac{1}{r}\partial_r\right) = \sum_{j=0}^{k}\binom{k}{j}\partial_r^{k-j}(1/r)\partial_r^{j+1} = (-1)^k k! \sum_{j=0}^{k}\frac{(-1)^j}{j!} r^{-k+j-1}\partial_r^{j+1}.$$

Therefore,

$$L\left(\frac{1}{r}\partial_r\right) = \sum_{k=0}^{m} a_k r^{k+1}\partial_r^k\left(\frac{1}{r}\partial_r\right)$$

$$= \sum_{k=0}^{m} a_k r^{k+1}\left((-1)^k k! \sum_{j=0}^{k}\frac{(-1)^j}{j!} r^{-k+j-1}\partial_r^{j+1}\right)$$

$$= \sum_{k=0}^{m}(-1)^k k! a_k \left(\sum_{j=0}^{k}\frac{(-1)^j}{j!} r^j \partial_r^{j+1}\right)$$

$$= \sum_{j=0}^{m}\frac{(-1)^j}{j!}\left(\sum_{k=j}^{m}(-1)^k k! a_k\right) r^j \partial_r^{j+1}. \qquad (11.2.10)$$

Now, using the hypothesis and comparing the coefficients in equations (11.2.9) and (11.2.10), we obtain

$$(k+1)\left(2a_k + (k+2)a_{k+1}\right) = (n-1)\frac{(-1)^k}{k!}\left(\sum_{j=k}^{m}(-1)^j j! a_j\right),$$

for $k = 0, 1, \ldots, m$; remember that $a_m = 1$ and $a_{m+1} = 0$. Taking $k = m$ in the above relation, we immediately see that $2(m+1) = n-1$. Therefore, n must be odd and $m = (n-3)/2$. This proves (a).

To obtain the recursive relations for the coefficients, we rewrite the above expression as

$$2(m-k)a_k = (k+1)(2m+k+4)a_{k+1}$$
$$- 2(m+1)\frac{(-1)^k}{k!}\left(\sum_{j=k+2}^{m}(-1)^j j! a_j\right), \qquad (11.2.11)$$

for $k = m-1, m-2, \ldots, 0$, the sum being absent when $k = m-1$. This answers (b). Incidentally, the recursion relations (11.2.11) prove that the operator L is *unique*, that is, any operator L with $a_m = 1$ that satisfies the specified commutator relation is unique.

To prove (c), write $M = \left(\frac{1}{r}\partial_r\right)^m r^{2m+1}$, $r \neq 0$. An easy induction argument shows that M can be written as $M = \sum_{k=0}^{m} b_k r^{k+1} \partial_r^k$, with $b_m = 1$. Thus, in order to show that $L = M$, it suffices to show that M also satisfies the stated commutation relation with $n = 2m + 3$. This can be accomplished by an inductive argument.

Exercise 11.6

Consider equation (11.1.14):
$$w_{tt} - w_{xx} + \alpha w_t + \beta w_x + \gamma w = 0, \ x \in \mathbb{R}, t > 0.$$

Put $v(x,t) = w(x,t) \exp(\alpha t - \beta x)/2)$. Then,

$$\left.\begin{aligned}
v_t &= \exp(\alpha t - \beta x)/2)\,(w_t + (\alpha/2)w) \\
v_x &= \exp(\alpha t - \beta x)/2))(w_x - (\beta/2)w) \\
v_{tt} &= \exp(\alpha t - \beta x)/2))(w_{tt} + 2(\alpha/2)w_t + (\alpha^2/4)w) \\
v_{xx} &= \exp(\alpha t - \beta x)/2)\,(w_{xx} - 2(\beta/2)w_x + (\beta^2/4)w).
\end{aligned}\right\}$$

Therefore,
$$v_{tt} - v_{xx} = \exp(\alpha t - \beta x)/2)\,((\alpha^2 - \beta^2)/4 + \gamma)w,$$

using equation (11.1.14). Depending on the sign of $(\alpha^2/4) - (\beta^2/4) + \gamma$, we put $\lambda^2 = \pm((\alpha^2 - \beta^2)/4 + \gamma)$, so that $\lambda > 0$. Thus, v satisfies the equation of the form (11.1.15).

Exercise 11.7

As observed in Section 11.1, the solution of equation (11.1.18), when f and ϕ are identically zero functions, is given by

$$u(x,t) = \frac{1}{2c} \int_{x-ct}^{x+ct} J_0(\lambda s/c)\, \psi(y)\, dy,$$

where $s^2 = c^2 t^2 - (x-y)^2$. Using the idea in Exercise 11.4, we now obtain the solution of the homogeneous equation (11.1.18) with $f \equiv 0$ as

$$u(x,t) = \frac{\partial}{\partial t}\left(\frac{1}{2c} \int_{x-ct}^{x+ct} J_0(\lambda s/c)\, \phi(y)\, dy\right) + \frac{1}{2c} \int_{x-ct}^{x+ct} J_0(\lambda s/c)\, \psi(y)\, dy$$

$$= \frac{1}{2}(\phi(x+ct) + \phi(x-ct)) - \frac{\lambda t}{2} \int_{x-ct}^{x+ct} J_1(\lambda s/c)\, \phi(y)\, \frac{dy}{s}$$

$$+ \frac{1}{2c} \int_{x-ct}^{x+ct} J_0(\lambda s/c)\, \psi(y)\, dy,$$

where s is as before and we have used the formula[1] $\dfrac{d}{d\eta}J_0(\eta) = -J_1(\eta)$. Finally, for the inhomogeneous equation, we use Duhamel's principle. Let $\phi = \psi = 0$ in equation (11.1.18). Consider IVP for any fixed $\tau \geqslant 0$:

$$v_{tt} - c^2 v_{xx} - \lambda^2 v = 0,\ x \in \mathbb{R},\ t > \tau,$$

$$v(x,\tau) = 0,\ v_t(x,\tau) = f(x,\tau),\ x \in \mathbb{R}.$$

Denoting the solution as $v(x,t;\tau)$, we have

$$v(x,t;\tau) = \dfrac{1}{2c}\int_{x-c(t-\tau)}^{x+c(t-\tau)} J_0\left(\lambda\xi/c\right) f(y,\tau))\,dy,$$

with $\xi^2 = c^2(t-\tau)^2 - (x-y)^2$ and for $t > \tau$. Thus, the solution of the inhomogeneous equation (11.1.18) with zero initial data is given by

$$u(x,t) = \dfrac{1}{2c}\int_0^t d\tau \int_{x-c(t-\tau)}^{x+c(t-\tau)} J_0\left(\lambda\xi/c\right) f(y,\tau))\,dy.$$

By linearity, we can now write the formula for the complete solution of equation (11.1.18) as $u(x,t) = I_1 + I_2 + I_3$, where

$$\left.\begin{aligned}
I_1 &= \dfrac{1}{2}(\phi(x+ct) + \phi(x-ct)) - \dfrac{\lambda t}{2}\int_{x-ct}^{x+ct} J_1\left(\lambda s/c\right)\phi(y)\dfrac{dy}{s}\\
I_2 &= \dfrac{1}{2c}\int_{x-ct}^{x+ct} J_0\left(\lambda s/c\right)\psi(y)\,dy\\
I_3 &= \dfrac{1}{2c}\int_0^t d\tau \int_{x-c(t-\tau)}^{x+c(t-\tau)} J_0\left(\lambda\xi/c\right) f(y,\tau))\,dy,
\end{aligned}\right\}$$

with s and ξ as before.

---

[1] $J_n(\eta) = \dfrac{1}{\pi}\int_0^\pi \cos(n\theta - \eta\sin\theta)\,d\theta$ for $n = 0, 1, \ldots$.

Exercise 11.8

First consider the case of odd $n = 2k+1$, $k \geq 1$. The main ingredients in the proof are Darboux equation (11.1.5) and the commutation relation (11.2.8) in Exercise 11.5. Let $h \in C^\ell(\mathbb{R}^n)$ for some $\ell \geq [n/2]+2$. We have

$$Q_h(x,t) = \frac{\sqrt{\pi}}{2\Gamma(k+\frac{1}{2})} \left(\frac{1}{2t}\frac{\partial}{\partial t}\right)^{k-1} (t^{2k-1} M_h(x,t)) = \frac{1}{c_n} L(M_h(x,t)),$$

for $t > 0$, where $L = L_m$, $m = (n-3)/2 = k-1$ is the operator defined in Exercise 11.5, with t replacing the variable r there and $c_n = 1 \cdot 3 \cdots (n-2)$. Thus, the coefficient of t term in $Q_h(x,t)$ equals 1 since $a_0 = c_n$. This immediately gives us $\lim_{t \to 0} Q_h(x,t) = 0$ and $\lim_{t \to 0} \partial_t Q_h(x,t) = h(x)$, using the relation $\lim_{t \to 0} M_h(x,t) = h(x)$. Next,

$$\frac{\partial^2}{\partial t^2} Q_h(x,t) = \frac{1}{c_n} \frac{\partial^2}{\partial t^2} (L(M_h(x,t)))$$

$$= \frac{1}{c_n} L \left(\frac{\partial^2}{\partial t^2} + \frac{n-1}{t}\frac{\partial}{\partial t}\right) M_h(x,t), \text{ using equation (11.2.8)}$$

$$= \frac{1}{c_n} L(\Delta_x M_h(x,t)), \text{ using the Darboux equation (11.1.5)}$$

$$= \Delta_x Q_h(x,t).$$

Thus, Q_h satisfies the wave equation. Finally,

$$\lim_{t \to 0} \partial_t^2 Q_h(x,t) = \lim_{t \to 0} \Delta_x Q_h(x,t) = 0.$$

This proves the function u given in the exercise is the solution of the IVP when n is odd.

Next, consider the case of even $n = 2k$, $k \geq 1$. The arguments in this case are similar to those used in the method of descent. For $\tilde{x} \in \mathbb{R}^{n+1}$, we write $\tilde{x} = (x, x_{n+1})$ with $x \in \mathbb{R}^n$ and $\tilde{x}_0 = (x, 0)$. For a function h defined in $\mathbb{R}^n$, we extend it to $\mathbb{R}^{n+1}$ by defining $h(\tilde{x}) = h(x)$. By extending the initial functions ψ and ϕ to $\mathbb{R}^{n+1}$ and using the first part of the exercise, we see that the function $\tilde{u}(\tilde{x}, t) = \frac{\partial}{\partial t} Q_\psi(\tilde{x}, t) + Q_\phi(\tilde{x}, t)$ is the solution of the IVP in $n+1$ dimensions. Put $u(x,t) = \tilde{u}(\tilde{x}_0, t)$. Since $\tilde{u}$ is a C^2 function, the function u, being its restriction to the lower-dimensional space $x_{n+1} = 0$, is also a C^2 function and it is the solution of the IVP in n dimensions. To conclude the proof, we show that u can be written in the form (11.1.8). Looking at the expressions in equations (11.1.8) and (11.1.9), we observe that this would be true if we can establish the following relation (remember $n = 2k$)

$$t^{2k-1} M_h(\tilde{x}_0, t) = \frac{2}{\sqrt{\pi}} \frac{\Gamma(k+\frac{1}{2})}{\Gamma(k)} \int_0^t \frac{r}{\sqrt{t^2-r^2}} (r^{2k-2} M_h(x,r)) \, dr, \quad (11.2.12)$$

for every smooth function h defined in $\mathbb{R}^n$ and extended to $\mathbb{R}^{n+1}$ as described above. Note that the spherical mean function on the left-hand side is $n+1$ dimensional, whereas it is n dimensional on the right-hand side.

To this end, let h be a smooth function defined in $\mathbb{R}^n$ and extended to $\mathbb{R}^{n+1}$ as described above. Then, for $t > 0$, we have

$$t^{2k-1} M_h(\tilde{x}_0, t) = t^{2k-1} \frac{1}{\sigma_{n+1} t^n} \int_{|\tilde{x}_0 - \tilde{y}| = t} h(\tilde{y}) \, d\sigma(\tilde{y})$$

$$= \frac{1}{\sigma_{n+1} t} \int_{|\tilde{x}_0 - \tilde{y}| = t} h(\tilde{y}) \, d\sigma(\tilde{y}).$$

Since h does not depend on y_{n+1}, we can perform integration with respect to y_{n+1} first. We have $d\sigma(\tilde{y}) = \frac{t}{|y_{n+1}|} dy$. Since the contributions of the regions $y_{n+1} > 0$ and $y_{n+1} < 0$ to the above integral are the same, we obtain

$$t^{2k-1} M_h(\tilde{x}_0, t) = \frac{2}{\sigma_{n+1}} \int_{|x-y|<t} \frac{h(y)}{|y_{n+1}|} dy$$

$$= \frac{1}{\sigma_{n+1}} \int_{|x-y|<t} \frac{h(y)}{\sqrt{t^2 - |x-y|^2}} dy$$

$$= \frac{2}{\sigma_{n+1}} \int_0^t dr \int_{|x-y|=r} \frac{h(y)}{\sqrt{t^2 - |x-y|^2}} d\sigma(y)$$

$$= \frac{2}{\sigma_{n+1}} \int_0^t dr \, \frac{\sigma_n r^{n-1}}{\sigma_n r^{n-1}} \int_{|x-y|=r} \frac{h(y)}{\sqrt{t^2 - |x-y|^2}} d\sigma(y)$$

$$= \frac{2\sigma_n}{\sigma_{n+1}} \int_0^t \frac{r}{\sqrt{t^2 - r^2}} \left(r^{2k-2} M_h(x, r) \right) dr.$$

Here, in the first and second lines, the integrals are n-dimensional volume integrals; the equality in the third line is obtained by changing to polar co-ordinates, and in the last line, M_h is the n-dimensional spherical mean function. Remember $n = 2k$. Thus,

$$\frac{\sigma_n}{\sigma_{n+1}} = \frac{2\pi^{n/2}/\Gamma(n/2)}{2\pi^{(n+1)/2}/\Gamma((n+1)/2)} = \frac{\Gamma\left(k + \frac{1}{2}\right)}{\sqrt{\pi}\Gamma(k)}.$$

Thus, equation (11.2.12) is true and this completes the proof.

Exercise 11.9

Let $A = [a_{ij}]$ be an $n \times n$ tridiagonal matrix. Consider the minor of the element a_{1n}, which is the product of all the subdiagonal elements, hence non-zero. By rank-nullity theorem, the rank of A is at least $n - 1$. Since for any eigenvalue λ of A, the nullity of $A - \lambda I$ is at least 1, it follows that it is equal to 1.

Now suppose A_1 and A_2 be real and symmetric $n \times n$ matrices, A_1 tridiagonal with all non-zero subdiagonal elements and A_2 a diagonal matrix with distinct diagonal elements. It follows that the matrix $\xi_1 A_1 + \xi_2 A_2$, for any real numbers ξ_1 and ξ_2, not both zero, is either a tridiagonal matrix with all non-zero subdiagonal elements or a diagonal matrix with distinct diagonal elements. In either case, all its eigenvalues are simple, as required.

Exercise 11.10

There is a non-singular real matrix $\mathbf{C}$ such that $\mathbf{C}^{-1}\mathbf{A}\mathbf{C} = \mathbf{J}$, where the matrix $\mathbf{J}$ is given by

$$\mathbf{J} = \begin{bmatrix} -\lambda_1 & 1 & 0 & \cdots & 0 \\ 0 & -\lambda_2 & 1 & \cdots & 0 \\ \cdots & \cdots & \cdots & \cdots & \cdots \\ 0 & 0 & \cdots & -\lambda_{N-1} & 1 \\ 0 & 0 & 0 & \cdots & -\lambda_N \end{bmatrix}$$

where $-\lambda_j$ are the eigenvalues of $\mathbf{A}$ with $\lambda_j > 0$ for all j. Thus, if we put $\mathbf{v} = \mathbf{C}^{-1}\mathbf{u}$, we see that the components of $\mathbf{v}$ satisfy the following (partially) decoupled system of equations:

$$v_{Ntt} - \lambda_N \Delta v_N = 0$$
$$v_{jtt} - \lambda_j \Delta v_j + v_{j+1} = 0, \; j = N-1, \ldots, 1.$$

These are homogeneous or inhomogeneous wave equations, with $\sqrt{\lambda_j}$ as speeds of propagation. The initial conditions prescribed on $\mathbf{u}$ get transferred to those on $\mathbf{v}$. Once the decoupled equations are solved for $\mathbf{v}$, we obtain the solution $\mathbf{u} = \mathbf{C}\mathbf{v}$.

Exercise 11.11

Let $(-\mathbf{A})^{1/2}$ denote the symmetric, positive definite square root of the symmetric, positive definite matrix $-\mathbf{A}$ and denote by $(-\mathbf{A})^{-1/2}$ the inverse of $(-\mathbf{A})^{1/2}$. As $\mathbf{A}$ commutes with $\mathbf{B}$, so are $(-\mathbf{A})^{1/2}$ and $(-\mathbf{A})^{-1/2}$. Consider the function $\mathbf{v}(x, x_{n+1}, t)$, $x \in \mathbb{R}^n$ defined by

$$\mathbf{v}(x, x_{n+1}, t) = \cosh\left(x_{n+1}(-\mathbf{A})^{-1/2}\mathbf{B}\right)\mathbf{u}(x,t),$$

where, for any square matrix $\mathbf{C}$, $\cosh(\mathbf{C})$ denotes the matrix $(1/2)(\exp(\mathbf{C}) + \exp(-\mathbf{C}))$. Now, a straightforward computation shows that $\mathbf{v}$ satisfies the following system of wave-like equations:

$$\mathbf{v}_{tt} + \mathbf{A}\Delta_{n+1}\mathbf{v} = \mathbf{0}, \; (x, x_{n+1}) \in \mathbb{R}^{n+1}, t > 0,$$

where Δ_{n+1} denotes the Laplacian in $n+1$ variables. Finally, by putting $x_{n+1} = 0$, we obtain the expression for $\mathbf{u}$ from that of $\mathbf{v}$.

Exercise 11.12

Put $\mathbf{v}(x,t) = \exp(t\mathbf{B})\mathbf{u}(x,t)$. Then,

$$\mathbf{v}_{tt} = \exp(t\mathbf{B})\mathbf{u}_{tt} + 2\mathbf{B}\exp(t\mathbf{B})\mathbf{u}_t + \mathbf{B}^2\exp(t\mathbf{B})\mathbf{u}$$
$$= \exp(t\mathbf{B})\left(\mathbf{u}_{tt} + 2\mathbf{B}\mathbf{u}_t\right) + \mathbf{B}^2\mathbf{v}$$

and $\mathbf{A}\Delta\mathbf{v} = \mathbf{A}\Delta(\exp(t\mathbf{B})\mathbf{u}) = \exp(t\mathbf{B})\mathbf{A}\Delta\mathbf{u}$, as $\mathbf{A}$ and $\mathbf{B}$ commute with each other. Thus, $\mathbf{v}$ satisfies the system

$$\mathbf{v}_{tt} + \mathbf{A}\Delta\mathbf{v} - \mathbf{B}^2\mathbf{v} = 0,$$

which is similar to equation (11.2.14). Following the procedure in Exercise 11.11, put

$$\mathbf{w}(x, x_{n+1}, t) = \cosh\left(x_{n+1}(-\mathbf{A})^{-1/2}\mathbf{B}\right)\mathbf{v}(x,t),$$

for $x \in \mathbb{R}^n$, $x_{n+1} \in \mathbb{R}$ and $t > 0$. Then, $\mathbf{w}$ satisfies the system

$$\mathbf{w}_{tt} + \mathbf{A}\Delta_{n+1}\mathbf{w} = 0, \; (x, x_{n+1}) \in \mathbb{R}^{n+1}, \; t > 0,$$

where Δ_{n+1} denotes the Laplacian in $n+1$ variables. Finally, we have

$$\mathbf{u}(x,t) = \exp(-t\mathbf{B})\mathbf{v}(x,t) = \exp(-t\mathbf{B})\mathbf{w}(x, 0, t).$$

Exercise 11.13

Comparing with the system in Exercise 11.11, we have $\mathbf{A} = -\mathbf{I}$ and $\mathbf{B}^2 = \begin{bmatrix} 0 & 1 \\ -1 & 0 \end{bmatrix}$. Since the eigenvalues of the matrix $\mathbf{J} = \begin{bmatrix} 0 & 1 \\ -1 & 0 \end{bmatrix}$ are purely imaginary, it is not possible to find a *real* matrix $\mathbf{B}$ such that $\mathbf{B}^2 = \mathbf{J}$. Nevertheless, we observe that $\mathbf{J}^2 = -\mathbf{I}$ and hence all the powers $\mathbf{B}^{2k}$, $k \geqslant 1$ are equal to either $\pm\mathbf{I}$ or $\pm\mathbf{J}$. Therefore, the matrix $\cosh(y\mathbf{B})$, $y \in \mathbb{R}$, is a linear combination of the matrices $\mathbf{I}$ and $\mathbf{J}$ with functions of y as coefficients. This observation also shows that the condition that the matrix $\mathbf{B}$ be real may be relaxed provided that the matrix generated through the cosh function is real. We thus put

$$v_1(x, y, t) = \varphi(y)\, u_1(x,t) + \psi(y)\, u_2(x,t),$$
$$v_2(x, y, t) = -\psi(y)\, u_1(x,t) + \varphi(y)\, u_2(x,t),$$

and attempt to find the functions φ and ψ so that the functions v_i satisfy the 2D wave equation, namely

$$(v_i)_{tt} - (v_i)_{xx} - (v_i)_{yy} = 0, \ i = 1, 2.$$

In order to recover u_i from v_i through the substitution $y = 0$, we also assume $\varphi(0) = 1$ and $\psi(0) = 0$. These requirements are met if φ and ψ satisfy the equations $\varphi'' = \psi$ and $\psi'' = -\varphi$, with $'$ denoting the differentiation with respect to y. We choose, among other possibilities,

$$\varphi(y) = \cosh(y/\sqrt{2}) \cos(y/\sqrt{2}) \text{ and } \psi(y) = -\sinh(y/\sqrt{2}) \sin(y/\sqrt{2}).$$

Thus,

$$v_1(x, y, t) = \cosh(y/\sqrt{2}) \cos(y/\sqrt{2}) u_1(x, t) - \sinh(y/\sqrt{2}) \sin(y/\sqrt{2}) u_2(x, t),$$
$$v_2(x, y, t) = \sinh(y/\sqrt{2}) \sin(y/\sqrt{2}) u_1(x, t) + \cosh(y/\sqrt{2}) \cos(y/\sqrt{2}) u_2(x, t),$$

and

$$v_1(x, y, 0) = v_2(x, y, 0) = 0,$$
$$(v_1)_t(x, y, 0) = \cosh(y/\sqrt{2}) \cos(y/\sqrt{2}) g_1(x) - \sinh(y/\sqrt{2}) \sin(y/\sqrt{2}) g_2(x),$$
$$(v_2)_t(x, y, 0) = \sinh(y/\sqrt{2}) \sin(y/\sqrt{2}) g_1(x) + \cosh(y/\sqrt{2}) \cos(y/\sqrt{2}) g_2(x).$$

Using the formula for the solution of the wave equation in two dimensions, we readily write down the expressions for v_1 and v_2. The substitution $y = 0$ then gives the expressions for u_1 and u_2. We have, for $t > 0$,

$$u_1(x, t) = \frac{1}{2\pi} \int_{x-t}^{x+t} g_1(\xi) \, d\xi \int_{-\sqrt{t^2-(x-\xi)^2}}^{\sqrt{t^2-(x-\xi)^2}} \frac{\cosh(\eta/\sqrt{2}) \cos(\eta/\sqrt{2})}{\sqrt{t^2 - (x-\xi)^2 - \eta^2}} \, d\eta$$

$$- \frac{1}{2\pi} \int_{x-t}^{x+t} g_2(\xi) \, d\xi \int_{-\sqrt{t^2-(x-\xi)^2}}^{\sqrt{t^2-(x-\xi)^2}} \frac{\sinh(\eta/\sqrt{2}) \sin(\eta/\sqrt{2})}{\sqrt{t^2 - (x-\xi)^2 - \eta^2}} \, d\eta$$

and a similar expression for u_2. We simplify the integral with respect to η by the substitution $\eta = s \sin \theta$, where $s^2 = t^2 - (x - \xi)^2$ and then using the following integral formula for the Bessel function of order 0:

$$J_0(z) = \frac{2}{\pi} \int_0^{\pi/2} \cos(z \sin \theta) \, d\theta, \ z \in \mathbb{C}.$$

See Refs. [1] and [27]. If $z = x + \mathrm{i}y$, we have

$$\cos(z) = \cos x \cosh y - \mathrm{i} \sin x \sinh y.$$

Putting all these expressions together, we obtain

$$u_1(x,t) = \frac{1}{2} \int_{x-t}^{x+t} \left[g_1(\xi) \Re\left(J_0((1+\mathrm{i})s/\sqrt{2})\right) + g_2(\xi) \Im\left(J_0((1+\mathrm{i})s/\sqrt{2})\right) \right] d\xi,$$

$$u_2(x,t) = \frac{1}{2} \int_{x-t}^{x+t} \left[g_2(\xi) \Re\left(J_0((1+\mathrm{i})s/\sqrt{2})\right) - g_1(\xi) \Im\left(J_0((1+\mathrm{i})s/\sqrt{2})\right) \right] d\xi,$$

where $\Re(z)$ and $\Im(z)$ denote the real and imaginary parts of a complex number z, respectively.

References

[1] Abramowitz, M. and Stegun A. (1972). *Handbook of Mathematical Functions.* New York: Dover.

[2] Apostol, T. M. (2011). *Calculus, vol. 1 and vol. 2.* New Delhi: Wiley India.

[3] Arnold, V. I. (1998). *Ordinary Differential Equations.* New Delhi: Prentice-Hall of India.

[4] Ascher, U. M., Mattheij, R. M. M. and Russel, R. D. (1995). *Numerical Solution of Boundary Value Problems for Ordinary Differential Equations.* Philadelphia: SIAM.

[5] Barták, J., Herrman, L., Lovicar, V. and Vejvoda, O. (1991). *Partial Differential Equations of Evolution.* New York: Ellis Horwood.

[6] Bellman, R. (1987). *Introduction to Matrix Analysis.* Philadelphia: SIAM.

[7] Bender, C. M. and Orszag, S. A. (1999). *Advanced Mathematical Methods for Scientists and Engineers I: Asymptotic Methods and Perturbation Theory.* New York: Springer.

[8] Birkhoff, G. and Rota, G.-C. (2003). *Ordinary Differential Equations.* New York: John Wiley and Sons.

[9] Bowman, F. (1958). *Introduction to Bessel Functions.* New York: Dover.

[10] Braun, M. (1978). *Differential Equations and Their Applications.* New Delhi: Springer International.

[11] Braun, M. (1975). *Differential Equations and Their Applications.* New York: Springer-Veralag.

[12] Brockett, R. W. (1970). *Finite Dimensional Linear Systems.* New York: Wiley.

[13] Callier, F. M. and Desoer, C. A. (1991). *Linear System Theory.* New Delhi: Narosa Publishing House.

[14] Chen, C. T. (1984). *Linear System Theory and Design.* New Delhi: Sauders College Publishing.

[15] Coddington, E. A. and Levinson, N. (1972). *Ordinary Differential Equations.* New Delhi: Tata McGraw-Hill.

[16] Courant, R. and Hilbert, D. (1989). *Methods of Mathematical Physics: Vol. II Partial Differential Equations.* New York: John Wiley and Sons.

[17] Davis, P. J. (1979). *Circulant Matrices.* New York: Wiley.

[18] Demidov, A. S. (2023). *Equations of Mathematical Physics.* New York: Springer.

[19] DiBenedetto, E. (2010). *Partial Differential Equations*, 2nd Edn. New Delhi: Springer International.

[20] Donnelly, H. (1987). Uniqueness of Positive Solutions to the Heat Equation. *Proc. Amer. Math. Soc.*, 99(2): 353–356.

[21] Dym, H. (2012). *Linear Algebra in Action.* First Indian edition. New Delhi: American Mathematical Society.

[22] Dym, H. and McKean, H. P. (1985). *Fourier Series and Integras.* Reprint edition. New York: Academic Press.

[23] Evans, L. C. (1998). *Partial Differential Equations*. Rhode Island: AMS, Providence.

[24] Friedman, A. (1964). *Partial Differential Equations of Parabolic Type*. New York: Prentice-Hall.

[25] Friedman, A. (1976). *Partial Differential Equations*. New York: Robert Krieger.

[26] Gantmacher, F. R. (1990). *The Theory of Matrices*, Providence Vol. 1. American Mathematical Society.

[27] Gradshteyn, I. S. and Ryzhik, I. M. (2007). *Table of Integrals, Series, and Products*, 7 Edn. New York: Academic Press.

[28] Halanay, A. and Samuel, J. (1997). *Differential Equations, Discrete Systems and Control - Economic Models*. London: Kluwer Acdemic Publishers.

[29] Ince, E. I. (1926). *Ordinary Differential Equations*. New York: Dover, Inc.

[30] Jackson, D. J. (1999). *Classical Electrodynamics*, 3rd Edn. New Delhi: Wiley India.

[31] John, F. (1978). *Partial Differential Equations*, 3rd Edn. New York: Springer Varlag.

[32] Jordan, D. W. and Smith, P. (2003). *Nonlinear Differential Equations*, 3rd (Reprint) Edn. Oxford: Oxford University Press.

[33] Keller, H. B. (1990). *Numerical Solution of Two-Point Boundary Value Problems*. New York: Dover.

[34] Kellogg, O. D. (1953). *Foundations of Potential Theory*. New York: Dover.

[35] Koshlyakov, N. S., Smirnov, M. M. and Gliner, E. B. (1964). *Differential Equations of Mathematical Physics*. Amsterdam: North-Holland.

[36] Lax, P. (1973). *Hyperbolic Systems of Conservation Laws and Mathematical Theory of Shock waves*. Philadelphia: SIAM.

[37] Lax, P. (2007). *Linear Algebra and Applications*, 2nd Edn. New York: Wiley.

[38] Lefschetz, S. (1977). *Differential Equations: Geometric Theory*. New York: Dover.

[39] Markushevich, A. I. (1977). *Theory of Functions of a Complex Variable*, 2nd Edn. New York: Chelsea Publishing Company.

[40] McOwen, R. C. (2005). *Partial Differential Equations: Methods and Applications*, 2nd Edn. Delhi: Pearson Education.

[41] Merkin, D. R. (1997). *Introduction to the Theory of Stability*. New York: Springer.

[42] Mikhailov, V. P. (1978). *Partial Differential Equations*. Moscow: Mir Publishers.

[43] Mischenko, A. and Fomenko, A. (1988). *A Course of Differential Geometry and Topology*. Moscow: Mir Publishers.

[44] Morawetz, C. S. (1981). *Lectures on Non-linear Waves and Shocks, TIFR Lecture Notes*. Springer-Verlag. New York.

[45] Nandakumaran, A. K. and Datti, P. S. (2020). *Partial Differential Equations:Classical Theory with a Modern Touch*. Cambridge: Cambridge University Press.

[46] Nandakumaran, A. K., Datti, P. S. and George, R. K. (2017). *Ordinary Differential Equations: Principles and Applications*. Cambridge: Cambridge University Press.

[47] Nishio, M. (1992). Uniqueness of Positive Solutions of the Heat Equation. *Osaka J. Math.*, 29: 531–538.

References

[48] Perko, L. (2001). *Differential Equations and Dynamical Systems*. New York: Springer.

[49] Prasad, P. and Ravindran, R. (1996). *Partial Differential Equations*, 3rd Edn. New Delhi: New Age International.

[50] Rudin, W. (1976). *Principles of Mathematical Analysis*. London: McGraw-Hill.

[51] Russel, D. L. (1979). *Mathematics of Finite Dimensional Control Systems*. New York and Basel: Marcel Dekker, Marcel Dekker Inc.

[52] Simmons, G. F. (1991). *Differential Equations With Applications and Historical Notes*. New York: McGraw-Hill Intenational.

[53] Sontag, E. D. (1998). *Mathematical Control Theory: Deterministic Finite Dimensional Systems*. New York: Springer.

[54] Taylor, M. E. (2011). *Introduction to Differential Equations*. Providence: American Mathematical Society, Indian (2013) edition.

[55] Vladimirov, V. S. (1986). (Ed.) *A Collection of Problems on the Equations of Mathematical Physics*. Moscow: Mir Publishers.

[56] Whitham, G. B. (1974). *Linear and Non-linear Waves*. New York: Wiley-Interscience.

[57] Widder, D. V. (1990). *The Heat Equation*. New York: Academic Press.

[58] Wiggins, S. (1990). *Introduction to Applied Nonlinear Dynamical Systems and Chaos*. New York: Springer.

Index

adjoint equation, 116
asymptotic series, 100
asymptotically stable, 143
autonomous system, 141

backward characteristic, 190
Brachistochrone problem, 197

canonical form, 238
canonical transformation, 148
catenary, 198
Cauchy problem, 185
    characteristic, 187, 203
    non-characteristic, 187
characteristic curve, 241
characteristic equations, 186
characteristic form, 242, 247
characteristic parallelogram property, 312
characteristic set, 242
characteristic variety, 242
coercive, 189
complete integral, 188
complex potential function, 150
conservation law, 190
control problem, 54
controllability, 53
controllability Grammian matrix, 55
cost
    optimal, 189
cycloid, 221

density, 261
Dirichlet principle, 198
domain of dependence, 308
domain of determinacy, 309
double layer potential, 261

eigenfunction
    boundary value problem, 111
eigenvalue
    boundary value problem, 111
ellipsoid, 238

energy functional, 148
entropy condition, 192
entropy solution, 192
envelope, 9
equation
    Bernoulli, 15
    Bessel's, 91, 96–98, 101
    Chebyshev, 92
    Darboux, 330
    diffusion, 235
    Duffing, 14
    Duffing's, 164
    elliptic, 235, 239
    Euler, 101
    exact, 4
    Hamilton-Jacobi, 188
    Hamilton-Jacobi-Bellman, 188
    heat, 235
    Hermite, 89
    hyperbolic, 235, 239
    hypergeometric, 92
    Jacobi, 15
    Klein–Gordon, 338
    Klein–Gordon–Fock, 338
    Laplace, 248
    Laplace, 235
        maximum principle, 247
        mean value formula, 247, 249
    Legendre, 91, 101
    parabolic, 235, 239
    Poisson, 248, 251
    Riccati, 15
    telegraph, 314, 332
    Tricomi, 239
    van der Pol, 14
    wave, 235
        multi-dimensional, 329
        one-dimensional, 307
        three-dimensional, 330
        two-dimensional, 331
equations
    Euler-Lagrange, 189, 198

equiangular spiral, 42
equilibrium point, 141, 143
   hyperbolic, 145
   isolated, 143
   non-isolated, 143
equipotential lines, 150
escape velocity, 19
exactly controllable, 54
exit time, 196

formula
   D'Alembert's, 307
   Fourier-Poisson, 285
   Green's, 117
   Green's representation, 252
   Hopf-Lax, 189
   Kirchhoff's, 330
   Lax-Oleinik, 192
   Picone, 120
   Poisson
      ball, 253
      upper half-space, 253
Frobenius series, 95
Frobenius theory, 85
function
   Hölder continuous, 254
   harmonic, 248
   sub-harmonic, 248
   super-harmonic, 248
fundamental solution, 236, 247

general solution, 188
generalized solution, 192
Gevrey class, 87
Green's function, 113
   dihedral angle, 257
   first octant, 257
   quarter-ball, 257
   strip, 257
   upper-half ball, 257
Green's identities, 255
Gronwall's inequality, 5, 16

Hamiltonian, 148, 188
Hamiltonian system, 148, 152, 154, 218
Harnack's inequality, 250
heat equation
   Dirichlet problem, 287
   Duhamel's principle, 286
   fundamental solution, 285
   inhomogeneous, 286
   mixed problem, 287
   Neumann problem, 287
heat kernel, 285
Hermite polynomial, 90
Hopf's lemma, 250
Huyghens principle, 332
hyperboloid, 238
hypoelliptic, 236

indicial equation, 94
initial value problem, 185
integrating factor, 3, 4
irregular singular point, 92
isogonal trajectory, 11

Joukowski airfoil, 153
Joukowski transformation, 153
jump condition, 192

Kalman's rank condition, 55
Kelvin transform, 259

Lagrange identity, 117
Lagrangian, 188
Legendre polynomial, 106
Legendre transformation, 189, 190
Liapunov
   function, 146
   instability theorem, 146
   stability, 143
   stability theorem, 146
Liapunov function, 145
limit cycle, 142, 171
linear stability analysis, 144
linear theory, 47
linearised system, 144
Lipschitz continuity, 6
Lipschitz continuous, 254
logarithmic spiral, 42
loss of regularity, 312

matrix
   diagonal, 48
   fundamental, 52
   nilpotent, 51
   transition, 48, 52
maximal interval of existence, 8
maximum principle
   strong, 249
   weak, 249, 254

method of characteristics, 185
method of descent, 331
minimum principle
   strong, 249
   weak, 249, 254

non-autonomous system, 141
non-hypoelliptic, 236
null controllable, 55
numerical range, 10

oblique trajectory, 11
observability, 53
observability inequality, 58
observable system, 56
operator
   Laplace, 247
orbit, 141
   closed, 142
   homoclinic, 164, 167, 181
   periodic, 142
ordinary point, 90
orthogonal trajectory, 11

paraboloid, 239
path, 142
periodic boundary condition, 112
Perron's method, 248
point
   characteristic, 236
Poisson kernel, 253
potential flow, 149
potential function, 150
potential theory
   local version, 261
Prüfer substitution, 120, 124
Prüfer substitution, 118
prey-predator system, 15
principal part, 242
principal symbol, 242, 247
propagation of singularities, 310

qualitative theory, 141

range of influence, 308
Rankine-Hugoniot condition, 192, 200
rarefaction, 209
rarefaction wave, 191, 226, 229, 231
real analytic function, 86
regular singular point, 92

self-adjoint system, 118
semi-concavity, 190
shock, 191, 192, 209
shooting method, 113
single layer potential, 261
singular point, 10, 87
singular point of first kind, 92
singular point of second kind, 92
singular solution, 216
solution
   optimal, 189
   periodic, 142
speed of propagation, 307
spherical mean function, 329
stream function, 150
streamline, 150
strip condition, 186
Sturm–Liouville problem
   regular, 111
   singular, 111
Sturm–Liouville theory, 111
subspace
   invariant, 50
surface
   characteristic, 236
   non-characteristic, 236
Sylvester's law of inertia, 238
symplectic matrix, 148

theorem
   Cauchy–Peano, 7
   center manifold, 145
   Chetaev, 146
   continuous dependence, 8
   Hartman–Grobman, 145
   Liouville's
      stronger form, 257
   Perron's, 144, 146
   Picard, 7
   Widder, 291
trajectory, 142
transversality condition, 187

uniformly elliptic, 247
unique continuation principle, 58

value function, 189
value function, 188

weakly hyperbolic, 241